SCIENCE, TECHNOLOGY, AND THE ART OF MEDICINE

Philosophy and Medicine

VOLUME 44

The titles published in this series are listed at the end of this volume.

SCIENCE, TECHNOLOGY, AND THE ART OF MEDICINE

European-American Dialogues

Edited by

CORINNA DELKESKAMP-HAYES

Director of European Programs, International Studies in Philosophy and Medicine, Germany

and

MARY ANN GARDELL CUTTER

Dept. of Philosophy, University of Colorado, Colorado Springs, U.S.A.

KLUWER ACADEMIC PUBLISHERS

DORDRECHT / BOSTON / LONDON

Library of Congress Cataloging-in-Publication Data

Science, technology, and the art of medicine : European-American dialogues / edited by Corinna Delkeskamp-Hayes and Mary Ann Gardell Cutter ; the contributions written in German for this volume were translated into English by Ruth M. Walker Moskop and S.G.M. Engelhardt.
p. cm. -- (Philosophy and medicine ; v. 44)
Includes index.
ISBN 0-7923-1869-2
1. Medicine--Philosophy. I. Delkeskamp-Hayes, Corinna. II. Cutter, Mary Ann Gardell. III. Series.
R723.S384 1993
610'.1--dc20 92-18579

ISBN 0-7923-1869-2

Published by Kluwer Academic Publishers,
P.O. Box 17, 3300 AA Dordrecht, The Netherlands.

Kluwer Academic Publishers incorporates
the publishing programmes of
D. Reidel, Martinus Nijhoff, Dr W. Junk and MTP Press.

Sold and distributed in the U.S.A. and Canada
by Kluwer Academic Publishers,
101 Philip Drive, Norwell, MA 02061, U.S.A.

In all other countries, sold and distributed
by Kluwer Academic Publishers Group,
P.O. Box 322, 3300 AH Dordrecht, The Netherlands.

The contributions written in German for this volume were translated into the English by Ruth M. Walker Moskop and S.G.M. Engelhardt.

Printed on acid-free paper

TABLE OF CONTENTS

EDITORIAL PREFACE

There are few more ambiguous, central, and puzzling concepts than art, science, and technology; they all have played important roles in accounts of medicine. The term 'art' has great antiquity. There is the Hippocratic work, "The Art", "Peri technê". The art is also mentioned with reverence in "The Oath" and elsewhere in the Hippocratic corpus. Much of what we associate with art is still true of medicine. The good artist at home with a medium and guided by talent is directed by insight and acts well. Often success is indisputable, yet when secured its exact origins remain obscure. Like good artists, good physicians have healed for millennia. The good physician diagnoses well and treats successfully, frequently without being able to provide an account in fully discursive terms of how the diagnosis was framed or the treatment selected. Experience can both provide guidance and serve as the justification for an action, especially when a more explicit account is unavailable.

Contemporary medicine was born of a science self-critical about its capacity to know. It was recast by the new sciences of anatomy, physiology, and bacteriology. Contemporary clinical medicine developed against emerging understandings of observer bias and the difficulties of overcoming it. Medicine was reshaped by a dialectic between clinical medicine and the sciences now called basic. This epistemologically self-critical medicine transformed the medical techniques of the experienced clinician into the technologies of scientific medicine. Art as *technê* provided the substantive (not only etymological) roots of technique and technology, which were themselves reshaped. Still, the difficulties of understanding medicine are disclosed by the fruitful ambiguities of *technê*, which compasses meanings such as art, skill, cunning of hand, craft, means whereby a thing is gained, a trade, and a set of rules or a method for doing something. This range of meanings includes pre-reflective skills, as well as explicit systems for acting. The history of medicine is in great measure the discursive appreciation of the meanings of *technê* or art. The well-trained art of the Hippocratic craftsman has become the examined technology tested in controlled trials. The *logos* of science has rendered the techniques of the clinician into the technologies of a scientific medicine.

These transformations have taken place against foundational developments in our understanding of the interplay of facts, theories, and values. They have also been driven by a moral concern to acquire knowledge responsibly and to act with prudence and care. The gaze of philosophy has come to focus on the *logos* of the art, on the meaning and significance of the knowings, doings, and makings that constitute the technologies of medicine. These moral and philosophical concerns about the arts, sciences, and technologies of medicine have given birth to the contemporary fields of bioethics and the philosophy of medicine, along with spurring the reflections that culminated in this volume.

The discussions regarding responsible knowing and medical action that led to this volume began in 1969 when one of the authors of this Preface (H.T.E.) was a Fulbright graduate fellow at Bonn University, where Corinna Delkeskamp was also studying. These discussions embraced the ideas that framed the first volume in this book series, *Evaluation and Explanation in the Biomedical Sciences*, published in 1975 (drawn from the First Transdisciplinary Symposium on Philosophy and Medicine, which was held in Galveston, Texas, in 1974). The concerns with medical explanations and evaluations articulated in that volume have continued as cardinal concerns of the *Philosophy and Medicine* book series. They anticipated this volume in a symposium held September 1 through 3, 1977, in Hamburg, Germany, while Corinna Delkeskamp (now "-Hayes"), who conceived of that conference, was preparing to give birth to her first child. The editors of this series take this opportunity to thank the Deutsche Forschungsgemeinschaft, the Freie und Hansestadt Hamburg, the Fritz-Thyssen Stiftung, the Harmannbund-Verband der Ärzte Deutschlands, the Reidel (Kluwer) Publishing Company, the Universität Hamburg, and the University of Connecticut School of Medicine for their support of that symposium: "Naturwissenschaft, Technik und die 'ärztliche Kunst'". In addition to these institutions, many individuals offered generously in time, energy, and insight culminating in the success of that symposium, which brought into being a set of conversations eventually leading to the present volume. We are as much indebted to them as to the scholars who contributed these essays.

Over the years, many persons sustained and enlivened discussions and explorations born of that symposium. These explorations continued at the Kennedy Institute of Ethics, Georgetown University, Washington, D.C., the Center for Ethics, Medicine, and Public Issues, Baylor College of Medicine, Houston, Texas, and the Institute for Advanced Study, Berlin. This volume took shape from this extended "traveling" colloquium. We wish to thank all who participated and encouraged us in the development of this book. We especially wish to acknowledge George Khushf and The Reverend Kevin Wm. Wildes, S.J., who had the art and possessed the techniques necessary to facilitate a dialogue over space and time. The series editors wish also to thank Ruth Walker Moskop and S.G.M. Engelhardt, who translated into English the articles by Dietrich von Engelhardt, Rudolf Gross, Lothar Schäfer, Nelly Tsouyopoulos, and Wolfgang Wieland.

Finally, the series editors owe a special debt to Corinna Delkeskamp-Hayes, who encouraged and informed these reflections, and to Mary Ann Gardell Cutter, who later joined this cluster of conversations. It is they who brought many voices and concepts together over the last two years and transformed them into this volume.

April 21, 1992

H. TRISTRAM ENGELHARDT, JR.
STUART F. SPICKER

CORINNA DELKESKAMP-HAYES

INTRODUCTION

This is a book about concepts: in what sense are "science", "technology" and "art" characteristic of medicine?[1] The complex web of their meanings will have to be explored.[2]

It is no easy task. The three concepts, while figuring quite commonly in present public discussions of medicine, are not often consistently used.[3]

Take for example "science": When physicians say that practitioners "apply" medicine ([19], p. 9), they sometimes imply that medicine is (essentially) a science which is then merely utilized in practice. A similar separation between medical science and medical practice is posited by health insurance agencies, as they restrict reimbursement to forms of treatment, the effectiveness of which has previously been scientifically established (for Germany, see [29], §135). These agencies thus exclude clinical trial treatments from standard (reimbursable) practice. Yet the prime method of improving that standard practice consists in conducting such trials, or in pursuing a practice-based scientific research, which denies the attempted separation.

Moreover, critical physicians and lay critics oppose "scientific medicine" altogether, because it disregards large areas of medical problems. Chronic illness, psychosomatic, psycho-social and psychic diseases, social deviance, handicaps, ageing and dying are thereby deprived of adequate attention (see for example [11], p. 197).[4]

Thus, neither in view of its juxtaposition with "practice" nor in view of its own limitations[5] is the concept of a (natural) "science of medicine" very helpful.

Given these difficulties, it is understandable that medicine is sometimes considered a "technology".[6] Medical practice is here construed teleologically[7] as producing "health goods", which are paid for — in more or less direct ways — by the consumer. Public policy-makers[8] employ its services as means for the realization of social goals, such as saving lives ([6], p. 135), securing an adequate level of health for society, encouraging responsible life styles, restricting ([43], p. XI) or encouraging ([3], p. 147, [10], p. 8) redistributive consequences, or, ultimately, increasing equality of social chances [17]. In a similar vein, statutory fee schedules put higher prices on the technical aspects of practice ([7], p. 8). They also distinguish minute aspects of

Corinna Delkeskamp-Hayes and Mary Ann Gardell Cutter (eds.), Science, Technology, and the Art of Medicine, 1–20.

medical activities in order to enhance the transparency of physicians' accounting. These schedules divide the care of patients into particular tasks, which are defined by procedural rules and whose success can be objectively controlled ([43], §§87, 294ff, p. xii).

Yet, on the other hand, physicians complain that patients prefer quick technological fixes to the burden of changing their unhealthy lifestyles, and both patients and physicians are accused of succumbing to the "magic" of the technical ([28], p. 12). Such observations, as well as the more general public criticism of medicine's recent "technologization" ([29], p. 205; H.D. Pohle in [20], p. 13; [16]), indicate discomfort with "medicine as a technology".

This discomfort sometimes motivates "art"-talk,[9] especially among physicians. The "art of medicine", aside from trivial uses (such as "state of the art"), is also invoked in a more encompassing sense. It is taken to imply the physician's special devotion to his task ([19], p. 15), and to preclude any primarily economic motivations. It is also closely associated with the person of the physician. (The question about to what extent it can be communicated through teaching is still disputed, see H.D. Pohle in [20], p. 12.) Physicians therefore refer to the "art"-character of their employment when defending their professional judgement against external regulation.[10]

Yet it is just this latter employment of the concept of an "art of medicine" which is hampered by a subtle paradox. Physicians agree about their obligation to further their art and to conduct scientific research. Through science medicine comes to conform to objective standards. Thus, medical practice is rendered objectively controllable[11] and in that sense impersonal. On the other hand, the "art" of medicine is thought to be incompatible with impersonalness.

This in itself poses no serious problem. Yet objective controls are presently also required of medicine for allowing policy-makers to supervise physicians' behavior. Hence, wherever art talk is used to oppose the objectifiability of medical practice (and thus to advocate a pre-modern traditionalism) [12] in order to discourage such external supervision ([18], [35]), it tends to conflict with physicians' own avowed (and modern) commitment to furthering their art.[13]

This difficulty is especially relevant in the context of present-day public policy attempts at cost containment. Here public policy not only regards medicine in view of its technological nature, it also transforms it along technological lines. The goal of cost containment is pursued either by permitting market forces to influence health care delivery ([43], §§105 (2), 121) or by enhancing bureaucratic controls ([43], §§106, 135ff, 141ff, 199ff, 275). In the first case, the physician is subjected to increased economic competition; in the second, he is made to function as fiduciary of societal resources. In each of these cases, the particular nature of medical transactions renders public controls regarding the economy or quality of medical services necessary. For this purpose, legal and organizational frameworks have been established which in turn encourage medicine's gradual technologization.[14]

In view of these difficulties, the question arises whether there is indeed anything so unique about medical services that a certain immunity from external regulation should be granted.[15] Most of the essays contained in this volume do not address policy decisions, nor do they intend to influence them. They pursue the more restricted goal of examining "what medicine is", or how the "art" of medicine can be understood in the context of medical science and technology. But in doing so, they also carry implications for the possibility of justifying professional claims, and for public policy decisions in general. In this sense, even a "book about concepts" may be seen to point beyond the merely theoretical.

The essays in this volume are grouped into four sections, with a concluding discussion at the end. These sections correspond to four ever more encompassing activities related to medicine: theory formation, explanation of disease, medical decision-making, and dealing with patients and the public. They can, in turn, be grouped under the more general headings of "activities relating to the establishment of medical knowledge" and "activities relating to medical practice". The concluding essay tries systematically to put it all together. It endeavors to show how, even if the uniqueness problem can be solved, the desired policy consequences are much harder to derive than might have been supposed.

The first section on theory formation in medicine focusses on medical research: How are "medical facts" conceived? And what constitutes "progress" in medical science?

Ever since the nineteenth century, medical theoreticians aspired to transform their merely personal and uncertain art into a scientific discipline. Thus at least the theory- (or knowledge-) part of medicine was deemed scientific, even if this meant restricting its scope to those auxiliary areas of biochemical research which fitted best into the established framework. In a similar vein, the more encompassing body of specifically medical knowledge that developed from ever expanding clinical research activities was modelled upon the natural sciences. The recent designation "biomedical sciences" illustrates this tendency.

Does such a model make sense? The three essays in the first section address this question.

Lothar Schäfer, in his exposition of Ludwik Fleck's 1935 treatise on scientific facts, takes up an early criticism of "medicine as a natural science". Fleck's arguments work on two levels: First, our understanding of natural science in general must be revised. Second, its new understanding should be modelled on medical knowledge (instead, as before, on physics).

Traditional theory (or philosophy) of science (which is still very much alive today) had rested on a positivist interpretation of what constitutes "medical facts" and on a deductive-nomological reconstruction of scientific statements about these facts. In Fleck's view, theoretical reflection on the sciences should not primarily concern the content of scientific laws and their logical interrelations. It should more fundamentally regard the formation process of these contents and relations as it is influenced by particular psychological and societal conditions. This idea was pursued by Thomas Kuhn, who cites (in the Preface to his *The Structure of Scientific Revolutions*) Fleck as one of his principal sources. Scientific observation is subject to constraints imposed by theoretical commitments. It is therefore not "objective". Different theories address different sorts of empirical evidence. There exists — so Fleck and Kuhn thought — no theory-neutral ground on which to evaluate scientific theories. Hence, actual choice between theories over time is not itself scientifically rational but depends on all sorts of culturally determined prejudices.

As to the manner of that dependence, Fleck goes beyond Kuhn in recommending the medical field as a new paradigm for understanding all the other supposedly "natural sciences". Using the example of syphilis research in the 1920s, Fleck exposes the particularly intense interdisciplinary cooperation as well as social and political involvement characterizing this area. He explains how the resulting communication problems led to external factors influencing theory formation. Schäfer points out that, while syphilis research provides merely the occasion for re-interpreting scientific research in general, the medical knowledge context of this research is offered as evidence for the adequacy of such a re-interpretation. Thus Schäfer's essay at the outset exposes the philosophy-of-science relevance of reflections on medical theorizing.

Nelly Tsouyopoulos exposes some weakness in the Gestalt Psychology foundation of Fleck's epistemology. Moreover, following Sadegh-zadeh ([39]; cf. also [40]), she proposes a different view of medicine's value-implications. Instead of changing our notion of "natural science" so as to make medical science its prime representative, one should distinguish between "theoretical" and "practical" science. The former is pursued for primarily epistemological reasons, the latter, while obviously encompassing natural science elements as well, is pursued for primarily non-epistemological reasons. The influence of social norms on medical research should not be taken to compromise the rationality of a supposedly theoretical endeavor. It can, instead, be justified in view of medicine's practical orientation.

Quite independently of the question whether medical science should be considered "practical" rather than "theoretical", *Reidar Lie* opposes Fleck's assumption that every influence of societal values on research must repudiate its scientific rationality. He concedes that scientific theories disregard "anomalies" for the sake of coherence. After all, coherence is a condition for the rationality of scientific theories. He also concedes that historical and social determinants influence the point at which observed "anomalies" are taken seriously and thus demand a replacement of thought patterns. Still, neither the internal validity of a theory's epistemological claims nor the progress involved in science's historical development is thereby refuted. Lie proposes that reference to such determinants could support this refutation only if they had caused scientists to neglect better substantiated alternative thought patterns. This, he argues, Fleck has not shown.

In the first section, three interpretations of medical knowledge have been suggested: as paradigm for an altered view of the natural sciences in general (Schäfer), as praxological theory (Tsouyopoulos), and (Lie, by implication) as natural science in the traditional sense.

The second section concerns explanation. Combining historical with present-day analyses, the authors explore the concept of "cause" in medical knowledge. Implicitly their arguments also point towards theoretical versus praxological explanations of conceptual change in medical theorizing, and thus towards the status of medical science which had been discussed in the first section.

José Luis Peset summarizes the development of medical causality from antiquity to modern times. He distinguishes a Galenic (and Aristotelian) from a modern (Hume-Kantian) model. The Greek essentialist causal pattern originally corresponded to a notion of disease as an imbalance of forces, or disorder in a cosmic sense. In the seventeenth and eighteenth centuries, this notion developed into the idea of classifiable disease entities. By contrast, the modern, phenomenalist causal pattern was at first fitted into a "clinical configuration"-reconstruction of these entities. These were, however, still reified, or supposed to constitute realities of their own. The notion of their causes was simply included in their descriptions. Only in the nineteenth century does the nosological system give way to physiological analysis; the notion of "cause" is now submerged in that of "pathological reactions". Bacteriology, however, again re-introduces the idea of external causation, which is later integrated into a newly Galenic multifactorial account.

Following the work of Michel Foucault, Peset concludes that not only changes in the philosophical interpretation of causes and essences influenced the thought of medical theorists, but that new practical problems in turn contributed to the shaping of medical concepts. Even though causal thinking was discouraged in view of contemporary phenomenalist reductions, the practical concern with contagious diseases and their bacteriological origins, as it were, forced the concept of efficient cause back into medical science. Similarly, epidemiological studies in the influence of environmental and social living conditions on the health status of populations kept alive the theoretically outdated multi-causal Galenic scheme.

If, as Peset has shown, in the nineteenth century medical theorizing considerably lagged behind the positivist reconstruction of science prevalent at that time, the question again arises as to whether this discrepancy should be interpreted on Fleckian or on praxological lines. In the first case, the historical development of medical thought is seen as determined by a merely passive reception of various (and sometimes contradictory) cultural influences. In the second case, the resulting anachronism originates in an active choice in favor of practical goals.

Dietrich von Engelhardt's account of causalism versus conditionalism in nineteenth century medical thought exposes that discrepancy in greater detail. In his view, the conceptual antagonism between medicine's (practical) interest in causal explanation and contemporary (theoretical) science denouncing such an interest as irrational has motivated a series of hotly disputed conceptual compromises. Cellular pathology depends for its explanations on the supposition of "dispositional properties", which was criticized as an (at that time) unscientific term. Theories integrating the cellular account of internal disease causes with the bacteriological account of external ones therefore treated "causes" more scientifically as "functional determiners" and replaced "explanations" with positivist "descriptions". In contrast to the still more radically positivist "conditionalists", the "causalists" retained the term "cause" as both intuitively and therapeutically useful, — and were accused of "mysticism". On the other hand, conditionalism accommodated positivistic determinism only by postulating an equivalence between events and the sum of their determiners. But therapeutic action requires that such disease-determiners be weighed. This scientifically "up-to-date" medical theory was therefore practically useless.

A still more penetrating case study is given by *Anne Fagot-Largeault*. Selecting as her prime example the development of Claude Bernard's medical thought, she shows how he must (first) choose between the positivistic restriction of laws to descriptions and medicine's need for etiological explanation. When facing the patient's unpredictable behavior towards disease-stimuli and therapeutic measures, he must (secondly) choose between maintaining the positivistic commitment to determinism and accepting the medical obligation to help, even while the determinants are still unknown. As Bernard opted for medicine, his alternatives reduced (thirdly) either to restricting practice to preventive endeavors on the ground of merely approximate causal knowledge (about what generally happens), or also including (individual) therapy at the price of facing an unlimited variety of unknowable idiosyncrasies. That is, for Bernard — and by Fagot-Largeault's implication also for his contemporaries — scientific positivism was only to be had at the expense of practice-orientation.

In contrast to Peset and von Engelhardt, Fagot-Largeault construes the conceptual muddle within nineteenth century medical theorizing neither as the result of (muddled) cultural influences (Peset) nor as the outcome of makeshift conceptual compromises (von Engelhardt). Instead, she proposes a conscious choice between the irreconcilably opposed interests of positivistic theory of science and of medicine. In her view, progress in medical science resulted from sticking with medicine's practical concerns. Thus, Bernard's ultimate opting for essentialism and his conceptual deviance (from theoretical science) was justified. It pointed, so she argues, in the direction of eventually sounder therapeutic progress in medicine. Nineteenth century medical knowledge did not become properly scientific insofar as it adjusted to the positivist pattern prevalent at that time. It became scientific by suffering the burden of anachronistic non-conformity and by conducting field medicine in that spirit. Her proof that this was the right decision is offered in terms of her own notion of modern medical scientificity: multicausal explanation of diseases, when applied to the epidemiological study of risk factors and subjected to statistical evaluation by means of mathematical models in a post-positivistic era, combines scientific respectability with practical effectiveness in both preventive and — so she claims — curative medicine.

Looking back at the issue of the first section, Fagot-Largeault's position can be taken to favor a praxological (instead of a Fleckian) interpretation of medical science's historical development. This line of thought is, however, not pursued by the two following authors.

Eric Juengst takes up the Aristotelian tradition, which Fagot-Largeault had recognized at the bottom of a praxis-oriented (albeit anachronistic) notion of medical causes. In Juengst's own (present-day) account of how to understand biological phenomena, this tradition is valued for its theoretical relevance. Moreover, the side-by-side existence of Aristotelian and positivist explanatory models in nineteenth century medical thought, which Fagot-Largeault had interpreted as irreconcilable opposition, appears to him as indicating a necessary complementarity. Both formal causes (merging into final) and efficient causes are needed for a full investigation of biological systems. Combining operational explanations (in view of goal-directed processes) and causal explanations (in view of mechanisms linking parts), such investigations have a legitimate place in medical theory as well.

In his view, the specific intertwinedness of heterogeneous concepts of causality in medicine, far from indicating any struggle between opposing theoretical and practical interests, presents an adequate conceptual framework for theorizing about the human body's functions and dysfunctions. Hence again, rather than arguing the need for a praxological interpretation of medical science, this intertwinedness suggests the particular aptness of medical theorizing to its natural subject matter. Nor is this account irresponsive to the concerns of medical practice.

Juengst's complementarity of explanatory models can be applied to different levels of a biological system's hierarchical organization. This permits the researcher to direct his inquiries towards his special purposes. Within Juengst's generally biological framework for functional analysis, specific pathological dysfunctions can be selected as the focus of functional research. Not only the practical interests typical of medicine as a whole, but also the more specific interests of researchers within the medical field are thus accounted for.

Quite obviously, Juengst's approach differs from that of Tsouyopoulos and Fagot-Largeault. All three authors examine how medicine's practical concerns enter into the shaping of medical knowledge. Juengst starts out with a biological theory primarily designed for doing theoretical justice to its objects of study. Only afterwards practical concerns are seen to guide further research at particular levels of the conceptual hierarchy. By contrast, Tsouyopoulos and Fagot-Largeault envisage a medical theory the very fundamental concepts and relational patterns of which are accepted or chosen with regard to such practical concerns. (See also Engelhardt in [14], to whom Juengst refers.) Let us call the former (i.e., Juengst's) understanding an "applied science" manner of conceiving medical theory's practical orientation, and reserve the term "praxology" for the latter understanding.

The final essay in this section seems again to favor the "applied science" sense. *Anne Marie Moulin* admits with Fagot-Largeault (and Juengst) the importance of multi-factorial analysis in medicine. Yet with Juengst (and unlike Fagot-Largeault), she wishes to salvage the scientific rationality (in the "theoretical" and "natural science" sense of the term) of multifactorial thinking. While Juengst had pursued this task by focussing on the heterogeneous notions of "cause" involved in such thinking, Moulin takes issue with the concern for individuality and the restriction to mere probabilism, both of which characterize Fagot's account of medicine's (praxological) scientificity. And while Juengst had attempted to salvage the theoretical fittingness of that heterogeneity to the biological object of study, Moulin attempts to salvage the nomological character of scientific laws for medical science. Immunology, in her view, will eventually permit physicians to "decode" individual idiosyncrasies in getting ill or responding to therapy. Its present rapid development entails the promise of explaining these idiosyncrasies in terms of general scientific laws. Moreover, what appears as a mere multiplicity of causes for disease today will increasingly become projected onto the plane of specific immune responses to specific external or internal conditions. It can thus be integrated into one comprehensive explanatory system.

The first two sections so far considered deal with medical theory and its either external or internal relationship to medical practice. But whether this relationship is (externally) derived from mere application or (internally) from that theory's very praxological orientation — in either case no support for any claimed uniqueness of the medical endeavor has so far been gained.

The next two sections focus on medical practice, as it in turn is guided by medical science[16]. Here the issue of medicine's unique status will be addressed in terms of the question to what extent medicine can be considered a technology, and whether the "art" of medicine is merely a preparatory stage in reaching its eventual technologization.

The third section concerns "art" and "intuition" as specific features entering into medical decision-making. Does medical practice, in resting on such decisions, differ from other forms of practice?

Wolfgang Wieland traces the historical roots of the concept "art" in medicine beyond its nineteenth century reduction to the aesthetic, and thereby merely subjective, realm. With the ancient notion of prudent practice, medicine, instead of occupying an esoteric rank, served as a model for all kinds of practice. Yet these practices, in depending on prudence (skill), were still eminently person-bound. By contrast, technology, the prime model of modern knowledgeable practice, is eminently non-person-bound.

Thus Wieland explores the present-day relevance of the "art of medicine" by examining at which points the physician is still "personally" needed, and where he will soon be replaced by technical staff. Till now, he argues, the physician's art is still required to bridge the gap between general laws and individual cases. Medical knowledge provides procedural rules which determine, for example, that if a patient exhibits certain symptoms, then certain diagnostic or therapeutic measures should be taken. But the physician must first determine whether the complaints offered by his patient and the results of his examination warrant the statement that he indeed has the required symptoms. So in order for science (and technology, we might add) to render medical practice rational in the sense of "open to objective evaluation and control", the (notwithstanding Moulin) persisting gap between universal laws and individual cases must be bridged by the physician's art. Thus the art of medicine furnishes the condition for the possibility of rendering practice scientific, or rational. It also determines for each stage in the development of medical science (and technology) the limits of that rationality. Wieland predicts, however, that further technological developments will render the physician's tasks increasingly trivial.

Rudolf Gross relativizes Wieland's prediction. Even the scientification of medical data does not render the physician's judgement superfluous. For Wieland, quantification of data had presented a model case for the ways in which the physician's art becomes dispensable, since here the subsumption problem disappears. Gross, however, shows how those data need again to be interpreted in qualitative terms such as "normal" or "abnormal", "significant" or "insignificant". Not only the application of general rules to specific cases, but already the interpretation of the rules themselves requires the physician's intuitive judgement.

Moreover, Gross challenges the view that science and technology only serve to make medical practice more rational. Wieland already had pointed to the fact that the amount of growth in medical knowledge far exceeds the amount of growth in knowledge that can be put to practical use. While absolutely speaking the rationality of medical practice increases all the time, proportionally it decreases. Gross points to an even weightier aspect of that problem. A new kind of uncontrolled empiricism is emerging from experimental use of new technical devices. The amount of know-how grows in this area as well, but as it exhibits no systematic order, the "rationality" it can confer to practice is at best atomic. The role of technology must thus be reassessed. On the one side, it both utilizes scientific knowledge and cultivates means-ends-adjustments, and these two rather different functions may cooperate in rendering practice rational. On the other side, in attending to very particular instrumental relationships, it also disrupts any existing coherence of knowledge and thus carries the risk of rendering practice less rational.

As a result of Gross' deliberations, the physician's judgement is needed at all levels of medical decision-making. Such judgements, Gross argues, can not be deduced according to any of the available decision procedures alone. They involve, instead, an "intuitive capacity" which physicians acquire in the course of extended clinical experience.

With this new term, Gross seems to posit a distinguishing characteristic of medical practice. He may, thus, be taken to open the possibility of physicians claiming privileged access to a somewhat esoteric knowledge. Outsiders would lack the training and experience necessary to evaluate medical decision-making. Perhaps this implication of Gross' own very circumspect remarks explains why both Spicker and Moseley have concentrated on opposing any irrationalist conclusions that an appeal to intuition might encourage.

Stuart F. Spicker adduces the work of A.R. Feinstein in demanding that "mere intuitions" should be rendered superfluous through the development of ever more specific scientific guidelines. Medical knowledge must be made to encompass ever more detailed and comprehensive aspects of practice. Even the language of patient complaints must become translatable according to objective criteria. In particular, Spicker rejects Gross' attempt at a psychology of knowledge justification for his claim that intuitions are intrinsically indispensable in medicine.

Ray Moseley supports this rejection by distinguishing three meanings of "intuition". He argues that only the third, which is really a very swift inference mistaken for an intuition, has a legitimate place in medicine. Hence, while there certainly is a need for further research into the justifiability of medical decisions, these may even now not be as inexplicable as is believed by those who make them.[17]

Whereas Gross has described the actual present state of scientific and technological knowledge relative to the physician's practice, Feinstein, as quoted by Spicker, takes the *desideratum* of improved scientific and technological rationality of practice as his starting point. From there, he deduces the changes in medical knowledge which are required if this *desideratum* is to be fulfilled. From this viewpoint, neither Gross' reliance on the intuition of the experienced practitioner, nor the "defensible sense" of intuitions as reconstructed by Moseley, can be tolerated in the long run. Reference to knowledge must be made conscious and explicitated in detail — if not in an emergency situation itself, then at least in retrospect for critical evaluation.

As a result, the discussion of medical decision-making in the third section finds the four authors so far considered to a certain extent unanimous as to the goal of an ever more invasive technologization of at least this aspect of medical practice.

Raphael Sassower takes up the philosophy of science issue discussed in the first two sections of this volume. He emphasizes technology's role in shaping not only medical practice, but also the very conceptual models underlying medical science. In his view, therefore, medicine presents a uniquely compelling example of that very intertwining of science with technology, which characterizes other supposedly "pure" science and technology fields as well. Pointing to recent developments in genetic engineering, Sassower shows how new technologies may influence even our view of medicine's proper goals. Is everything that can be technologically accomplished also desirable in view of medicine's traditional value commitments?

Of the four authors previously considered, only Wieland introduces a *caveat*. At the end of his paper, he intimates that patients might not accept a "medicine without physicians". This implies that the question as to the person-boundness of medicine, which he had raised in the context of the reliability of medical judgement, and thus the question of medicine's special status, should be pursued in other quarters.

This double question lies at the center of the fourth section, in which the grounds and limits of the medical profession's legitimate authority and its obligation towards humaneness are examined.

That medicine is a changing field was affirmed already in the first two sections, where progress in medicine was examined as to its theoretical or praxological conditions. Similarly in the third section, the growing technologization of medicine was exposed. In all three cases, such progress was taken to be desirable. Two of the authors in the fourth section, by contrast, take up present public criticism of medicine. They show how the purely scientific and technological progress in this field has introduced other changes which are not desirable in view of medicine's proper goal, the care of patients.

Both authors, *Stephen Toulmin* and *Hans-Martin Sass*, agree in blaming these undesirable side effects at least in part on medicine's all too pervasive orientation towards science. To be sure, both proceed from different points of attack. For Toulmin, physicians' preoccupation with medical science must be corrected by a regard for historiology (as defined in the works of Giambattista Vico). For Sass, medical science's hidden value implications must be recognized and subjected to public discussion and a resulting adjustment. But as they both agree in taking the definition of medicine's proper goal from medicine's ethos, they both call for a thoroughly deontological rendering of medicine's (in their view) properly praxological orientation.

Toulmin frames his criticism in terms of an *argumentum ad hominem.* If medicine is understood as a technology today, and if the notion of technology implies that goals are realized through knowledgeable practice, then the goal of proper patient care requires that practice should go beyond the narrowly technological. Diagnostic and therapeutic decisions are adequate only if they rest on an adequate evaluation of the "medical situation". The only way to understand that situation is to attend to patients' wishes and complaints, that is, to enter into interpersonal sympathetic communication. Hence, physicians should rediscover history-taking as their central responsibility, and should consider personal involvement as the necessary condition for meeting that responsibility. Toulmin acknowledges the difficulty of this task. The Cartesian tradition which underlies scientific thinking encourages a radical separation of the physical (as the domain of scientific examination and control) and the mental (as remaining outside). When accepted within medicine, this tradition is responsible for the prevalent restriction of physicians' attention to the somatic aspects of what in reality are psychosomatic disease-complexes.[18]

The insight that such restriction renders medical services less effective in view of overall patient well-being is of course shared by writers who have limited their view to the "matter of fact" side of medicine. Thus Juengst's method for investigating biological phenomena is also claimed to permit inclusion of the psychic domain. He sees this domain as a simply more encompassing realm again accessible in terms of causal and functional laws. And thus Moulin indicates that immunological responses are comprehensible only in the context of psychological data. Yet in each of these cases, regard for the psychic is added only as an afterthought to an otherwise natural science approach. This leaves the manner in which this regard should be integrated into explanation and practice again — as Gross indeed claims — to the intuition of researchers and practitioners. Fagot-Largeault's account of the epidemological determination of risk factors is also neutral as to whether these factors are sought within the physical, environmental or psychic domain. Yet for psychic factors to figure within that determination, they must be somehow classified and enumerated — which is difficult to achieve in generally acceptable terms.

Toulmin is not satisfied with such makeshift adjustments of "somatic science". His objection to physicians' understanding their knowledge as "scientific" is also directed against the segmentation of patient care into a multiplicity of medical specialties, each dealing with only limited aspects of the patients' well-being. Moreover, this objection has a more fundamentally moral motivation as well. Cartesian science's ideal of and claim to objectivity fosters a mental attitude which leads physicians to regard their patients as "cases" (lack of sympathy and respect) and to employ them as research material (abuse of the moral privilege of science). The only way both to render medical services more effective and to safeguard more efficiently against unfair treatment is to dismiss the Cartesian standard of rational knowledge altogether.

Toulmin therefore, first, contrasts scientific with technical knowledge and specifies this, second, by having the latter include a double regard for (again) science and for history. The first contrast concerns intentions (regarding theory versus practice); the second duplicity concerns modes of knowing (the general versus the particular). Toulmin thus determines an epistemological framework for understanding what is special about medicine. This special nature regards medicine's "human side", which is, however, integrated into the notion of the technical "matter of fact" side itself. Moreover, this notion of "technical", as it is bound up with medicine's commitment to the patient's "good", realizes not merely instrumental but also ethical values.

Mary Ann Gardell Cutter, a third author of this section, extends Toulmin's thought in a more systematic manner. She investigates the role values play in medicine, through a framework involving an interplay among science, technology, and art, one that is not reducible to any one of the three dimensions. As a result, she is able to support and amplify Toulmin's view of the grounds and limits of the medical professions' legitimate authority (both epistemically and non-epistemically speaking) and its obligation toward patients.

For Sass, the shortcomings of present-day medicine derive from the fact that traditional theory of science restricts objectivity and hence rationality to statements about facts. Value issues are thereby allotted to merely subjective whims, and thought unresponsive to reasonable treatment. In the case of medicine, such a misconception has particularly harmful consequences. After all, so Sass argues, medical theory as well as practice rests on notions like "health" and "disease", "adequate care" and "life", all of which are value-laden. In addition, medicine as a whole is a social institution and is therefore vulnerable to legal, political, and bureaucratic influences. These influences in turn depend on societal value choices and thus partake in society's historical change. Failure to reflect on this dependency, in Sass' view, has led to the present-day decay in medicine's ethos. Neither the obligation to further the art of medicine, nor the obligation toward the individual patient is properly met any more. Bureaucratic and legal intrusions have ruined the profession's moral integrity. Secularization and value pluralism, together with a pervasive scientification and technization of life and the existence of a publicly funded health care system, have encouraged consumer attitudes and a shunning of responsibility on all levels of society.

Looking back at the first section of this book from this perspective, the Fleckian account of medicine's merely passive receptiveness with respect to cultural change represents just an undesirable state of affairs. As a remedy, Sass demands a dialogue between society and the medical profession, aiming at more conscious and more generally accepted value choices. The profession should actively influence the social setting within which it must function. The traditional narrow "fine art of healing", which is defined in terms of medicine's ethos, must be expanded. It also needs to cover a commitment to scientific and technological progress in medicine, a commitment to freeing the physician-patient relationship from external intrusions, and a commitment to educating the public about lifestyle health risks.

Whereas the first three sections of this volume focussed mostly on the "matter of fact" side of medical theorizing and decision-making, the fourth section, in attending to medicine's ethos, emphasizes its "humane" side. Thus, while the first three sections left undecided the question whether medicine should be understood as a science, which gets merely applied in practice, or whether already its theoretical part should be interpreted in a praxological sense, the authors of the fourth section clearly favor the latter view.

The concluding essay by *Corinna Delkeskamp-Hayes* attempts to describe this result in greater detail. Taking up the initial question concerning medicine's place between "science", "technology", and "art", she devises a rational reconstruction of the very heterogeneous arguments put forth in this volume. In the course of this reconstruction, she derives a number of statements about various senses in which different sides of the three title terms' meanings permit an understanding of the various aspects of the medical endeavor. Taken together, these statements establish a more comprehensive "praxological" account of medicine's special nature. Thus, the theory-of-science discussions occupying the first sections of this volume are systematized in such a way that the political implications drawn out in the volume's last section can also be accommodated.

International Studies in Philosophy and Medicine
Freigericht, Germany

NOTES

[1] Given the vast array of modern medical specialties and institutional settings for practice (not even to speak of agencies for preventive or epidemological issues), one hesitates to use a single term to cover the whole ground. Nevertheless it will be supposed here that the term "medicine" has an identifiable meaning despite the heterogeneity of objects it denotes, and it will be attempted to map the complexity of that meaning.
[2] There are two journals which discuss particular aspects of related problems: The *Journal of Medicine and Philosophy*, and *Metamedicine* (formerly *Metamed*). But systematic investigations about that which distinguishes medical theorizing and practicing from other ways of theorizing and practicing are scarce. A good summary of the literature on epistemological issues in medicine up to 1980 is found in [5]: see also [42].
[3] This inconsistency also bears witness to a fundamental cultural cleavage which hampers present-day attitudes toward science and technology in general. The "modern" belief in scientific and technological progress and its societal utilizations as a sure road for serving humanity is shaken as the ambivalence of that progress reveals itself. (See, for example, [49]; [36]; [40], p. 70ff; and [67]). Relief is sought in several rather incompatible ways: Some recommend modifying the technological model itself in the direction of an oeco-technology (as in [37]). Others endorse an individual rights foundation for social frameworks in which scientific and technological progress can be linked to persons' autonomous decisions [13]. Still others hope that the revival of some pre-modern wholistic worldview will permit humans to feel once more securely embedded in society and nature [24]. As for medicine, certain aspects of these different construals of "post-modernity" are reflected in the contrast between contractarian [14] and value-ontological accounts of the physician-patient encounter [47]).
[4] There is also, especially in Germany, a rich tradition of anthroposophical criticism of scientific medicine. See, for example, [9].
[5] Physicians sometimes accuse medical experts of exaggerating the scientific standing of medicine when testifying in courts (M. Arnold in [20], p. 17). They also claim that unreasonable expectations concerning what science in medicine "can do" are responsible for much of the public's dissatisfaction ([19], p. 11).
[6] A still worthwhile discussion of technology and related concepts along with their relationship to science can be found in [27]. It is conceptually superior to the more recent and thematically more pertinent essays on technology in medicine in [44]. On the interrelation of technology with science, see [1].

In the English language, "technology" is used to cover both German terms "Technik" and "Technologie". The definition by Mitcham (in [31], p. 282), where technology is rendered as "the making and using of artifacts", merely addresses one particular aspect of "Technik". In the context of this introduction, it makes sense to understand by "technology" both a "theory related to practice" and a "practice related to theory". I shall reserve the more modest term "technical" to cases where the theoretical aspect is negligible.
[7] See the definition "knowledge of how to fulfill certain human purposes in a specifiable and reproducible way" in [8].
[8] Among the "customers" of the health professions, one finds not only patients but also employers, unions, insurances, and the state ([45], p. 76).

[9] See, for example ([7], p. 8). An example for exceptions is [34].
[10] See R. Kulhorn in ([20], p. 15) and H. Heuser-Schreiber in ([20], pp. 19, 23).
[11] Persuasive arguments for the already medical necessity of cost-benefit analyses are given in ([33], p. 284f) and ([41], p. 294).
[12] The matter is actually a bit more complicated. The corporate organization of the medical profession can be interpreted in various manners as either representing "modern" or "pre- (post-) modern" principles. "Professional independence" in the sense of Freidson indicates an anachronism for any modern view about power requiring democratic justification (see [46], [50]). On the other hand, "professional organization" in present-day West European countries also constitutes an organizational resource which not only — in an immediate sense — encourages superior professional performance and thus serves the interests of society in good medicine, but which also — in a mediate sense — can be used by the (modern) state for serving additional, extra-medical societal goals.

Whereas in the first case "the profession" is seen as an agent in its own right defending the interests of its members against the interests of society, in the latter case, it appears as an agent of the state who uses the threat of further legal intrusion for enforcing that collective self-control which serves its own (state) interest against that of individual physicians (see [26]).

But even this juxtaposition does not go unchallenged. Baier ([3], [2]) has argued against the acceptability of this second interpretation: Corporate bargaining policies are so difficult to unveil, that non-medical individuals have no means of defending their interests. Hence, Baier argues, no manner in which the state may try to take advantage of such organizational resources is democratically legitimate. In his view, then, even the modern employment of pre-modern structures is unacceptable.
[13] Accordingly, it is frequently observed that corporate structures in general and the corporate organization of medicine in particular discourage innovation. (See, for example, [23], p. 290f.)
[14] Accordingly, one of the most powerful German professional organizations (the Hartmannbund) interprets this transformation as resulting from an unwelcome infringement on professional independence [7]. Such infringements are perceived not only in increased regulatory activity, but also in exposure to market forces. Professional freedom is demanded both through independence from outside scrutiny, and in security from the more invasive economic risks. An area sheltered from social, legal, political, and also monetary pressures is thus demanded (cf. [5]).
[15] There exist of course economic analyses of the economic uniqueness of medical services (for example, [4]). But their relevance for policy decisions is disputed. (See, for example, [32].)
[16] Cf. the term "cognitive art" in [47], but a much more sophisticated analysis is sketched in terms of a "science of action" in [21] and [22].
[17] It is interesting to observe that whereas "intuition" is little regarded by conceptual analysts of the "knowledgeable practice" of medicine, its function for achieving holistic assessments concerning the situational context of economic practice presently enjoys a more favorable reception (see, for example, [48]). But then, among the traditionally down-to-earth managers, there may not be as much reason for fighting obscurantism as there is among physicians.

[18] Cf. the concern about "psychosomatic speculations" compromising the empirical status of medical science in [5]; [28], p. 7; [42], p. 100ff.; and [30].

BIBLIOGRAPHY

1. Agassi, J.: 1980, 'Between Science and Technology', *Philosophy of Science* **47**, 82ff.
2. Baier, H.: 1987, 'Benötigen wir eine Ethik der Medizin? Der Freiraum der Ärzte zwischen Moral, Politik und Recht', in L. Bress (ed.), *Medizin und Gesellschaft. Ethik, Oekonomie, Oekologie*, Springer Verlag, Berlin, p. 131ff.
3. Baier, H.: 1988, 'Gibt es eine Ethik des Sozialstaats?', in G. Gäfgen, (ed.), *Neokorporatismus im Gesundheitswesen*, Nomos Verlagsgesellschaft, Baden-Baden, p. 231ff.
4. Balthasar, R.: 1977, 'Ökonomische Aspekte zur Kostenexplosion im Gesundheitswesen', *Metamed* **1** (1,2), 43–54.
5. Bauch, J.: 1979, 'Psychosomatik als Paradigma', *Medizin, Mensch, Gesellschaft* **4** (4), 199ff.
6 Bauch, J. and Ropohl, G.: 1985, *Die unvollkommene Technik*, Suhrkamp Verlag, Frankfurt.
7. Bourmer, H.: 1975, 'Zur Einführung, Gedanken zum Selbstverständnis des Arztes am 75. Jahrestag des Hartmannbundes', in H. Schadewaldt, *et al.* (eds.), *75 Jahre Hartmannbund*, Verband der Ärzte Deutschlands, Bad Godesberg.
8. Brooks, H.: 1981,'Technology, Evolution and Purpose', in T.J. Kuehn and A.L. Porter (eds.), *Science, Technology and National Policy*, Cornell University Press, Ithaca, New York, p. 35ff.
9. Buchleitner, K.: 1977, 'Die Medizin in der Krise', *Die Kommenden* **3**, p. 5ff, 4, p. 5ff.
10. Cassel, D. and Henke, K.-D.: 1988, 'Reform der gesetzlichen Krankenversicherung in der Bundesrepublik Deutschland zwischen Utopie und Pragmatik: Kostendämpfung als Strukturreform?' in H.-M. Sass (ed.), *Ethik und öffentliches Gesundheitswesen*, Springer Verlag, Berlin. p. 13ff.
11. Deneke, J.F.V.: 1989, 'Sind die Freiberufler noch frei?', *Hessisches Ärzteblatt* **6**, 330.
12. Deppe, H.-U.: 1987, *Krankheit ist ohne Politik nicht heilbar*, Suhrkamp Verlag, Frankfurt.
13. Engelhardt, H.T., Jr.: 1991, *Bioethics and Secular Humanism: The Search for a Common Morality*, Trinity Press International, Philadelphia.
14. Engelhardt, H.T., Jr.: 1986, *The Foundations of Bioethics*, Oxford University Press, New York.
15. Engelhardt, H.T. Jr. and Erde, E.L.: 1980, 'Philosophy of Medicine', in P.T. Durbin (ed.), *The Culture of Science, Technology, and Medicine*, The Free Press, New York, p. 367ff.
16. Ferber, C. von: 1988, 'Gesundheitsverantwortung und Gesundheitsfinanzierung', in H.-M. Sass (ed.), *Ethik und öffentliches Gesundheitswesen*, Springer Verlag, Berlin, pp. 113–134.
17. Gäfgen, G.: 1988, 'Gesundheitspolitik als Verteilungspolitik', *Medizin, Mensch, Gesellschaft* **13**, 95ff.

18. Goeckenjan, G.: 1988, 'Wandlungen im Selbstbild des Arztes', *Medizin, Mensch, Gesellschaft* **13**, 41ff.
19. Gross, R.: 1976, *Zur klinischen Dimension der Medizin*, Hippokrates Verlag, Stuttgart.
20. Hrycyk, J. (ed.): 1988, *Medizin und Gesellschaft: Naturwissenschaft, Technik und die ärztliche Kunst; Verantwortung von Arzt und Pharma-Industrie für das Wohl des Patienten*, Schering Aktiengesellschaft, Berlin.
21. Hucklenbroich, P.: 1981, 'Action Theory as a Source for Philosophy of Science', *Metamedicine* **2** (1), 55–73.
22. Hucklenbroich, P.: 1983, 'Therapie und Handlungstheorie', in R. Töllner and K. Sadegh-zadeh (eds.), *Anamnese, Diagnose und Therapie*, Burgverlag, Tecklenburg, p. 201ff.
23. Knappe, E.: 1988, 'Neokorporatistische Ordnungsformen als Leitbild einer Strukturreform im Gesundheitswesen?', in G. Gäfgen (ed.), *Neokorporatismus im Gesundheitswesen*, Nomos Verlagsgesellschaft, Baden Baden, p. 271ff.
24. Koslowski, P.: 1988, *Die postmoderne Kultur*, C.H. Beck'sche Verlagsbuchhandlung, München.
25. Kurzrock, R.: 1978, 'Vorwort', in R. Kurzrock (ed.), *Medizin, Ethos und soziale Verantwortung*, Colloquium Verlag, Berlin.
26. Lehmbruch, G.: 1988, 'Der Neokorporatismus der BRD im internationalen Vergleich und die "Konzertierte Aktion" im Gesundheitswesen', in G. Gäfgen (ed.), *Neokorporatismus im Gesundheitswesen*, Nomos Verlagsgesellschaft, Baden Baden, p. 11ff.
27. Lenk, H. and Moser, S. (eds.): 1973, *Techne, Technik, Technologie, Philosophische Perspektiven*, Verlag Dokumentation, Pullach.
28. Lüst, R. *et al.* (eds.): 1986, *Beobachtung, Experiment und Theorie in Naturwissenschaft und Medizin*, Verband der Gesellschaft Deutscher Naturforscher und Ärzte, Stuttgart.
29. Mattheis, R.: 1988, 'Macro-Allocation in Health Care in the Federal Republic of Germany', in H.-M. Sass and R.U. Massey (ed.), *Health Care Systems*, Kluwer Academic Publishers, Dordrecht, p. 201ff.
30. *Medizin, Mensch, Gesellschaft* **4** (4), 1979.
31. Mitcham, C.: 1980, 'Philosophy of Technology', in P.T. Durbin (ed.), *The Culture of Science, Technology and Medicine*, The Free Press, New York, p. 282ff.
32. Oberender, P.: 1988, 'Zielvorstellungen zum Gesundheitswesen', in H.-M. Sass (ed.), *Ethik und öffentliches Gesundheitswesen*, Springer Verlag, Berlin, p. 267ff.
33. Pfaff, M.: 1988, 'Micro-Allocation in the Health-Care-System: Fiscal Consolidation with Structural Reforms?', in H.-M. Sass and R.U. Massey (eds.), *Health Care Systems*, Kluwer Academic Publishers, Dordrecht, p. 267ff.
34. Rie, M.: 1988, 'The American and West German Health Care Systems: A Physician's Reflection', in H.-M. Sass and R.U. Massey (eds.), *Health Care Systems*, Kluwer Academic Publishers, Dordrecht, p. 75ff.
35. Rohde, J.J.: 1968, 'Die Selbsteinschätzung des Arztes und seine Einschätzung in der modernen Gesellschaft', in H. Kampen-Haas (ed.), *Soziologische Probleme medizinischer Berufe*, Westdeutscher Verlag, Köln.
36. Rohrmoser, G.: 1984, 'Zur neuren Problematisierung des Fortschritts', in H. Kleinsorge and C.E. Zoeckler (eds.), *Fortschritt in der Medizin, Versuchung oder Herausforderung?*, TM Verlag, Hameln, p. 11ff.

37. Ropohl, G.: 1983, 'Der Mensch und die Wissenschaften vom Menschen', *Die Beiträge des 12. Kongresses für Philosophie in Innsbruck 1981*, Solaris Verlag, Innsbruck, p. 99ff.
38. Sadegh-zadeh, K.: 1980, 'Wissenschaftstheoretische Probleme der Medizin', in J. Speck (ed.), *Handbuch wissenschaftstheoretischer Begriffe*, Vandenhoeck und Ruprecht, Göttingen, p. 406ff.
39. Sadegh-zadeh, K.: 1983, 'Medizin als Ethik und konstruktive Utopie', *Medizin, Ethik, Philosophie* **1**, 1ff.
40. Schipperges, H.: 1980, 'Entwicklungstendenzen der modernen Medizin und neue Prioritäten', *Medizin, Mensch, Gesellschaft* **5**, 68ff.
41. Schoene-Seifert, B.: 1988, 'Micro-Allocation in the Health Care System: Is and Ought', in H.-M. Sass and R.U. Massey (eds.), *Health Care Systems*, Kluwer Academic Publishers, Dordrecht, p. 293ff.
42. Schulte, F.: 1978, 'Die ärztliche Entscheidung', in R. Kurzrock (ed.), *Medizin, Ethos und soziale Verantwortung*, Colloquium Verlag, Berlin, p. 97ff.
43. *SGB V, Gesetzliche Krankenversicherung*, 1989, Deutscher Taschenbuch Verlag, München.
44. Silomon, H. (ed.): 1983, *Technologie in der Medizin, Folgen und Probleme*, Hippokrates Verlag, Stuttgart.
45. Spicker, S.F.: 1988, 'Rechte, Ansprüche und Rationalisierungen im Gesundheitswesen', in H.-M. Sass (ed.), *Ethik und öffentliches Gesundheitswesen*, Springer Verlag, Berlin, p. 65ff.
46. Streit, M.E.: 1988, 'Neokorporatismus und marktwirtschaftliche Ordnung', in G. Gäfgen (ed.), *Neokorporatismus im Gesundheitswesen*, Nomos Verlagsgesellschaft, Baden Baden, p. 33ff.
47. Thomasma, D.C. and Pellegrino, E.D.: 1981, 'Philosophy of Medicine as the Source for Medical Ethics', *Metamedicine* **2** (1), 5ff.
48. Ulrich, H. and Probst, G.J.B.: 1988, *Anleitung zum ganzheitlichen Denken, Ein Brevier für Führungskräfte*, Verlag Paul Haupt, Bern.
49. Zimmerli, W.C.: 1982, 'Prognose und Wert: Grenzen einer Philosophie des Technology Assessment', in F. Rapp and P.T. Durbin (eds.), *Technikphilosophie in der Diskussion*, Friedrich Vieweg und Sohn, Braunschweig, p. 139ff.
50. Zweifel, P. and Eichenberger, R.: 1988, 'Der Korporatismus der niedergelassenen Ärzte', in G. Gäfgen (ed.), *Neokorporatismus im Gesundheitswesen*, Nomos Verlagsgesellschaft, Baden Baden, p. 159ff.

PART I

MEDICAL FACTS AND SCIENTIFIC PROGRESS: THE SCIENTIFIC STATUS OF MEDICAL KNOWLEDGE

LOTHAR SCHÄFER

ON THE SCIENTIFIC STATUS OF MEDICAL RESEARCH: CASE STUDY AND INTERPRETATION ACCORDING TO LUDWIK FLECK

This essay devotes itself exclusively to the presentation and analysis of Ludwik Fleck's *Genesis and Development of a Scientific Fact* [4]. Although this work has been given little recognition, Fleck's work has significantly influenced discussions in the philosophy of science over the past two decades. Such influence is evident, for example, in the writings of Thomas S. Kuhn ([9], [10]).[1]

The writings of Ludwik Fleck deserve attention for numerous reasons. The traditional philosophy of science represented by Carnap, Hempel, and Reichenbach was primarily concerned with the natural sciences. Further, it was concerned with the formal aspect of these sciences in such a way that the empirical sciences were treated more or less on the model of formal systems like mathematics and logic. Occasional illustrative examples were drawn from medicine to substantiate a point.[2] A systematic study of the special characteristics of medical science had not been undertaken until Fleck.

Fleck challenges the thesis that the deductive-nomological model of explanation sufficiently accounts for the nature of the scientific method. All sciences, including medicine, were thought to conform to this model of rationality. Particular kinds of medical research, e.g., those concerned with the infectious diseases or the acts of diagnosis and prognosis in connection with well-known diseases, may fulfill such standards of scientific rationality. However, what concerns Fleck is the implied assumption that medical research is 'scientific' only to the extent that it appears merely as a non-specific case of a scientific stereotype — not when it is typically medical in its techniques and procedures.

Fleck challenges the validity of this traditional understanding of scientific activity. In his view, traditional philosophy of science misinterprets the nature of science in two ways: (1) it limits itself to the consideration of finished results and does not take into account the process-character of scientific work; and (2) it considers science the business of single individuals. In criticizing these points, Fleck fashions a systematic analysis of science taken from the standpoint of medicine, an area in which he himself was

Corinna Delkeskamp-Hayes and Mary Ann Gardell Cutter (eds.), Science, Technology, and the Art of Medicine, 23–38.

trained and had worked for years. The following essay presents a summary of Fleck's discussion of the nature of medical 'facts', focussed on a case study of syphilis.

I. SYPHILIS AS A MEDICAL 'FACT'

The first major objection Fleck makes against the accepted interpretation of science involves its interpretation of the concept of 'fact'. The medical context does not easily lend itself to the positivist concept of a 'fact' as developed, for example, by Mach, Carnap, and Schlick. How, then, do facts of experience come about in medicine? In order to answer this question, Fleck investigates a phenomemon that was generally accepted as a fact in medicine, namely, the disease called 'syphilis', whose symptoms and development were generally known. Fleck shows how four lines of thought beginning in the fifteenth century are clearly distinguishable in the notion of 'syphilis'.

The first reference to syphilis can be traced to the early fifteenth century when it was understood as an "ethical-mystical" disease entity, the *Lustseuche* (disease of lust). This notion includes not only syphilis, but diseases such as gonorrhea, *ulcus molle* (a group of skin diseases that even today are viewed as non-specific), and non-venereal diseases such as gout. The venereal nature of syphilis was explained astrologically. The sidereal cause, a particular relation of Jupiter and Saturn in the house of Mars on October 25, 1494, was held responsible for the appearance of venereal disease. "Benevolent love surrendered to the evil planets Saturn and Mars and the sign of scorpio explains why the genitals presented the first point of attack for the new disease" ([4], p. 2). Astrology thus assisted in establishing the venereal character of syphilis as its first *differentia specifica.* The first basic understanding of syphilis as a venereal disease, i.e., *Lustseuche katexochen,* was influential during the centuries that followed.

The second important line of thought came from medical practitioners who learned that certain diseases could be separated from the larger group of chronic skin diseases according to how mercury ointments affected them. Diseases in this group were favorably influenced, and sometimes completely cured, by mercury treatment. As a result, a particular scabies — *scabies grossa* (one of the old names for syphilis) — was separated from the domain of skin diseases. With Fleck, one might call this conception of syphilis the "empirical-therapeutic" notion of syphilis. However, even this notion did not yet determine the specific disease entity. Until the nineteenth century, the claim was made that "sometimes mercury does not cure the carnal scourge, i.e., syphilis, but makes it even worse" ([4], p. 5). Mercury was actually effective only in the treatment of so-called "constitutional syphilis" (i.e., that which was in the stage of the generalized disease). The primary, properly venereal stage of syphilis, localized in the genitals, remained tied to the *Lustseuche* idea. These two approaches to the disease, which together described the "ethical-mystical" disease entity, developed alongside each other. As Fleck puts it: "Theoretical and practical elements, the a priori-magical and the purely empirical, mingled with one another" ([4], p. 5).

The so-called "experimental-pathological" — or, as Fleck would also say, "pathogenic" — disease entity emerged as the third major line of thought in the development of the notion of syphilis. This understanding resulted from the efforts of a group of physicians to demonstrate on the basis of an identity theory that almost all venereal diseases (e.g., gonorrhea, chancres) can at any time cause syphilis. They sought to show experimentally that gonorrheal infection could cause chancres, and that the latter could also cause the former. Another group of research pathologists working with a dualistic theory distinguished gonorrhea and the soft chancres from syphilis. The separation of the different pathologies proved to be extremely difficult. The general doctrine of the mixing of humors and the notion of the "corrupted blood of the syphilitic" became relevant in this context. The notion of the corrupted blood had been commonly appealed to in all cases of general illness. But while its significance decreased in the explanation of several other diseases, it increased in the case of syphilis.

It was the so-called "Wasserman reaction" that first brought scientific specificity to the notion of syphilis. Its findings fashioned a fourth major line of thought in the understanding of syphilis. The development of this line of thought — referred to by Fleck as the "serological-etiological" disease entity — is more complex than is usually thought. Out of the field of bacteriology, which had developed quite independently from concern with syphilis, a view of the causal agent, *Spirochaeta pallida*, emerged. The Wasserman reaction, though tied specifically to this disease, subsequently ushered in the development of a different science, namely, serology ([4], p. 15).

These four lines of thought, and their conflicting interaction over time, have produced the current specific concept of syphilis and have fostered its acceptance as a medical 'fact'. Only with this concept did it become possible to differentiate among the various stages of syphilis, and to distinguish this disease entity from other ones (e.g., gonorrhea) having similar signs and symptoms. As Fleck puts it: "*Spirochaeta pallida*, together with the Wasserman reaction, helped to classify the *tabes dorsalis* and progressive paralysis characterizing syphilis" ([4], p. 17). Since spirochaetes were found in the lymph system soon after infection, even the first stage of syphilis came to be viewed no longer as a localized disease but already as a stage of the general disease ([4], p. 17).

In response to this analysis by Fleck, one is inclined to advance the following objection: In past centuries, no medically useful concept of a fact that is free of mystical, religious, and sociological notions existed. Yet, with the full development of serology and bacteriology, the concept of infectious disease became clearly determined so as to be consistent with the positivist concept of 'fact'. Fleck can show, however, that this is not the case. Even after the discovery of *Spirochaeta pallida*, syphilis could not be *defined* as the disease caused by this particular agent. After all, some organisms have the agent in their blood without contracting the disease. The agent must therefore be seen as one symptom among others that characterizes the disease. With this subsumption of the agent under the symptoms, Fleck's opposition to the positivist preoccupation with distinguishing between disease-causing factors and disease-specific indicators becomes clear. His concern is to point out how bacteriological theories as well as particular presumptions regarding the nature of disease and health enter into determinations of a medical 'fact'. In short, medical facts are "theory-laden".

II. THE COLLECTIVE NATURE OF RESEARCH

Fleck's principle concern extends beyond the thesis of the theory-laden character of perceptions in general, and of medical observations in particular. He aims at establishing an anti-individualistic model of our claims to knowledge. Using his case study, Fleck demonstrates that the establishment of a scientific 'fact' involves factors not primarily based in the thought of individual researchers. The persistence of ideas bound to a society, an historical situation, or a culture, appears to Fleck to be as important as, or more important than, the intentions, experimental techniques, and actual work of the individual researcher. In short, Fleck concludes from his case study of syphilis that science must be understood as a collective process. The Wasserman reaction, the domain in which research on syphilis finally achieved its medical scientific level, is for Fleck an instance illustrating the collaborative nature of research. Alternatively, the traditional philosophy of science errs in presenting science as an accomplishment of individuals, of great single scientific personalities.

That Schaudinn's recognition of *Spirochaeta pallida* as the causal agent of syphilis was heralded as a simple discovery constitutes, in Fleck's view, a misrepresentation. His recognition of the agent should have been seen as a collective process. According to Fleck's revised description, Schaudinn, on the basis of contemporary views about syphilis and about causative agents, suggested that *Spirochaeta pallida* should be recognized as the causative agent of syphilis. The significance of *Spirochaeta pallida* was duly accepted and used for further developing syphilology ([4], p. 40). Finally, the development of the Wasserman reaction appears as a complicated "intercollective" process. The very point of departure of Wasserman's work came from the outside. As Wasserman reports:

> The head of the Ministry, Friedrich Althoff, asked me to his office when Neisser had returned from his first expedition, and the French were far ahead in experimental biological research on syphilis. He suggested that I work on this disease to assure that German experimental research has a share in this field ([4], p. 68).

The appearance of the Wasserman reaction was, therefore, from the very beginning steered by external factors. In particular, the national contest, in an area viewed as very important politically, cannot be overlooked. In such a situation, the social pressure pushing scientific investigation forward is especially strong. That the effort belongs more properly to a community of thought than to an individual is illustrated by Fleck when he says:

> Owing to the controversy with Ehrlich, the instrument was supplied by Bordet and Gengou. Wasserman and Bruck perfected and expanded it. Because of rivalry with the French, Althoff mapped out the new territory and applied the necessary pressure. Neisser offers the pathological material and his experience as a physician. Wasserman as director of the laboratory was responsible for the plan, and Bruck as a colleague executed it. Siebert prepared the sera. Schucht, an assistant of Neisser's, produced the organ extracts. These are the ones whose names we know. But there certainly were many suggestions concerning technical manipulations, modifications, and combinations from others whom it is impossible to list. Citron subsequently decisively improved the dosing. Landsteiner, Marie, and Levaditi, among others, published the first practical method of preparing the extracts. Skills, experience in the field, and ideas whether "wrong" or "right" passed from hand to hand and from brain to brain. These ideas certainly underwent substantive change in passing through any one person's mind, as well as from person to person, because of the difficulty of fully understanding transmitted knowledge. In the end an edifice of knowledge was erected that nobody had really foreseen or intended. Indeed, it stood in opposition to the anticipations and intentions of the individuals who had helped build it ([4], p. 69).

Fleck refers in this context to the accomplishment of Columbus, who also set out with the completely different goal of discovering a sea passage to India, but achieved something much greater — the discovery of a new continent. Similarly, the goal that Wasserman and his colleagues eventually reached was by no means the original goal, but rather the fulfillment of an ancient communal desire: the proof of "syphilis blood".

The first efforts of Wasserman, Bruck, and Neisser in 1906 show quite clearly that they were concerned with the proof of a specific antigen. The proof of a syphilis *antigen* was actually the established goal. The proof of a syphilis *antibody*, though it produced only 15–20% positive results in the early studies and is mentioned only in passing, eventually became the actual discovery and accomplishment of the Wasserman reaction. Understanding how this initial 15–20% became 70–90% would not be possible without returning to all of the unnamed co-researchers who achieved the optimal adjustment of test results through technical tricks and expedients — e.g., a little more or a little fewer reagents, a little more or a little less reaction time, etc.

In Fleck's view, the fact that the Wasserman reaction unleashed great research activity may be understood in light of deep-rooted social needs and conditions. The prescientific notion of syphilis as an ethically-weighted *Lustseuche* may have had its strongest after-effects here. The effect of this special moral significance of syphilis on research activity cannot be overestimated. Fleck notes that tuberculosis, although it caused more harm than syphilis over time, did not receive great attention. In part, this was due to perception of it not as a cursed disease, but rather as a "romantic" one ([4], p. 78). A national competition over a socially unimportant disease is inconceivable; no ministry director would be able to arouse the best scientists of his nation to combat it.

Quite clearly, the "social atmosphere" — to use a Fleckian term — first motivates the more focused research group in its investigation. Subsequently, through a continually growing number of co-workers and their interaction, this group produces something like a collective experiment and then a communal, anonymous development of the reaction that raises the initially meager results to their eventual usefulness. Particularly important factors in this process were congresses, press reports about their results, and finally the legal ordinances concerning the application of the Wasserman reaction. Indeed, one must acknowledge the avalanche of serological work on syphilis. In a 1927 summary report on the serological diagnosis of syphilis, Laubenheimer cites 1,500 articles on this theme, although only newer articles were included. Adding these to the foreign language and lesser-known articles and the clinically-oriented essays that were not wholly taken into account, Fleck estimates the total number of articles at 10,000. Obviously, there are not many problems that have been so extensively treated. This indicates how inadequate any explanation of syphilis in terms of individual research interests would have been. Consider as well how primitive cancer research would be if not for the collective nature of medical research.

III. HOW DOES COLLECTIVE RESEARCH ESTABLISH ITSELF?

Fleck calls the community in which medical research on the etiology of syphilis was undertaken the "serological collective". Clues about its structure, who belongs to it, and what problems it pursues, may be gained by looking at the manner in which new members are introduced in this research field. Fleck speaks, for example, of the "initiation into the domain of the Wasserman reaction ... according to German ritual" ([4], p. 80) with examples drawn from the 1910 edition of what he calls the "catechism" by Citron, a student of Wasserman [1]. This book was used as a textbook even during the 1920s. Fleck identifies five factors in this introductory work that both characterize the community of serologists and permanently determine their activities. These five factors have nothing to do with any experimentally established results or methods. First of all, they include (1) a particular concept of infectious disease, inappropriately applied. Fleck refers here to the fact that even this supposedly experimental notion still involved the old demon of sickness, forcing itself — quite irrespective of rational grounds — upon scientists' ways of thinking. The underlying notion that an organism is a separate, self-sufficient unit with fixed boundaries that can become sick or ill only by invading, disease-causing factors from outside is exposed as fictitious. It cannot be sustained and was not a part of the better biological research of the time. Here after all, the idea of a dynamic system without any strict separation from the environment had already begun to take hold. Even in the first two decades of this century, infection was thought to be less the invasion of infectious agents into a "healthy" body than a very complex interaction of states within the organism. Thus, among the serologists, the concepts of disease and health remain on prescientific grounds.

Another concept Fleck criticizes in Citron's introductory work concerns (2) immunity, conceived in its classical sense. Many notions of the doctrine of immunity derive from what he calls the "epoch of chemical madness". Under the influence of great chemical successes in physiology, it was hoped that all of biology could be explained through the action of chemically defined substances. One appealed to "toxins, amboceptors, complements as chemical individuals, and to their opposites as antitoxins, anticomplements, etc." ([4], p. 63). But in using such expressions, the young doctrine of immunity undoubtedly employed a very primitive and defective model. Citron's introduction also exhibits (3) many other habits of thought Fleck considers no longer scientifically defensible — for example, the separation and respective weighing of *humoral and cellular factors*, or the notion of *specificity*. It also contains (4) methodological initiations — as, for example, into the role of the control method whose status as an independent authority is vastly overestimated. Finally, (5) there are some objectionable general doctrines, such as the view a) that understanding proceeds not through intuition, but through observations; b) that such observations aim at a diagnosis, that is, insertion into a system of distinct disease entities; and c) that therefore such entities exist and must be identifiable through analytic methods.

Fleck calls these five factors "constituents of the thought style of the serological collective" ([4], p. 64). They determine the direction of research and connect it with its tradition. Fleck does not consider such factors unnecessary. They are in fact viewed as the necessary conditions of research. *They have, however, little to do with truth or falsehood*: they were encouraging; they gratified; and they were generally accepted. They were then superseded, not because they were discovered to be false, but rather because ideas had changed and developed ([4], p. 64).

What appears here is a very complicated interaction among three factors: first, the individual researcher; second, what Fleck calls the community of thought (*Denkkollektiv*); and third, reality. These three factors are not independent. According to Fleck, the dependence of the so-called scientific 'fact' on the style of thought is particularly important. To this dependency, the discussion now turns.

IV. PATTERN-SEEING AS BOUND TO THOUGHT STYLE

Fleck takes Carnap's understanding of protocol sentences and of underlying concepts of strict acceptance of immediate experience to task. Based on bacteriological studies, Fleck demonstrates that such a thing as a presuppositionless observation simply does not exist. There are always arbitrary decisions at work as well as habitual thought styles (*Denkstile*) that predetermine the very properties available for description. Presuppositionless observations are psychologically impossible and logically nonsensical. Rather, observations enter into the understanding in two different ways, many intermediate stages being omitted. They enter initially as unclear visions and subsequently as developed, immediate pattern-seeing ([4], p. 93).

Immediate pattern-seeing doubtlessly presupposes experience in the particular realm of knowledge: only after many experiences, themselves preceded by theoretical introduction and practical instruction, will one have the ability simply to perceive something such as a univocal phenomemon.[3] Improvement of this ability, however, is accompanied by a decreasing capacity to realize falsifying instances. The disposition for "directed perception" is accompanied by the loss of an ability to perceive the heterogeneous. Therefore, Fleck holds, *the readiness for directed perception lies at the root of each mode of thought*. Pattern-seeing comes to be a matter of thought-styles.[4] Accordingly, the notion of having experience acquires an entirely new sense by entering into the analysis of modes of thought. The initial diffuse seeing, on the other hand, is still shapeless, undirected, and chaotic. Jumbled partial elements are grasped in it, and contradictory moods push the unfocussed view in different directions. The 'fact', i.e., the fixed object, is lacking at this stage. For Fleck, 'fact' means 'resistance' to the arbitrariness of thought.[5] Its relation to collective thought is threefold:

1. *Every fact must be in line with the intellectual interests of its thought collective*, since resistance (i.e., factuality) is possible only where there is striving toward a goal. Irrelevant states of affairs, which may correspond to some other (e.g., an aesthetic) interest, are not experienced as providing resistance and are thus not considered as (e.g., medical) facts.
2. *The resistance must be effective as such within the thought collective. It must be brought home to each member as both a thought constraint and a pattern open to immediate experience.*
3. *The fact must be expressed in the style of the thought collective*... ([4], pp. 101–102).

Consider an example taken from bacteriology. The doctrine of the "unchangeability of characteristics of the species" was deemed a simple truth, a thought constraint for the collective of bacteriologists after Pasteur and Koch. Instances of variability were simply not recognized. Two conditions are in effect here: (1) a denial of the old doctrine of variability, whereby a universal *coccobacteria septica* was said to transform itself into every possible shape (Billroth); and (2) the successful exercise of the methods used in the collective: namely, to innoculate only 24-hour-old cultures; very fresh and very old cultures were simply not investigated. Secondary changes in cultures were therefore not observed at all. Wherever any variation was occasionally observed, it was viewed as a mistake of the experimenter. Only when Neisser and Massini (1906) investigated cultures not only after 24 hours, but also on the following days, were variations established that could not be suppressed or concealed.

This example shows that the renewed appearance of variable thinking did not bring about a return to pre-Koch times. Rather it represents yet another change of modes of thought. The "charm of the harmony of illusions", as Fleck calls it, was broken, and further discoveries became possible that were inconceivable in the classical phase with its assumptions of invariability.

V. THE STABILITY OF THOUGHT COLLECTIVES OR THOUGHT STYLES

Another issue that Fleck confronts concerns the question: How can we understand the observed stability and persistence of "thought style"—or in Thomas S. Kuhn's words, a "paradigm" or "disciplinary matrix"? The hermetic compactness of the thought style and its "tendency to persist" are for Fleck two sides of the same sociologically-describable circumstance.[6] The separation of specialists from the general public, that is, the establishment of an esoteric circle distinguished from the uninitiated, provides the first identifying factor. Within such circles, graduated relationships and corresponding forms of communication come into play—e.g., between teachers and students. Fleck sees the teacher-student relationship as a reflection—on a personal level— of the relationship between the elite and the masses. It is borne by trust on the one side, and by a feeling of dependence on public opinion on the other. A similar intellectual dependence also characterizes the relationship among peers in the scientific community, for each is dedicated to "solidarity of thought in the service of a superindividual idea" ([4], p. 106). In what follows, Fleck confirms that each "intracommunal exchange of ideas" is governed by a special feeling of dependence.

> The general structure of a thought collective entails that *the internal communication of thoughts, irrespective of content and logical justification, should lead for sociological reasons to the strengthening of a thought structure.* Trust in the initiated, their dependence upon public opinion, intellectual solidarity between equals in the service of the same idea, are parallel social forces which create a social shared mood and, to an ever-increasing extent, impart solidarity and conformity of style to these thought structures ([4], p. 106).

Non-epistemic factors are seen then to play central roles in the fashioning and persistence of thought collectives.

VI. PROSPECTS FOR CHANGE IN THOUGHT STYLE

Given this structural tendency toward persistence and stabilization, how do change and progress in scientific research, the particular marks of science, come to pass? How do basic changes in the thought style — or as Kuhn calls them, 'paradigm shifts' — occur?

Every scientist belongs, in addition to his special intellectual group, to the public domain of everyday life. Generally, he will also be a member of some narrower or broader scientific community. *Tendencies toward change* follow, according to Fleck, from the modes of communication and dissemination of knowledge among individuals themselves, especially where they belong to different groups. As Fleck says: "The intercollective communication of ideas always results in a shift or a change in the currency of thought" ([4], p. 109).

This statement contains a hypothesis about the function of language and the relation of linguistic utterances to their meanings. *Fleck views language as an institution that not only makes possible the communication and thereby the reproduction of scientific knowledge, but also contributes to the development of that knowledge through the "misunderstanding" or shift of meaning that occurs in every communication.* The ideal language — or 'invariance of meaning' postulate — of the logical positivists was intended to prevent shifts of meaning. For Fleck, the violation of this postulate is important not only for ordinary language, but also for scientific language. Terms used in different thought collectives assume distinctions in meaning such that mutual understanding between members of different groups is difficult, if not impossible.[7] Fleck finds in the different linguistic representations of science an index both to its collective structure and to factors related to its dynamic character.

Public demand for information about the results of scientific investigation implies a tendency toward exoteric or publicly understandable presentation. Popularization even belongs, in a sense, to (specialized) science, and Fleck assigns to it an important function, no matter how critically it might be viewed by the scientist. Fleck here distinguishes three types of scientific literature (all of which serve the goal of popularization): journal literature, manual literature, and textbook literature; he correspondingly speaks of three different sciences: journal, manual, and textbook science ([4], p. 119).

Popular exposition for Fleck has an important function insofar as the common sense position presents itself in it. He calls this position the "personification of the everyday thought", which must be considered a "universal source for many particular thought collectives" ([4], p. 112). Fleck's expression, "popular exposition of science", is admittedly not an entirely happy one, since it is often taken to mean the watered-down, third-hand presentation of scientific results. However, for Fleck, it is a realm where common notions assume their very important role. So that notions from the everyday world will be needed in situations where a shift of theory occurs in order to identify objects and object domains for which different interpretations are available. And so, new discoveries become possible only when a thought constraint loosens, that is, when the meanings of terms change with their use in different thought collectives. In such situations, recourse to exoteric popular notions come strongly into play.

Fleck also assigns to popular knowledge the establishment of the ultimate norms, or as he calls them, the ideal values of certainty, simplicity, and vividness. In any case, the specialist gets his belief in these values as ideals of knowledge from the beliefs of popular knowledge ([4], p. 115). It is almost an essential characteristic of explicative presentation that the knowledge explicated becomes more and more exoteric or popular. After all, the institution of science as a societal activity quasi-necessitates maximal generalization, that is, general communicability.

Journal literature, according to Fleck, is characterized by a temporary and personal tenor. The fragmentariness of the problems, the randomness of the materials, and the particular technical set-up indicate this; the language of journal articles expresses this feature. Provisional and personal expressions dominate. The 'we' often found here is not plural *majestatis*, but rather a plural *modestiae*.

In contrast, the task of "manual science" is presentation in an ordered, complete system in which individuality disappears in favor of the solemnity of an impersonal, certain declaration. "Manual knowledge" does not consist in a mere summary of the individual findings published in journals. Rather, it is constructed out of these like a mosaic in which individual efforts constitute the building stones ([4], p. 128).

Between journal science and manual science, there exists a tension that makes scientific change tangible. On the one hand, the journal science strives to be accepted into the manual, as shown by its reference to the ruling thought style manifested in manuals. Thus, journal literature bears witness to the fact that only "intracollective exchange of ideas" can lead out of the temporary and tentative phase of research into that of objective certainty ([4], p. 121). "Each journal article contains in the introduction or in the conclusion such a reference to the manual science as proof that it seeks to enter the manual and views its present position as temporary" ([4], pp. 119–120).

Manual knowledge, on the other hand, remains dependent on the productive accomplishments of research; it always limps behind the protagonists of discovery. Within the tension between journal literature and manual science is the point where psychological and sociological factors effect changes in institutionalized research. According to Fleck, the plan of systemization (i.e., manual knowledge) takes shape

> through esoteric communication of thought — during discussion among the experts, through mutual agreement and mutual misunderstanding, through mutual concessions and mutual incitement to obstinancy. When two ideas conflict with each other, all the forces of demagogy are activated. And it is almost always a third idea that emerges triumphant: one woven from exoteric alien-collective, as well as the two controversial strands ([4], p. 120).

In manual knowledge, a thought style solidifies into a thought-constraint. In it, the normative force of science expresses itself. In it is determined

> what cannot be thought in any other way, what is to be neglected or ignored, and where, inversely, redoubled effort of investigation is required. The readiness for directed perception becomes consolidated and assumes a definite form ([4], p. 123).

The dynamism of science is thus indebted to the fact that science is an institutionalized, compartmentalized process that embraces the difference between the pinnacle of research and the standard communication of its results. Since there exists a need for communication among the various groups of scientists, this communication inevitably results in a shift of meanings that can assume a knowledge-broadening function. The origin of theoretical alternatives thus becomes explicable in view of the institutionalized forms of scientific research and their modes of exposition. The productive researcher must adhere to the standards of manual knowledge, and yet he also knows that the information in the manuals is always already surpassed by research. He bows to what is generally accepted, although he knows that the official doctrine of tomorrow will have to be devised from the primary research of today's journal literature.

VII. CLOSING REMARKS

As the foregoing illustrates, Fleck has provided a view of medical science that is different from the traditional one advanced by logical positivists. Fleck's concern is to show the inadequacy of those accounts of medical science that address the process of science and its collective nature. Kuhn's work underscores Fleck's contributions. Unlike Fleck, however, Kuhn works in the context of a general philosophy of science. Since the mid-sixties (at the latest), this new way of understanding the genesis and development of scientific knowledge had been widely discussed and hailed as the successor to the Hempel-Oppenheim model. Through Kuhn's work, Fleck's ideas join Popper's methodology of critical rationalism, providing the basis for "theory-dynamic reconstruction".

The influence of Fleck's views on the way in which scientific knowledge is construed should come as no surprise considering that Fleck concerns himself primarily with the nature of medical research. Medical research may be seen as a type of natural science research which endows medicine with its scientific status. Beyond this, however, it seems to be no accident that Fleck developed this new concept as a medical scientist concerned with understanding the nature of medical knowledge *per se*. Nowhere else is the interdependence of different scientific communities so fundamental and obvious as in medicine. From its anatomical and physiological investigations to research in microbiology or neurophysiology — almost all realms of natural science mingle with one another. In this interdependence among heterogeneous disciplines, medical research provides an excellent paradigm for the collective structure of knowledge.

Fleck's work is ony a starting point. Medicine's epistemological complexity along with its social relevance offers many stimulating opportunities for study in the philosophy of medicine and science.

Universität Hamburg
Hamburg, Germany

NOTES

[1] Kuhn mentions Fleck in the preface of his well-known *The Structure of Scientific Revolutions*, where he introduces the "Society of Fellows of Harvard University" as the sustainers of his research. There he writes: "That is the sort of random exploration that the Society of Fellows permits, and only through it could I have encountered Fleck's almost unknown monograph, *Entstehung und Entwicklung einer wissenschaftlichen Tatsache*, an essay that anticipates many of my own ideas ... Fleck's work made me realize that those ideas might require to be set in the sociology of the scientific community. Though readers will find few references to either these works or conversations below, I am indebted to them in more ways than I can now reconstruct or evaluate" ([9], preface, p. ix). Since *Kuhn does not make a single reference to Fleck's work* in his whole book, this relationship has until now not been investigated. This I wrote in 1977. For recent work see the introductions in [5] and [7], and esp. see [6].

[2] Semmelweiss's battle against childbed fever is also an example of the systematic testing of hypotheses in a context of scientific explanation in Hempel [8].

[3] "Experience" may not be identified with a knowledge of the relevant concepts, theories, and methods. It has to do with the acquisition and exercise of a practical skill and dexterity. To know how one maintains one's balance and how to walk does not yet

mean that one can walk a tightrope. An unmistakable element or individuality enters into science here that was found in earlier methodologies reserved to the creative researcher, or even the genius.

The ordinary routine of science appeared to be explainable in an "objective" way. Fleck's work makes clear that the training of experts requires not only the communication of "knowledge", but also a practical exercise of procedures that formerly seemed reserved for the arts and trades ([4]; see also [6]). Kuhn follows Fleck not only in his sociological perspective, but also in this pragmatic turn of the concept of science (see also note 5 below).

[4] Fleck is influenced by Gestalt psychology in his theory of perception. The process of discovery is explained by a sudden change of Gestalt. "Suddenly, one simply doesn't understand how the earlier Gestalt was possible and how the incompatible one could remain unnoticed" ([4], p. 118n).

[5] Fleck here employs the pragmatic conception of reality developed by C.S. Peirce and William James. Presumably Fleck was introduced to these ideas through Wilhelm Jerusalem, who made American pragmatism known in Vienna. In any case, Jerusalem is more often cited as the precursor of Fleck's own ideas than anyone else. Jerusalem, who with Durkheim, Levy-Bruhl, and others at the turn of the century, aided the breakthrough of the sociological perspective, translated and edited the Lowell lectures of James in 1908 that contained a long chapter on the pragmatic concept of truth ([11], p. xxviiff).

It seems to me a strange phenomenon that Kuhn can be celebrated as a metascientific revolution in the United States of the sixties and seventies, without seeing that in the critical points it is a question of the reimportation of Peirce's original ideas via James, Jerusalem, and Fleck.

[6] It is not hard to recognize here Kuhn's claim that "crises" are necessary, that paradigm destruction is a condition for innovation. Fleck's terminology certainly suggests a much stronger polemic and appraisal. This is not really Fleck's addition. Rather, Fleck accepts in his terminology those values that are ratified by the scientists themselves. When the habits of thought, the universally accepted assumptions, are outstripped and replaced by a new theory, there arises out of the belief in the progress and truth of the innovation, the claim that the supposed truths of its predecessor may have been a coherent set of ideas, but did not represent an unsurpassed truth. This Fleck expressed in his talk of a "charm of the harmony of illusions".

[7] Here Fleck addressed all of the problems pursued under the title of the incommensurability of theories by Feyerabend ([2], [3]) and then forcefully by Kuhn ([9], [10]).

BIBLIOGRAPHY

1. Citron, J.: 1907, *Die Methoden der Immunodiagnostik und Immunotherapie*, Leipzig.
2. Feyerabend, P.K.: 1970, 'Against Method. Outline of an Anarchistic Theory of Knowledge', *Minnesota Studies in the Philosophy of Science* **4**, 17–130.
3. Feyerabend, P.K.: 1965, 'Reply to Criticism. Comments on Smart, Sellars, and Putnam', in I. Cohen and M. Wartofsky (eds.), *Boston Studies in the Philosophy of Science* **2**, pp. 223–261.

4. Fleck, L.: 1979, *Genesis and Development of a Scientific Fact* (ed. by T.J. Trenn and R.K. Merton), University of Chicago Press, Chicago and London.
5. Fleck, L.: 1935, *Entstehung und Entwicklung einer wissenschaftlichen Tatsache*, Basel, Switzerland (new edition by L. Schäfer and T. Schnelle, Suhrkamp Verlag, Frankfurt/M., 1980).
6. Cohen, R.S., and Schnelle, T. (eds.): 1986, *Cognition and Fact: Materials on Ludwik Fleck* (Boston Studies in the Philosophy of Science, Vol. 87), Dordrecht.
7. Fleck, L.: 1983, *Erfahrung und Tatsache: Gesammelte Aufsätze* (ed. by L. Schäfer and T. Schnelle), Suhrkamp Verlag, Frankfurt/M.
8. Hempel, C: 1966, *Philosophy of a Natural Science*, Prentice-Hall, Englewood Cliffs, New Jersey.
9. Kuhn, T.S.: 1970, *The Structure of Scientific Revolutions*, 2nd edition, University of Chicago Press, Chicago, Illinois.
10. Kuhn, T.S.: 1977, *The Essential Tension: Selected Studies in Scientific Tradition and Change*, University of Chicago Press, Chicago, Illinois.
11. Oehler, K.: 1977 'Zur deutschen Rezeption des amerikanischen Pragmatismus', in W. James, *Pragmatismus*, F. Meiner Verlag, Hamburg.

NELLY TSOUYOPOULUS

THE SCIENTIFIC STATUS OF MEDICAL RESEARCH: A REPLY TO SCHÄFER

The paper by Lothar Schäfer [16] provides a critical approach to the subject, "The Scientific Status of Medical Research", which is based on a study of Ludwik Fleck ([2], [18]). This study offers an alternative to traditional philosophy of science, all the more since it has influenced, as Schäfer underlines, Thomas Kuhn's work on the structure of scientific revolutions [10]. Kuhn's work has been discussed in recent years as the most interesting alternative to the positivistic philosophy of science. The traditional philosophy of science fails above all to explain the peculiar character of medicine. As Schäfer argues, it has never been able to analyze medicine as a science different from physics. The work of Fleck provides, then, a hopeful approach to a fundamentally new concept of science.

Fleck criticizes two essential points of traditional philosophy of science. The first point concerns the understanding of 'facts'; the second point concerns the interpretation of knowledge and research as products of individuals. Fleck explains his view by analyzing a special case: syphilis. The traditional philosophy of science, he says, constructs the 'fact' syphilis as being something definite and static, as a result of a final solution. Such a construction fails to explain the complexity of this phenomenon. Fleck suggests a new construction containing components that could show the dynamic character of medical facts. According to this construction, the following components belong to the 'fact' syphilis: a) the ethical-mystical component, which concerns the venereal character of the disease; b) the empirical-therapeutic component which derives from methods of practical treatment; c) the experimental-pathological component which focuses on the infectious character of the disease; and finally, d) the serological-etiological component.

An interesting point in Fleck's analysis is the consideration of the historical and sociological aspects of science, which were overlooked by the traditional philosophy of science. Fleck worked out the historical components of the 'fact' syphilis and then used them as arguments to support his thesis about the collective character of scientific research [18]. Thus Fleck's analysis of science implies a question about the mutual relevance of history and philosophy of science, which has often been an object of discussion in recent years ([6], [13]). The main question that arises here is: can the modus of interaction

Corinna Delkeskamp-Hayes and Mary Ann Gardell Cutter (eds.), Science, Technology, and the Art of Medicine, 39–46.

between the philosophy and history of science, implied in the models of Kuhn and Fleck, provide a new non-positivistic aspect of the history of science?

The main difficulty in this respect is not the model as such but the application of its categories to the historical investigation. Rothschuh has already pointed out this difficulty in his discussion of Kuhn's model ([13], p. 84). I will try to make it clear by discussing an example from Fleck's model, as Schäfer has presented it.

The venereal character of syphilis was shown in Fleck's study to constitute the first ethical-mystical component in the whole complex of the phenomenon; it is also called an 'a priori magic' element, which provides, together with astrology, the first 'differentia specifica' of the concept to this disease. Now, a critical research of the historical sources shows, on the contrary, that the venereal nature of this disease (that is to say, the specification of the disease as one that affected, above all, the genitals, and that was contracted while having intercourse with an affected person) was at first a result of observation, and of the efforts toward empirical treatment. Moreover, according to Fracastoro's famous work on that subject, it was discovered while discussing the contagious nature of the disease ([3], [7]). The moral idea of pleasure did not play an essential role in this context. The ethical idea of this disease — the consideration of it as a product of immorality — belongs much more to the mentality of the eighteenth and nineteenth centuries; through the historiography of the nineteenth century, this idea was projected back to the Renaissance, even unto antiquity. In the early history of epidemic syphilis, it was unusual to disapprove of the affected persons or consider them as scandalous. During the first epidemic of syphilis, in which the virulence of the disease was very great and the process extremely painful, "it began to be said that this was the disease of St. Job"[1] ([15], p. 646). Even the names 'venereal disease' — "*Lues venerea*", "*Venusseuche*" and "*Lustseuche*" themselves became prevalent only in the eighteenth century.[2] In the nineteenth century, an attempt was made to show its immoral character through absurd etymologies of the former names.[3]

Fleck supports this idea of the "ethical-mystical" component with a single quotation about the conjunction of the planets from the year 1484. But this quotation belongs to the poem "Vaticinium in epidemicam scabiem", which does not seem to have been any important work about syphilis. The historiography of the late nineteenth century has for the first time considered this poem an essential source for the history of this disease. Haeser, who was an influential historian of his time, suggested that this dark poem contains mythological and astrological considerations about the conjunctions of Jupiter and Saturn in the year 1484 as the almost generally accepted cause of epidemic *lues venerea*, or venereal disease[4] ([5], pp. 238–239). But Haeser's interpretation manifests nineteenth century ideology, a view according to which the whole medical tradition had, up until this time, been irrational and non-scientific.

Thus, Fleck's view of the history of medicine reflects views of the positivistic history of medicine, although his general aim is to criticize positivism [18]. In short, Fleck did not demonstrate what the real function of history could be in a new or modified concept of science. While keeping former ways of thinking current would indeed prevent medical science from considering its latest point of view as an everlasting truth, this fact does not provide any grounds for using several steps from the history of medicine as "components" of a concept or of a fact. Historical events cannot be used as arguments to either found or support categories that belong to the theory of knowledge, or to the philosophy of science.

Further, a central point of Fleck's analysis concerns the role of observation in scientific research. According to Fleck, observations are neither immediate (i.e., direct, free of limitations), nor are they true or false. At first, we see only an indistinct something, which then becomes a *Gestalt*. The *Gestalt* will then be seen through a kind of directed perception. This kind of "seeing" provides the "reality" but what is real does not apparently coincide with what was just given. Every fact is connected with the intellectual environment (the intellectual community's way of thinking) through some quasi-axiomatic rules such as "every fact must be on the line of mental interests of the intellectual group" ([2], p. 207).

Fleck criticizes the positivistic thesis of observation from the point of view of psychological theories that arose in the 1920s. He obviously accepts the 'theory of seeing' suggested by Kurt Koffka,[5] but he does not mention the work of Claude Bernard, who discussed the problem of observation in great detail ([1], pp. 11–27).

Fleck is expressing here something similar to the main idea of the 'Gestalt theory', a movement in psychology which began in Germany early in the twentieth century. The central idea of this movement is that the characteristic qualities and modifications of the whole (in this instance, of the intellectual community) cannot be composed from similar qualities and actions of its parts ([8], p. ix).

Fleck speaks about the so-called "Denkstilgebundenes Gestaltsehen", which Schäfer considers as an important category of the philosophy of science. I would underline with Schäfer the correctness of Fleck's elaboration on the collective character of scientific research: The aims of scientific research are indeed suggested by the scientific community. Individuals who want to participate in research programs must be accepted and must be given by the scientific community a place and function at the existing institutions. Every step of their work will then be examined, judged and evaluated according to the principles, rules and aims of the scientific community. Such rules, aims and principles again are not autonomous but depend on more general economic, cultural, political and social factors. It is therefore a central demand of the history and philosophy of science to consider all levels of interaction between those several factors. If we want to consider the dynamical character of scientific research, it is also important to have a new category, since we no longer accept the positivistic idea of linear progress. Nevertheless, I do not agree with Schäfer and Fleck that the 'Gestalt' Theory would help to solve these problems.

The central critical point in Schäfer's paper concerns the practical character of medicine. Schäfer is right, in saying that the traditional philosophy of science fails in explaining this peculiar character of medicine. However, Schäfer believes that the problem can be solved within the framework of Fleck's suggestions.

In considering the practical character of medicine we are confronted with several difficulties. The action-character of medicine has in recent years been the focus of further inquiry ([12], [19], [20]). In research, however, it is more difficult than in clinical medicine to discover its peculiar practical character. It has always been argued that research is the domain in which new medical knowledge is obtained — thus widening the theoretical horizon of medical science and providing new theoretical matter — which is then applied to clinical medicine. Also, it has been argued that the research of pure natural science is often motivated, and even manipulated, by a view to practical application. From these arguments emerges the following question regarding the philosophy of medical research: How can the primarily practical character of medical research be defined in order to ascertain a substantial difference between medical research and research in natural science?

At first, it must be mentioned that there are also many theoretical questions which must be answered in medical research. As a rule, theoretical issues are to be decided often through experiments which function as falsificators in the 'Popperian' sense ([11], pp. 8, 49, 51, 54, 160, 195, 224f). Such theoretical questions, however, even if necessary and fruitful, do not reveal the special character of medical research.

The special character of medical research is, first of all, to be found in the *form* of setting the problem. This form is: "*A* must be changed." Such a form never occurs in theoretical science. Similar forms expressing theoretical problems are: "What is *A*?" "What is the truth about *A*?" "How does *A* come to be not-*A*?" "How does A^1 change to A^2?" "How can *A* be explained, measured, etc.?" All these forms can be relevant for medical research as well, but they are not identical, nor can they replace the main form: "*A* must be changed." For stating the content of *A* it is not necessary to ask again a theoretical question such as "What is *A*?" which demands an exact definition of *A*. It can be empirically, and also, with certain limitations, freely estimated. Consider, for example, suffering, invalidity, indisposition, high mortality, deformity, and the relative health of a class of people. It is not the essence of *A* which interests us, but its relation to the special formulation of the practical problem. It must be considered as "something" which must be changed. Its character depends upon the 'deontic form', not upon generality.

The difficulty in finding a general concept of disease with real currency originates from confounding the practical form of setting the question with a theoretical one. In connection with this difficulty, I would like to refer to the critical essay of Kazem Sadegh-zadeh about the general concept of disease. Sadegh-zadeh has shown that expressions such as "Disease is G" never have the status of real statements; they can only be pure definitions [14].

Thus, the practical character of medicine may not be understood only as an application of knowledge; applicability as such cannot provide a real distinction between theoretical and practical science. "Practical" in this instance is to be explained as a plan according to some ideal implying rules of realization through certain actions. Such rules must be considered as essential for medical research, and also as an important subject for the theory of scientific research. The plan, as the final objective of medical research, differs not only from the goal of theoretical science, but also from that of art: The work of art comes into being without preconditions and is planned quite freely. On the other hand, the practical project of medicine cannot be planned in total freedom, for it has to take real needs into consideration, so that it is, to a certain extent, determined through the situation which must be changed. Still, the practical plan is not necessarily derived from the knowledge of real needs. The relation between the knowledge of real needs and the practical plan is not the same relation as the one between a theoretical problem and its solution, because in practical planning there is always a free space of alternative solutions of the same problem. By a certain valuation of "*A* which must be changed", there are several plans which could claim equal validity as objects of scientific research. Some examples of these are: "relief from pain"; "back-formation of pathological states (processes)"; "eradication of an endemical disease"; and "improvement of the general average health of a certain population". Even by a very special determination of *A*, such as "Many people suffer from nephritic disorder", the practical plan is not necessarily bound to the real needs; the free space of decision still exists. Thus the corresponding aim in this case can be a concrete plan which demands: a) "The number of individuals suffering from nephritic disorder within a certain population must be diminished", or an equivalent plan; b) "No matter how many individuals suffer from kidney disease, modern medicine must be able, by means of artificial kidneys, to help all those individuals whose kidneys are irreparably damaged." Each of these plans gives a different orientation to scientific research. The orientation of scientific research then determines the direction and relevance of clinical action, and according to that orientation, one can judge "success", "efficacy", and "desired result", and thus manipulate the day-to-day efficiency of physicians.

The discussion above makes way for some critical remarks about the scientific status of contemporary medical research. Modern medicine conceives its practical character in a rather trivial way, that is, as an application of theories. As a result, the whole of scientific research in medicine considers the intermediate goals of solving theoretical problems — at which it has been successful — as its principal and actual characteristic. Thus, its aims are only directed towards efficiency in theoretical problem-solving, and for this reason, modern medical research has a latent tendency towards a "therapeutic nihilism"[19].

Finally, the critical view proposed by Schäfer is an appropriate point of departure for an analysis of medical research. That is, current philosophy of science fails to discover the very character of medical science. Instead, it considers only *that* as "scientific" in medical research which is not typical for medicine. Therefore, the criteria of this current positivistic philosophy of science have to be extended and the concept of science modified. Towards this end, it is necessary to reflect upon the dimensions of the "practical" and those of "history".

Westfälische Wilhelms-Universität Münster
Münster, Germany

NOTES

[1] "It began to be said that this was the disease of St. Job, and for this reason a picture of that saint was painted in San Lorenzo" (from Mattarazzo, "Cronaca della Città di Perugia", ca. 1500, [15], p. 646).

As the historian Haeser says, "Wir lesen von Königen, Kaisern, ja von Päpsten, die an der Krankheit zu leiden hatten. In Folge dessen verlor dieselbe für manche Kreise sogar fast ihren anstössigen Charakter; ja man rühmte sich ihrer, und betrachtete sie beinahe als Attribut der Vornehmheit" ([5], p. 315).

[2] The usual name of the disease during the 16th century was *morbus gallicus*. Toward the end of this century, the term *morbus venereus* was introduced but this last term had not been generally accepted before the 18th century. The work of J. Astruc, *De morbis venereis* (Paris 1736), has been very influential for its generalization ([7], pp. 51, 542).

[3] For example, as Haeser says, "Gewiß liegt nichts so nahe, als morbus gallicus durch französische Krankheit zu übersetzen. Dennoch bleibt... die Möglichkeit nicht ausgeschlossen, daß gallische Krankheit ursprünglich keine geographische Bedeutung hatte... Gale ist angelsächsisch geradezu Geilheit, fleischliche Lust" ([5], p. 251). Consider the following citation from Ruy Diaz de Ysla: "Tractado contra el mal serpentino," (1539) shows the different mentality of the Renaissance by discussing the same subject: "Die Italiener und Neapolitaner aber, die niemals von dieser Krankheit Kunde gehabt hatten, nannten dieselbe mal frances... und im portugiesischen Indien nannten sie die Inder mal de los Portugeses;... Aber, wie Galen sagt, durch die Namen werde ich nicht geheilt: die Heilversuche seien richtig und gut" ([7], pp. 652–655).

[4] Astrological events in the 16th century are often considered as remote causes of diseases, but in the sense of natural phenomena producing mechanically some effects and as a rule in connection with the 'theory of contagion' ([7], p. 544).

[5] Koffka's objections are directed against the so-called "positivism" as the prevailing scientific attitude of his time; in opposition to this attitude, Koffka suggests that the word 'see' should be used in a "phenomenological sense", that is, in such a way that the words are pure descriptions of our experience. Moreover, Koffka's proposal implies that we can never be sure or unsure about what we see, nor can we be right or wrong, for everything that we see is part of the behavioral environment [9].

BIBLIOGRAPHY

1. Bernard, C.: 1912, *Introduction a l'étude de la Médecine Expérimentale*, 3rd ed., Librairie, Paris.
2. Fleck, L.: 1935, *Entstehung und Entwicklung einer wissenschaftlichen Tatsache. Einführung in die Lehre vom Denkstil und Denkkollektiv*, Basel; F. Bradley and T.J. Trenn (trs.) and T.J. Trenn and R.K. Merton (eds.) in 1979, *Genesis and Development of a Scientific Fact*, University of Chicago Press, Chicago, Illinois.
3. Fracastoro, H.: 1968 (1546), 'Drei Bücher von den Kontagien, den kontagiösen Krankheiten und deren Behandlung' (von V. Fossel übersetzt und eingeleitet), in K. Sudhoff (ed.), *Klassiker der Medizin*, Vol. 5, Joh. Am. Barth Leipzig.

4. Fuchs, C.H.: 1850, *Theodorici Ulsenii Phrisii Vaticinium in epidemicam scabiem, quae passim toto orbe grassatur, nebst einigen Nachträgen zur Sammlung der ältesten Schriftsteller über die Lustseuche in Deutschland*, Göttingen.
5. Haeser, H.: 1882, *Lehrbuch der Geschichte der Medizin und der epidemischen Krankheiten*, 3rd ed., Vol. 3, G. Fischer, Jena.
6. Hanson, N.R.: 1971, 'The Irrelevance of History of Science to Philosophy of Science', in S. Toulmin and H. Woolf (eds.), *What I Do Not Believe and Other Essays*, D. Reidel Publishing Company, Dordrecht, pp. 274–287.
7. Hendickson, G.L.: 1934, 'The "Syphilis" of Girolamo Fracastoro. With some Observations on the Origin and History of the Word "Syphilis" ', *Bulletin of the Institute of the History of Medicine* **4**, 515–546.
8. Köhler, W.: 1924, *Die psychischen Gestalten in Ruhe und im stationären Zustand*, Verlag der Philosophischen Akademie, Erlangen.
9. Koffka, K.: 1935, *Principles of Gestalt Psychology*, Harcourt-Brace Jovanovich, Inc., New York.
10. Kuhn, T.S.: 1970, *The Structure of Scientific Revolutions*, 2nd ed., University of Chicago Press, Chicago, Illinois.
11. Popper, K.R.: 1969, *Logik der Forschung*, 3rd ed., J.C.B. Mohr Tübingen.
12. Rothschuh, K.E.: 1965, 'Prinzipien der Medizin', Urban and Schwarzenberg, München and Berlin.
13. Rothschuh, K.E.: 1977, 'Ist das Kuhnsche Erklärungsmodell wissenschaftlicher Wandlungen mit Gewinn auf die Konzepte der klinischen Medizin anwendbar?', in A. Diemer (ed.), *Die Struktur wissenschaftlicher Revolutionen und die Geschichte der Wissenschaften*, Verlag Anton Hain, Meisenheim am Glan, pp. 73–80.
14. Sadegh-zadeh, K.: 1977, 'Krankheitsbegriffe und nosologische Systeme', *Metamed* **1**, 4–41.
15. Sandison Brock, G.: 1901, 'An Early Account of Syphilis and of the Use of Mercury in its Treatment', *Janus* **6**, 592–595, 645–647.
16. Schäfer, L.: 1992, 'On the Scientific Status of Medical Research; Case Study and Interpretation According to Ludwig Fleck', in this volume, pp. 23–38.
17. Scheube, B.: 1901, 'Über den Ursprung der Syphilis', *Janus* **6**, 648–655.
18. Tsouyopoulos, N.: 1982, 'Auf der Suche nach einer adäquaten Methode für die Geschichte und Theorie der Medizin: Auseinandersetzung mit Ludwik Fleck's "Entstehung und Entwicklung einer wissenschatlichen Tatsache"', *Medizinhistorisches Journal* **17**, 20–36.
19. Tsouyopoulos, N.: 1984, 'German Philosophy and the Rise of Modern Clinical Medicine', *Theoretical Medicine* **5**, 345–357.
20. Wieland, W.: 1975, *Diagnose, Überlegungen zur Medizintheorie*', Walther de Gruyter, Berlin, New York.

REIDAR KRUMMRADT LIE

LUDWIK FLECK AND THE PHILOSOPHY OF MEDICINE: A COMMENTARY ON SCHÄFER AND TSOUYOPOULOS

The logical empiricists, or logical positivists, believed that it would be possible for any scientific field to find a set of observation statements or descriptions of facts that one could use to support the more problematic theoretical statements that assert something about unobservables. The support relation between observation statements and theoretical statements in turn was supposed to be theory neutral and based purely on the logical relations between these two types of statements. Ludwik Fleck criticized this conception of science, arguing that it would be impossible to isolate a set of statements that are supposed to be descriptions of facts independent of one's particular theoretical viewpoint. In their commentary on Fleck, both Lothar Schäfer and Nelly Tsouyopoulos point out that his account of theory structure and dynamics represents a critique of the standard empiricist conception of science.

It is clear that there are important insights in Fleck's work. As Schäfer points out, there is an emphasis on the collective nature of scientific research: scientists do not carry out their investigations in isolation from their colleagues, or in isolation from the concerns of the rest of society. Fleck is also sensitive to the historical development of a field of research and to the different types of activities within a given field (such as in his description of the differences between journal science and textbook science). It is also significant that Fleck used examples from biomedical research, instead of focussing on the physical sciences that were, and still are, the basis for most of the work in philosophy of science. Since, however, many of the topics introduced by Fleck have been extensively discussed in philosophy of science after his book was published, Fleck's own contribution may not be as illuminating for present-day philosophy of science and medicine as Schäfer and perhaps also Tsouyopoulos may want. For example, Fleck places great emphasis on the theory-ladenness of observations and the incommensurability of theories (see [2], pp. 36, 90–92, 100). There is now available an extensive literature on these topics; in order, therefore, to advance our understanding of scientific change today, one cannot simply point out, as significant, the fact that scientists tend to find what they look for, or that the meaning of central terms in a particular theory can only be understood within the framework of that theory. Regarding the so-called theory-ladenness of

Corinna Delkeskamp-Hayes and Mary Ann Gardell Cutter (eds.), Science, Technology, and the Art of Medicine, 47–54.

observations, one may, for example, maintain that although our observations are always colored by some theory or another, there are some observations which can be understood within a theory accepted by all researchers, and which therefore can be used to test rival theories. The 'incommensurability thesis' has also been extensively discussed in recent years, and new theories of meaning have been introduced, which will have to be considered in any adequate philosophy of science concerned about the problem of the meaning of theoretical terms (see [6], [7], [8], [9] and [10] for a discussion of the incommensurability thesis, and [3] and [5] for new meaning theories).

Therefore, I do not think, contra Schãfer and Tsouyopoulos, that the importance of Fleck's contribution lies in this general criticism of traditional philosophy of science. Rather, I believe that it is Fleck's analysis of examples from bacteriology and immunology that potentially can pose a formidable challenge to anyone who would defend an account of rational theory change in the biomedical sciences. In the traditional model, theories are supposed to be abandoned because they can no longer explain the available experimental evidence, and new theories are accepted because they can better account for that experimental evidence. There is an accumulation of knowledge, in the sense that subsequent theories can explain everything previous theories could explain and something more in addition, or in the sense that successive theories are continuously getting closer to the truth. In this way, science arrives at an ever more adequate picture of the world. If it can be shown that scientists as a matter of fact do not behave in this way, it would indeed be difficult to argue that scientific change can be rationally reconstructed. In his book, Fleck argues that scientists do not choose their theories because they are true, and reject theories because they are false:

> It is altogether unwise to proclaim any such stylized viewpoint acknowledged and used to advantage by an entire thought collective, as "*truth or error*". Some views advanced knowledge and gave satisfaction. These were overtaken not because they were wrong but because thought develops. Nor will our opinions last forever, because there is probably no end to the possible development of knowledge just as there is probably no limit to the development of other biological forms ([2], p. 64).

Fleck's examples are introduced in order to support this thesis. Let me illustrate this by examining his discussion of the introduction of the Wassermann reaction.

Fleck discusses the establishment of the connection between syphilis and the Wassermann reaction. According to the traditional picture of science, one would have to be able to point to certain observational statements, or results of experiments, by which one can establish this connection. Fleck argues that it is impossible to single out any particular experiment or experiments that can serve that purpose. Rather, there is a process of trial and error where a group of scientists attempts to modify the procedures used, and only after a period of such extensive modification (with "a little more" reagent, or "a little less"), can one say that the *fact* of the association between the Wassermann reaction and syphilis became established. That is, initially there was only a loose association, with many exceptions, and it is only when the scientists involved became conditioned a certain way, by the fact that they were involved in this process of trial and error, that they could arrive at the perceived connection between these two phenomena.

The significance of this account from the point of view of philosophy of science is not that the research involved a community of scientists rather than a single individual, as suggested by Schäfer. It is also not significant that it was "steered by external factors", such as national pride, or that the initial goals of the investigators were different from what they finally were able to achieve. A philosopher of science inclined to empiricism would simply argue that as long as you can show that there is an association between the results of a procedure, which can be unambiguously characterized, and certain symptoms or signs, which can also be unambiguously characterized, it is irrelevant how that knowledge was acquired. According to the empiricist, all that one can legitimately say in this case anyway is that the result obtained from a specified test is a reliable indicator of the presence of certain other observable phenomena, or even that the test is a partial definition of the theoretical term 'syphilis'. What makes Fleck interesting for present-day philosophy of science is his account of how the scientists in a field of research actually go about deciding which observations to believe and which to discard.

In the initial stages of a new field of research, there is chaos, no apparent order, and many exceptions to suggested generalizations. Then the scientists start their experiments, often with an idea about what they want to find, such as the postulated association between blood and syphilis, and attempt to discover some order, or systematic connections between the elements. As they do this, they discard some observations, and accept others. Based on this accepted knowledge, and new experiments, more systematic connections are established, until the development of the field has reached a stage where the relationships are almost necessarily true, rather than contingent. Fleck argues:

> The more developed and detailed a branch of knowledge becomes, the smaller are the differences of opinion. In the history of the concept of syphilis we encountered very divergent views. There were far fewer differences during the history of the

Wassermann reaction, and as the reaction develops further, they will become even rarer. It is as if with the increase of the number of junction points,... free space were reduced. It is as if more resistance were generated, and the free unfolding of ideas were restricted ([2], pp. 83–84).

If this is an accurate description of the development of a typical field of research, theories are not chosen because they fit the experimental evidence, but are chosen if they can bring some order into apparent contradictions. Scientists do not bother about developing a theory that will explain the diversity of the phenomena observed, but they simply disregard certain experiments in order to be able to suggest a coherent theory which will explain 'all' the evidence. Because the theories have been made to fit the evidence in this way, established generalizations become almost necessarily true after some time. Such a picture of scientific activity is of course in stark contrast with traditional ideas.

Once such a system has been established, it becomes the basis for the work of all scientists in the field, that is, it becomes part of their 'thought style'. The scientists belonging to this 'thought collective' cannot see the world otherwise. They think that the accepted facts have been objectively established, despite the fact that they are the result of the particular past history of the field:

Once a field has been sufficiently worked over so that the possible conclusions are more or less limited to existence, or nonexistence, and perhaps to quantitative determination, the experiments will become increasingly better defined. But they will no longer be independent, because *they are carried along by a system of earlier experiments and decisions*, which is generally the situation in physics and chemistry today ([2], p. 86).

As mentioned above, I do not think that Fleck's thesis is significant if it is interpreted to mean that scientists who belong to different thought collectives are not able to understand each other's language, or that their observations are colored by the particular thought style which they accept. Rather, I think that Fleck's thesis is important because it accurately describes the way scientists as a matter of fact choose theories, and the way a scientific field actually develops, and because that description does not fit our intuitive notions about how a scientific field should rationally change.

One should note, however, that if Fleck's account is to serve as a challenge to the notion of scientific rationality, two additional theses need to be established. First, one needs to show that the examples chosen by Fleck are more or less representative of the way science in general develops. If these examples are atypical in some way or another, one cannot use them to establish a general point about science. This point is also made by

Tsouyopoulos.

Second, and more important, it is not sufficient to show that scientists have a tendency to disregard anomalous observations. If one is able to give good reasons for why the scientists in a particular case disregarded some observations, and decided to accept others, it is hard to see why a theory that does not explain the 'anomalies' cannot be rationally defended. It is, of course, difficult to suggest criteria by which one can judge whether or not there were good reasons for a particular choice. Without defending my position, I shall simply maintain that there are no good reasons for rejecting a piece of experimental evidence, if there are scientists who in fact disagree with that judgment, and who think that any adequate theory should be able to explain that particular piece of experimental evidence. That is, in order for Fleck's account to pose a challenge to the traditional notion of scientific theory change, we need to show that, again as a matter of fact, there were scientists who believed that those facts that were disregarded by the others should not have been ignored, but should have been used to support other, rival theories. In addition, we need to show that there is no theory-neutral methodology that can decide this controversy between rival groups.

If this can be established for a number of historical cases, it means that the acceptance of a theory is inextricably linked to the acceptance of certain standards of evidential support; each theory is therefore regarded as the best theory according to its own criteria, and there are no neutral criteria by which to judge the competing theory. According to Gerald Doppelt, this is the most fundamental of the theses advanced by Thomas Kuhn, and the thesis most fundamental, and most damaging to the traditional idea of science as a continuous process of cumulative development, or development leading to ever closer approximations to truth [1]. Rather, as Kuhn argued, scientific development is a process marked by radical discontinuities. Let me now briefly explain why such an account poses a challenge to traditional notions of scientific rationality.

Kuhn thinks that there is a sense in which subsequent theories can be regarded as progressive compared with previous theories. The old paradigm is abandoned because it is confronted with an increased number of 'anomalies', and the new paradigm has been able to solve some of these problems, and in addition is very effective at solving new problems. Kuhn maintains: "Later scientific theories are better than earlier ones for solving puzzles in the often quite different environments to which they are applied" ([4], p. 206). Although the new paradigm in this way may be able to solve more problems than the old, there will always be some facts that are not explained by the new paradigm (cf. Fleck's point that scientists choose to disregard certain facts in order to accept a coherent thought style). This, according to Kuhn, means that there is no cumulative development, even in the sense in which it was maintained in the instrumentalist versions of logical empiricism, and certainly not in the sense that subsequent theories are a better representation of reality than previous theories.

This, then, is a historical thesis: During episodes of scientific change, there is, as a matter of fact, always, or very often, a loss, as well as a gain in explanatory power of successive theories. This particular thesis is then used to argue that there can be no rationally compelling reasons for choosing one particular theory over another. The reason why there cannot be compelling reasons for theory choice is that although the increased problem-solving effectiveness of the new theory may be a reason in favor of the new paradigm, it can never be a *compelling* reason, because each paradigm incorporates certain standards about how to weigh the importance of problems, and consequently each theory will be judged superior by its own standards. Doppelt argues

> Relative to the standards of old established paradigms, the observational explicanda lost in a new paradigm constitute decisive or very strong reasons against the latter's adequacy and in favor of the former's — even if the latter explains far more data than the former ([1], p. 58).

If it therefore is the case that, as a matter of historical fact, subsequent theories do not explain everything previous theories explained, and the scientific methodologies by which theories are judged pick out these 'lost facts' as especially important, then this creates a serious problem for anyone who may want to argue for the rationality of scientific change. It has been exceedingly difficult to propose theories of confirmation in terms of a purely logical relationship between theories and evidential statements. The alternative is, of course, to refer to some methodological standard that employs empirical knowledge of the world. If, as Kuhn maintains, acceptance of these standards is associated with the acceptance of the rival theories, then we cannot make theory-neutral judgments concerning the adequacy of a particular theory.

Has Fleck, then, been able to show, in the analysis of his examples, that scientific development can be characterized as described above? From his account of the introduction of the Wassermann reaction, it is actually not clear whether the development of this diagnostic procedure is problematic. A case could very well be made for the position that no alternative, or rival, system of thought was available, challenging the particular choices made by

Wassermann and co-workers. In order for Fleck to have made his case that this development in some sense cannot be justified, he would have had to show that there were certain facts that were disregarded by these investigators, and that should not, according to other scientists, have been disregarded. He should have pointed out that there were other, rival, scientists who developed a different approach to syphilis based precisely on these disregarded facts. Since Fleck has not done this, we simply do not know whether this particular example can be used to support his theses.

I have taken Fleck's important and interesting thesis to be the claim that actual case histories of scientific change will show that certain approaches were abandoned by scientists despite the fact that rival theories did not regard that abandonment as defensible. Since he did not demonstrate this in the cases discussed, Fleck has not sufficiently defended this thesis. However, there is still a possibility that we may use Fleck's account to challenge the traditional notion of scientific rationality.

First, since a scientific field develops when scientists make choices that although not controversial may not be fully justified, we do not know whether an alternative system of thought could have been developed given a different starting point, and a different set of facts to disregard. That is, since the 'thought style', which happened to be chosen, depended on certain, more or less, arbitrary choices, we do not know how adequate the actual choice is compared with other potential rivals. This is, however, quite a modest claim, because as long as no actual rivals have been proposed, we do not know whether any of the potential rivals can pose a real challenge to the actual choices which were made.

Second, a full analysis of the case studies introduced by Fleck may in fact reveal that there were rival approaches that were not pursued. If that position can be maintained, then Fleck's examples can be used to argue against a traditional notion of scientific change. Or one may, inspired by Fleck's first attempt, try to analyze different case studies using his perspective. These studies may then reveal that Fleck's position can be defended after all. Given that so little work has been done on the history of medicine from the perspective of philosophy of science, this should be a fruitful endeavor, and should result in important insight into the structure and dynamics of biomedical theories. It is Fleck's achievement that, after fifty years, his work is still one of the few available systematic investigations in this field of research.

University of Oslo
Oslo, Norway

BIBLIOGRAPHY

1. Doppelt, G.: 1978, 'Kuhn's Epistemological Relativism: An Interpretation and Defense', *Inquiry* **21**, 33–86.
2. Fleck, L.: 1935, *Entstehung und Entwicklung einer wissenschaftlichen Tatsache. Einführung in die Lehre vom Denkstil und Denkkollektiv*, Basel; English translation by F. Bradley and T.J. Trenn (trs.) and T.J. Trenn and R.K. Merton (eds.) in 1979, *Genesis and Development of a Scientific Fact*, University of Chicago Press, Chicago.
3. Kripke, S.: 1972, *Naming and Necessity*, Basil Blackwell, Oxford.
4. Kuhn, T.: 1970, *The Structure of Scientific Revolutions*, 2nd ed., University of Chicago Press, Chicago.
5. Putnam, H.: 1975, 'The Meaning of Meaning', in H. Putnam (ed.), *Philosophical Papers II: Mind, Language and Reality*, Cambridge University Press, Cambridge, pp. 304–324.
6. Scheffler, I.: 1967, *Science and Subjectivity*, Bobbs-Merill, Indianapolis.
7. Scheffler, I.: 1972, 'Vision and Revolution; A Postscript on Kuhn', *Philosophy of Science* **39**, 366–374.
8. Shapere, D.: 1964, 'The Structure of Scientific Revolutions', *Philosophical Review* **73**, 383–394.
9. Shapere, D.: 1966, 'Meaning and Scientific Change', in R. Colodny (ed.), *Mind and Cosmos: Essays in Contemporary Science and Philosophy*, University of Pittsburgh Press, Pittsburgh, pp. 41–85.
10. Shapere, D.: 1971, 'The Paradigm Concept', *Science* **172**, 706–709.

PART II

CAUSALITY AND EXPLANATION IN MEDICINE: THE REGARD FOR PRACTICE IN MEDICAL KNOWLEDGE

JOSÉ LUIS PESET

ON THE HISTORY OF MEDICAL CAUSALITY*

I. INTRODUCTION

Throughout this essay it will be my intention to show how current approaches to the concept of medical causality[1] derive from man's historical struggle with disease, the struggle of "human groups" with "pathogenous agents". It is only within the framework of this historical dialectic that the problem of medical causality attains its true meaning. The question of the cause of an illness is not univocal, but rather possesses different meanings, depending on the social and cultural coordinates in which the question arises. In this sense, it is necessary to distinguish between at least two clearly differentiated periods in the history of the problem of medical causality. The first period, whose paradigm is the "Galenic", commences with the Hippocratic writings and does not come to an end before the nineteenth century. The other, which slowly developed in medicine during the nineteenth and twentieth centuries, derives from Locke's and especially Hume's empirical criticism of the Aristotelian doctrine of causality.

II. THE CLASSICAL PARADIGM

The classical idea of causality is based on the primary and fundamental concept of *physis*, Nature. For the Greeks, Nature is order or the principle of order, harmony, beauty, and justice. This natural order, conformity to which constitutes justice, is structured on three levels, individual, social, and cosmic. It follows from this that health for the Greek physician consists of the balance of order, not only individual but also social and cosmic order, while disease is inbalance or lack of harmony in this order. Disease is a disorder of Nature. And as Nature is, according to the pre-Socratics, "the divine", the study of the order and disorder of nature, the science of health and disease, cannot in the last analysis be separated from knowledge of the divine, that is, *sophía*, or more exactly philosophy: "The Physician who is at the same time a philosopher is like unto the gods", as the famous Hippocratic dictum goes.

Health consists of the harmony of the *physis* of man. Yet, it would be wrong to think that the Greek physician understood *physis* in individual

Corinna Delkeskamp-Hayes and Mary Ann Gardell Cutter (eds.), Science, Technology, and the Art of Medicine, 57–74.

categories. It is rather the opposite. The Greek always considered the individual *physis* as an abstraction from and concretization of the general and only *physis*. In this respect it is important to point out the basic role played by socio-political ideas in the constitution of the Greek *tékhnê* or art of medicine. In the same way as the Athenian democracy is harmonic, just, and reasonable, so also is the nature of man in all its manifestations. Hence, the first physician, Alcmaeon of Crotona, in giving a rational definition of health and disease, has recourse to the political world for his basic terms. Health is, to use the Greek work, the *isonomía* of the potencies, that is, a balance, an "equality of rights", while the *monarkhía* of one over the others gives rise to disease (Diels-Kranz B4). This principle also underlies the explanation in the Hippocratic text, *On Airs, Waters and Places,* of the soft and not very bellicose character of the Asiatics in contrast with Europeans. This character is attributed to their subjugation to a tyrannical government as well as to the climatic differences from Europe.

It is no accident that Greek interest in the *aitía* begins in the pre-Socratic period, specifically with Democritus, who affirms that it is preferable to discover an etiology than to possess the kingdom of Persia, while Leucippus writes that nothing happens by chance but everything occurs according to motive and of necessity. To know the reality of things is identical to knowing their causes. Thinking in a scientific way is synonymous with thinking in a causal way. Hence Alcmaeon's *monarkhía* of the potencies is not only the "definition" of disease, but the "cause" of disease. As will be affirmed from Aristotle onwards, "causal definitions are essential definitions". "Causal" thinking is synonymous with "essential" thinking and with "scientific" thinking. To know the cause of a phenomenon is the same as knowing its essence. As Aristotle says: "For men of experience know that the thing is so, but do not know why, while the others (i.e., the technicians) know the 'why' and the 'cause'" (Metaphysics 981a28–30). And medicine is a *tékhnê*, the *tékhnê iatrikê*. Consequently, what health and disease are is something that cannot be known except by studying their causes. This is the aim of the Hippocratics and Galen: "It is not sufficient to know the localization of a disease, nor should we be satisfied with doing just this, it is necessary to go further, to get to the cause which produces this damage" (K., VIII, 131).

The classical paradigm of this causal medical thinking comes from Galen of Pergamum. He applies the Aristotelian etiological doctrine to medicine. For Aristotle there are four causes: material, formal, efficient, and final. The first two rather than being independent causes are really co-causes of any reality. They never exist in separation from each other. Matter and form are co-principles of any *physis* or nature. Hence, the material cause and the formal cause constitute the *physis* of man, his *physiología*, the first great chapter of Galenic medicine. Human disease would not be possible if man had not a determinate constitutional and physiological structure: this is precisely what Galen called *aitía proêgoumenê*, that is to say, internal or predisposing cause. Without health, disease would be impossible. Hence health is, in a certain sense, the cause of disease. The Aristotelian material and formal causes, in this way, give rise to what Galen called *aitía proêgoumenê*.

In order to have a disease, another cause, the cause that converts the *physiología* to *pathología*, is necessary. It is therefore a cause that is no longer natural nor physiological, but contra-natural and anti-physiological, which is the real efficient cause of disease according to Aristotle. Galen called this cause *aitía prokatarktikê*, the exciting and external cause. Together with cause as *res naturales* there is, therefore, cause as *res contranaturales* or *praeternaturales*. Causal thinking is the origin not only of *physiología*, but also, and to no lesser extent, of *pathología*. The origin of the latter is to be found in the efficient, or *prokatarktikê*, cause of disease. At the same time disease, as simply that, is not identifiable with its efficient cause, since it consists in the result of the action of the efficient cause on the material and formal causes, that is to say, it consists of the action of the *prokatarktikê* cause on the *proêgoumenê* one. Disease, *nósos*, is the end of all such causal processes, what Aristotle called the final cause and Galen the *aitía synektikê*, the conjunct or proximate cause.

One last and historically important problem about the Galenic doctrine of causes remains to be treated, namely, the theme of the *res non naturales* (i.e., "the non-naturals"), those things that lie between the *res naturales* of the *physiología* and the *res contranaturales* of the *pathología*. What are these *res non naturales*[2] from the "essentialist" and "causal" point of view? The question has a more satisfactory negative answer than a positive one, as the very denomination of the *res non naturales* indicates. The so-called *res non naturales* are really *res naturales* realities, nature, and as such are material and formal causes of disease. However, they also have the characteristic of being potential efficient causes of disease. Thus, they can act as *aitía prokatarktikê* and as *aitía proêgoumenê*, as physiological things and as pathological things. And so, in the last analysis, they are neither physiological nor pathological things, but according to Galen "neutral" things. Their scope is therefore properly neither physiological nor pathological, but that of hygiene. This is what Galen deals with in his book *De sanitate tuenda*. He systematizes these ideas in another book, *Ars medica*, along lines that are ever afterwards to become classical: air, motion and rest, sleep and watch, that which is taken in, that which is excreted or retained, passions of the mind. It is very curious that in modern times, after medicine has gotten rid of the Galenic causal system of physiology and pathology, the Galenic doctrine of hygiene still prevails with even renewed force. The ultimate reason for this very important phenomenon lies in its peculiar characteristics which permitted Galenic hygiene to escape the clutches of physiological and pathological rationalism. The new hygiene had to be developed on the basis of more empirical criteria, according to a more modern point of view.

III. THE MODERN PARADIGM

The formal characteristic of classic etiology is its conceptualism or rationalism. It is an *a priori*, deductive etiology. It is not *a posteriori* or inductive. Is an *a posteriori* and empirical doctrine of causality possible? This is the great question that philosophy and modern medicine tries to answer.

This question was one of the great disputed focal points of English empirical philosophy. It is in particular well captured by Hume in his *Treatise on Human Nature*, where he sets out to attack the rationalist concept of causality and to lay the foundations for an empirical causality. His arguments are clear and decisive. Only efficient causes are true causes from an empirical point of view. It is only possible to speak about efficient causes when the empirical and "analytic" analysis shows that an effect is due to a determinate cause. However, to prove this is not an easy task, as a multiplicity of forces, some known and some unknown, intervene. This obliges the new philosopher or scientist to proceed with extreme care.

Hume's critique of the rationalist doctrine of causality was based on the progress of modern science: he endeavored to give a philosophical grounding to the new idea of causality derived from modern need. But at the philosophical level the project posed insurmountable problems, because the new method had destroyed the whole edifice of classical metaphysics. Could metaphysics be possible afterwards? This is the difficult problem that confronted eighteenth-century philosophy. In France the problem was posed in all its radicality by Condillac, who was to have immediate influence on medicine, especially among the circle of the "ideologues" and "nosographists" and later, though in a markedly different way, among the "positivists". The common feature of all these groups lay in their "phenomenalism", that is, in their total rejection of the possibility of knowing the "thing in itself" and in their limiting of knowledge to pure appearance, the phenomenon. Naturally, this destroys the ancient theory of causality based on the "essence" and "substance" of things. This old kind of causality no longer interested philosophers and scientists. Now only the fixed rules of phenomena are relevant. Even if a modern scientist uses the word causality, he uses it in a phenomenalistic sense.

In Germany Kant elaborated Hume's theories. Hume's work left room for "analytical propositions" of a formal character and also room for "scientific hypotheses". Causality is not an analytic proposition; therefore, it must be an hypothesis. Kant endeavored, however, to avoid this drastic consequence by setting up a compromise. For him causality is not a simple hypothesis but an *a priori* category of human understanding. So, after Kant, it was once again possible to talk about causality. However, the knowledge of reality permitted by the *a priori* categories was no longer of the essence or substance of things, their "noumenon" as it was for ancient rationalism, but only of their "phenomenon" perceived in experience, their appearance. The causality recovered by Kant was not the ancient "substantialist causality", but instead a new phenomenalistic causality. Essence or the substance of things cannot be known through pure natural reason, but rather through other means, such as feeling, faith, etc., which in the last analysis are supranatural. This was the position of philosophical idealism, especially Fichte and the German *Naturphilo-sophen*. It is also the position held by the so-called "romantic medicine". But Romanticism faded, especially during the period of revival of Kantian thought. The great German science of the latter half of the nineteenth century and the first decades of the twentieth century had its epistemological foundations in its "return to Kant". So Neo-Kantianism and other more or less related philosophies (Vitalism, Neo-Hegelianism, Historicism, Pragmatism, etc.) were born.

These are also the epistemological bases on which the medicine of the contemporary period is founded and has developed. No longer is medicine able to talk about "cause", at least if the term is understood in its classical meaning. Medical etiology since the second half of the nineteenth century is no longer an "essentialistic" etiology, but rather a "phenomenalistic" one.

IV. THE EVOLUTION OF THE NEW CAUSALITY

Within medical thought, the Galenic causal scheme had lasted for centuries. The famous authors of the middle or modern ages made few innovations, at least in general causal outlines. Sydenham applied his energies to the study of the telluric or miasmatic causes, but this had only a minor influence. With

Boerhaave or Stahl, the scheme is the same. The fact that the master of Leiden reduces the *res non naturales* to four (*ingesta, gesta, retenta* and *applicata externa corpori*) does not add a great deal of theoretic innovation either ([29], [30]). However, while the theoretic scheme was maintained, innovations that tended to expand or contradict it appeared. Other causes were empirically observed. They accumulated and prepared the way for a future theoretic revision of the concept and theory of the causes of disease. As the feudal system came to an end, social, political, and economic innovations revealed a change in man's relationship with and dominion over nature. The birth of the bourgeois city, trade, geographical discoveries, work in factories and mines gave rise to a change in disease processes, and particularly in the way man confronted them. The appearance or the marked increase of certain infectious or epidemic diseases are of great importance, for example plague, syphilis, yellow fever, and, later on, cholera. The modern physician's attention had to be turned to the study of how these diseases were transmitted. This resulted in the discovery or supposition that certain live microorganisms were responsible for them. The theory of *contagium vivum* and the *generatio aequivoca* was used to explain these new facts or suppositions ([5], [18], [24], [37], [45], [57], [61], [62]). Microscopy, new exploratory techniques, and careful observations justified them, and at least gave them empirical validity.

On the other hand, while the cosmos was seen as causing disease by means of nosogenous *constitutiones*, *miasmata*, and *animalcula*, the social order was also seen to produce disease. Following Paracelsus's description in his book *Von der Bergsucht und andern Bergkrankheiten* (1533–1534), observations were made regarding the nosogenous potential of occupational risks to health. In 1700, these developments were reviewed by the Italian Bernardino Ramazzini in his *De morbis artificum diatriba*, a truly encyclopedia on workers' hygiene and occupational diseases. Until then, concern with hygiene was restricted exclusively to the nobility and royalty. But Ramazzini's studies spanned from the working class to the nobility. In 1710, he published *De principum valitudine tuenda*, and from then onwards physicians insisted that social conditions differentially caused disease, depending on one's membership in a particular social class. Thus in the eighteenth century the popular author of hygiene treatises, Tissot, makes a clear distinction between some forms of disease to which the higher or lower classes are respectively more prone. In the former, the disease is more frequent and more sophisticated; in the latter, more infrequent and more natural. Without being fully right, at least with regard to the frequency of many diseases, this account is nonetheless important because it showed that many physicians recognized that different social classes had different propensities for different diseases. There was also the view that there was a natural expression of disease, a "savage" disease in Rousseau's sense. Hence, Tissot affirms that as one goes up on the social scale "one's health seems to go down by degrees", diseases become diversified and complicated, and increase. Their number is large "in the upper bourgeois classes... and it is the greatest possible among the *gens du monde*". The same ideas can be found in Cabanis: "the industrious and active population" leads a more moral and healthy life, is less prone to illness, and in case of illness requires less specialized physicians to cure them ([19], p. 15). Finally, the economic, political, and social bases for the responsibility of governments towards the diseases and the requirements of their more needy subjects were affirmed by J.P. Frank in his well-known lecture *Akademische Rede vom Volkselend als der Mutter der Krankheiten* (1790). His work *Medizinische Polizei* (1784) can be considered the beginning of the new public hygiene.

At the level of *res naturales* and *non naturales*, the Galenic theories likewise began to be dismantled. This can be appreciated in light of the change of view regarding the role of diet in maintaining good health or as being the cause of illness. Certain authors, the strictest followers of Paracelsus, rejected the important role played heretofore by food in medical theory, replacing it by polypharmacy. Others, on the contrary, accepted it and tried to renew it. Thus, S. Blankaart spoke of "acidifying" and "alkalinizer" diets and considered illness to be a thickening and acidosis of the blood. A stable diet began to be investigated in terms of different nosological and nosogenical criteria. Similar investigations were undertaken by Tissot and J.P. Frank, who correlated food and the social class with different levels of health ([2], [11], [15]).

However, a new causal theory was not possible in medicine without first having a new and comprehensive theory of disease. Galenic essentialism resolved nothing, nor did attempts at explanations based merely on physics and chemistry. Sydenham's new ideas were merely a step forward. His rejection of philosophical and even physiological explanations had delayed a forthcoming solution to the problem. Moreover, this allowed Galenism to persist and it even permitted various other systems which were later proven untenable. Moreover, Sydenham's most faithful followers, the nosographists, initiated a new reification of the processes of illness, which Foucault was to call a "medicine of species". Let us consider briefly this phase of nosographic thinking.

In seventeenth- and eighteenth-century medicine, the primordial clinical rule was to fix the species. "Never treat an illness without first making sure of the species" ([20], p. 2), Gillbert used to say. From the outset of the new science and the new philosophy (Locke, Hume), an essentialist study of the nature of illness was no longer possible. However, a new "essence" was found in the clinical configurations of diseases and these were again converted into things. The "essences" of species were defined by their historical configurations; analogies of the species shaped their relations and their laws of production. Also, as L.S. King affirms, "The disease is that which it is defined to be" ([18], p. 23). Causes, understood as attendant circumstances in terms of the Galenic multicausal scheme, formed part of this same definition as components of the clinical table. Causes, likewise, entered as definitional elements sought in order to classify and identify the disease. Thus, Boerhaave eagerly looked for the remote causes, the whole set of which forms the proximate cause. For Boerhaave, this cause necessarily acts in determining and defining the disease. "The presence of this cause makes and continues the disease, and the absence of it removes the disease". In other words, "The cause of disease is the same thing as the disease itself" ([28], p. 23). The individual, with these attendant circumstances, is still an accident, a modulation of those fixed species.

The difficulties did not frighten off the "nosographists" from this task of "tracking down" these species in their most natural state. As zoologists, they were attempting to make museums in their clinical hospitals. The "savage" disease sought shelter in the family, attended by a non-denaturalizing, wait-and-see medicine. These authors followed Sydenham, who introduced classificatory thought into medicine. They also followed his firm belief in a natural order, similar to the one shown by scientists regarding the vegetable and animal world [36].

Clinical teaching presents what is already known. It moves from established knowledge towards the unknown. Since the truth is still synthetic, everything is given and the manifestations are merely its consequences. Moreover, it is a science or knowledge complete in itself. Clinical hospitals, just like scientific-medical treatises and books, collect and classify the organized body of nosology. Clinics were neither open, nor specialized, as they will be later. The choice of patients was made qualitatively according to those pre-established species. The patient was merely a case, who in his clinical examination or autopsy confirmed what was already known. Teaching and the clinic in this context were really "décryptements" [19].

V. MODERN TIMES

The teaching, scientific, and hospital institutions had to change, but this only became possible at the end of the eighteenth century, when the bourgeoisie came to power. Physicians effected a new system of alliances with the newly constituted power and a new dominion of medical experiences was established. Free enterprise and political liberalism were reflected in the medical world. Institutions were reformed; the traditional medical corporations were abolished; dogmatic teaching in universities and clinical schools ceased; the ancient hospitals considered as asylums or prisons were transformed.

The French example is obvious. The Société Royale de Médicine or the revolutionary reform projects illustrate the new medical aspirations. The Societe, as opposed to the Faculté, began its activities during the last quarter of the eighteenth century. It inaugurated a new and more all-embracing system, and likewise a more Hippocratic point of view. Its members were in charge of studying diseases with a threefold task; survey, elaboration, and control of data. The pathological world was explored on three classic levels: meteorological observations (pressure, temperature, wind), observations of human life (air, water, places, land, temperaments, societies)[3] and common and unusual diseases.

In these complications, the individual, the society, and the cosmos were taken into account [19]. Hospitals were made public and open, as was the teaching in them. They were open to all diseases, without prior selection. The new hospital became a neutral, homogeneous field of observation, parallel to the political domain now inaugurated. The new clinical observation no longer had institutional barriers; these also had been wiped away, as had the epistemological ones. Following Condillac and Cabanis there was a search of a universal element to be attained through analysis. Likewise, there was a search for an operational logic to be applied to that element in order to secure new knowledge. This new synthesis was not yet sure whether to follow the path of language or that of calculation. Clinical observation, understood analogically as a beautiful sensitivity, functioned according to the scientific standard of chemistry, seeking a chemical type of analysis, working like fire which achieves elemental purity in chemistry. The existence of such elements was supposed or postulated, as also was a series of principal and limited pathological phenomena that would in addition assure the alphabetic structure of the disease. Thus, an authentic nominalist reduction, which had been sought over the centuries, was obtained by converting a disease into a name, while at the same time supposing a verbal structure for it. Barthez was the first to use the term "éléments morbides" [19].

It was recognized that disease (or death, as was shortly to be said) itself caused decomposition of the body in a natural way. Disease itself was localized not only in a logical or natural classification, but in the space offered it by the human body. Hippocrates sought the localization in individuals; Morgagni found it in organs; Bichat saw that the bearers of disease were simpler elements, namely, the tissues of his *Anatomie Générale* (1801); Virchow would soon be able to refer to cells. Disease would no longer be a Galenic essence or a Sydenhamian species, but rather a suffering of organs and of functions. The suffering of the human body between life and death was no longer considered a struggle between the natural and the non-natural. Disease was a resistance to death, a resistance that life opposes to death (Bichat). In this way disease acquires a new meaning no longer derived from a natural essence. Disease is an alteration of the vital processes, leading to death, and the vital processes are anatomical and physiological properties ([32], [33], [52]). The study of disease and of medicine generally carries out this analysis, initially of tissues and functions, then of cellular and of physiological properties. This new concept of disease arose in these attempts to locate disease (Morgagni, Bichat), analyze it (Bichat, Broussais), and name it (Fourcroy, Alibert).

Following the anatomo-clinical way of thinking in medicine, localization of disease becomes universal and causal. Perhaps Broussais is a good example of this attempt with his investigation of the localization and causes of fevers. With Broussais, "... the seat of the disease is no more than the link-up point of the irritant cause; a point which in turn is determined by the irritability of the tissue and the irritation force of the agent. The local space of the disease is in turn and immediately a causal space." The disease is: "An organic reaction to an irritant agent; the pathological phenomenon ... is prisoner within an organic framework where the structures are spatial, the determinations are causal, and the phenomenon are anatomic and physiological" ([19], pp. 191–192). A medicine of suffering organs begins, where the physician has to determine which organ is suffering, how it has come to suffer and how this suffering can be avoided. Shortly afterwards, the physiopathological physicians undertook a similar task with regard to suffering and functions. When the physiological function and its alterations are objectified, reduced to quantities and laws, and, in short, studied scientifically, the physiopathological medical scientist will be able to approach the problem of which function is suffering, why this happens, and how to avoid it. And as in the case of the anatomo-clinical researchers, the new concept of life underlies the new science. "The state of illness is no more than physiological life in altered conditions," said Bernard and Frerichs. Laín Entralgo comments on this phrase as follows: "The fundamental principle of physiopathological nosology affirms that there is no essential difference between normal life and ill life" ([34], p. 347).

Anatomo-clinical and physiopathological scientists appeared aloof from the causal problem, but they provided the key for a new approach to the problem. "Medicine of diseases has ended; a medicine of pathological reactions is starting, a structure of experience which has dominated the 19th century and to a certain extent the 20th century, since the medicine of pathogenous agents, not without methodological modifications, will be inserted there" ([19], p. 194).

There were, however, certain types of diseases which traditionally lay outside the scope of classificatory medicine, e.g., infectious diseases, especially the epidemic ones.[4] They were not easy to fit into a classification, as they were defined as collective morbid phenomena often considered unrepeatable. Moreover, they demanded or imposed an obligatory causal consideration, which varied between climatic theories, aerial theories, or live contagiousness. Their behavior, fulminant reproduction, and diffusion along the routes of communication gradually encouraged physicians to opt for the theory of the *contagium vivum*, in other words, to consider epidemic diseases as transmissible by human beings. But this marked causal consideration would take time to become established, so much so that in both Europe and America it was not at all rare to consider cholera in 1832 as a non-specific and non-contagious disease. There are many reasons for this, among them being the following: on a medical theoretic level, the medicine of epidemics had been forgotten during the eighteenth century. On the one hand, the most feared epidemic, the plague, had not reappeared; on the other hand, the medicine of species or the different medical systems (vitalism, and the followers of Brown and Broussais) dominated the medical scene. Moreover, the infectious disease in vogue at the time, smallpox, showed very specific and fixed pathological behavioral norms. Diseases that did not exactly fit these norms were considered neither specific nor contagious. Thus, cholera did not admit of vaccination; it was not transmitted directly between human beings; it was greatly influenced by seasonal variations. Moreover, it affected people who had already suffered from it. All this led to the strangest theories about the new epidemic. It was held to be either not a species at all or else a subspecies together with other choleriform diseases. It was either non-transmissible or else it was transmissible by air, by *miasmata* of the ancient type, or by ferments based on the new chemistry (Berzelius, Liebig).

Underlying these medical difficulties, there were other social, scientific, and economic ones. Generally speaking, the new economic liberalism feared quarantines and the interruptions in trade which the contagionist theories involved. The same occurred in Europe when yellow fever arrived from America. These measures, moreover, were costly and difficult to establish, especially for the new liberal governments. The Sanitary Movement of the Anglo-Saxon countries on the contrary worked well as an empirical matter. But their hygienic rules did not seem to necessitate a contagionist theory. All this was accompanied by the lack of laboratories and microscopes, especially in the Anglo-Saxon area.

But everything would soon change, because after the first outbreak of cholera many innovations were to appear. A large number of microscopic causal agents were discovered; laboratories and technical advances multiplied (achromatic microscope, tinctures, cultures, inoculations). The research rules of the new bacterial pathology were laid down by Henle, Pasteur, and Koch. On the other hand, the bourgeoisie had firmly secured power and had realized that it was more profitable to impose quarantines than suffer the consequences of an epidemic. It is a curious fact that in the United States of America, the consideration of cholera as a contagious disease requiring isolation was favored by the success of the Northern protectionist states. Thus after the War, in 1866 the New York Academy of Medicine declared the need to act as if one were confronted with a transmissible disease, "in accordance with the hypothesis (or the fact) that the cholera diarrhea and rice-water discharges of cholera patients are capable, in connection with well-known localizing conditions, of propagating the cholera poison" ([50], p. 348). And so the Metropolitan Board of Health of New York, acting accordingly, managed to achieve a notable decrease in its mortality statistics. On the other hand, and parallel to this, the Sanitary Movement failed in the case of many diseases because the new labor and urban conditions could not be resolved without a new theory of infectious diseases.

A new and comprehensive nosological edifice was being established. The new bacteriology — and with it toxicology (Orfila, Erhlich) — would provide a mechanism for causal explanation which would complete the clinical and pathological discoveries. It was now possible to fit a syndrome, a cause, and a nosological explanation into an adequate framework. But there was still a risk here, another possible reification of disease, identifying the *causa morbi* with the *ens morbi*, as Klebs tried to do. Rudolf Virchow violently reacted against this, establishing the difference between an irritant agent (microbe) and the bearer of the disease (cells), thus overcoming this danger. Thanks to the work by Robert Koch and the enumeration of his rules, a new causal way of thinking was established, directly influenced by induction and positivist determinism ([34], pp. 389–390; [27]). The works of Auguste Comte and J.S. Mill supported the validity of these causal inferences.

But it would not only be in the field of environmental pathogenic agents that a causal account of disease could be developed. Investigations continued to be elaborated at other levels, in the social and individual sphere. The nosogenic potentiality of the new industrial society is shown in surveys and statistics. The works of Tissot and J.P. Frank are closely followed by Thackray, Chadwick, and Villermé. All modern hygiene has its starting point here. On the other hand, not only was it seen that social conditions injured the lower and working classes due to their harsh life, but it was also recognized that mental health was compromised by the new socio-industrial structure. As a result, both higher and lower classes suffered. The clinics of Charcot and Freud were filled with hysterical people of very diverse backgrounds: some from the laboring classes whose work in Paris had damaged their health: others from the bourgeoisie, rendered ill in their leisure by sophisticated psychic traumas. This double lesional possibility of society, socio-material and socio-psychic, adds a new nosogenic level to human life. Towards the middle of the nineteenth century, the necessity of an authentic social medicine was outlined by Virchow and S. Neumann. At the beginning of the twentieth century this project became a true scientific reality, thanks to the work of Gottstein and Grotjahn.[5]

Finally, with the birth of modern medicine, the nosogenic potential of individual causes was not forgotten. In this way, the classic causal scheme was maintained, although with quite a different content and interpretation. At any rate, the importance of the causes internal to the ill person himself were by no means overlooked by the new medicine. The discovery of the laws of heredity and genetics was a fundamental breakthrough (Galton, Mendel). Similarly fundamental were the various attempts at objectifying the role played by the individual constitution as a nosogenic predisposition (Sigaud, Viola, Kretschmer, Lombroso). The study of gametogenesis and embryogenesis, on the other hand, enabled many diseases to be traced to congenital factors. Also, acquired living conditions — formerly known as *sex res non naturales* — gave rise to the new hygiene, particularly personal hygiene. C.W. Hufeland, with his work *Makrobiotik* (1823), was one of the pioneers in this field.[6]

As the modern age dawned, the very ancient causal scheme appeared as follows:

A. *Environmental causes.*
1. The cosmic or telluric environment: this led to an ample literature on topography, geography, and medical cosmology.
2. Lesions and aggressions by material agents of organic or inorganic nature, highly variable in size and activity.
3. Chemical lesions: toxic and poisonous, organic or inorganic.
4. Contagion through live agents: parasites, bacteria, and viruses.

B. *Causes arising from the social environment.*
1. The socio-material origin of disease in a class structured society.
2. The psycho-social origin of disease: psychic traumas of very diverse natures deriving from affective, family, work, or social relations.

C. *Causes due to the individual's biological environment.*
1. Heredity and constitution.
2. Fecundation, gametogenesis, or embriogenesis.
3. Acquired: biological or psycho-social.

VI. CONCLUSIONS

In spite of the positivist-determinist optimism, the traditional causal way of thinking came to be acknowledged as indefensible. Neither sufficient nor sole cause could be found to exist. Perhaps in a few diseases, as Koch showed in the case of tuberculosis, it is possible to demonstrate the need for some particular cause, but not that the cause is unique or sufficient. The cause-effect necessity disappears, just as it did in physics and philosophy. Physicians followed suit, because for them as well neither sufficient nor sole causes of disease were any longer possible. That is, neither strict identifications between syndromes and their causes, nor determinist explanations of signs and symptoms were possible. In fact, in the medical and human struggle against disease, a completely new era of thinking had been ushered in where the Aristotelian cause-effect scheme had collapsed. Replacing this scheme were systems of relations, with more or less logical validity, permitting various levels of abstraction in the quest for models that permit the disease to be apprehended, understood, and controlled. These systems attempt to relate signs and symptoms with diverse causal factors and with theoretic explanations on an anatomo-physiological level. Thus, the different causal factors fit into a new relational formulation: "a relation identification or pattern-relation analysis — a relation of variables is organized in a model of explanation with a nosological structure (i.e., according to the laws of pathology) to account for the pattern of signs and symptoms constituting a syndrome" ([16], p. 233).[7]

Consejo Superior de Investigationes Científicas
Madrid, Spain

NOTES

* The editors wish to thank Michael C. White for the original translation of this essay.
[1] I am speaking about causality in the first of the possible meanings established by M. Wartofsky ([63], p. 292).
[2] L.J. Rather defines the *sex res non naturales* (i.e., air, food and drink, motion and rest, sleep and wakefulness, emotions, secreta and retenta) as follows: "The concept of doctrine may be stated briefly as follows: *there are six categories or factors that operatively determine health or disease, depending on the circumstances of their use or abuse, to which human beings are unavoidably exposed in the course of daily life*" ([48], p. 337).
For further studies of Hippocrates, Galen, and their followers, see: [4], [9], [21], [22], [25], [35], [38], [39], [44], [60], [64].
[3] For the influence of traveling, maps, and climate on the medical thinking, see: [26], [40], [59].
[4] For further studies of the history of epidemics and contagious agents, see: [1], [7], [8], [12], [13], [42], [47], [50], [56], [58].
[5] For recent analysis of the use of social and statistics studies, see: [6], [20], [23], [53].
[6] Concerning hygiene, see [54], [55]. About Cesare Lombroso [46]. Also Peset, J. L.: 1983, *Ciencia J Marginación*, Critica, Barcelona.
[7] This essay is dedicated to Agustín Albarracín, Diego Gracia Guillén, Pedro Laín Entralgo, and Tristram Engelhardt.

BIBLIOGRAPHY

1. Ackernecht, E.H.: 1948, 'Anticontagionism Between 1821 and 1867', *Bulletin of the History of Medicine* **22**, 562–593.
2. Ackernecht, E.H.: 1971, 'The End of Greek Diet', *Bulletin of the History of Medicine* **45**, 242–249.
3. Agassi, J.: 1976, 'Causality and Medicine', *The Journal of Medicine and Philosophy* **1**, 301–317.
4. Albarracín-Teulón, A.: 1970, *Homero y la Medicina*, Prensa Española, Madrid.
5. Allen, P.: 1947, 'Early American Animalcular Hypotheses', *Bulletin of the History of Medicine* **21**, 734–743.
6. Bariety, M.: 1972, 'Louis et la méthode numérique', *Clio Medica* **7**, 177–183.

7. Bates, D.G.: 1965, 'Thomas Willis and the Epidemic Fever of 1661: A Commentary', *Bulletin of the History of Medicine* **39**, 393–414.
8. Biraben, J.N.: 1975–1976, *Les hommes et la peste en France et dans les pays européens et méditerranéens*, Mouton, Paris.
9. Bylebyl, J.J.: 1971, 'Galen on the Non-natural Causes of Variation in the Pulse', *Bulletin of the History of Medicine* **45**, 482–485.
10. Dagi, T.F.: 1976, 'Cause and Culpability', *The Journal of Medicine and Philosophy* **1**, 349–371.
11. Debus, A.G.: 1972, 'The Paracelsians and the Chemists: The Chemical Dilemma in Renaissance Medicine', *Clio Medica* **7**, 185–199.
12. Doetsch, R.N.: 1962, 'Early American Experiments on "spontaneous generation" by Jeffries Wyman (1814–1874)', *Journal of the History of Medicine* **17**, 325–332.
13. Doetsch, R.N.: 1964, 'Mitchell on the Cause of Fevers', *Bulletin of the History of Medicine* **38**, 241–259.
14. Eisenberg, L.: 1976, 'The Outcome as Cause: Predestination and Human Cloning', *The Journal of Medicine and Philosophy* **1**, 318–331.
15. Elaut, L.: 1968, 'Les règles d'une gastronomie hygiénique, exposées par le médecin-humaniste Georgius Pictorius', *Clio Medica* **3**, 349–359.
16. Engelhardt, H.T., Jr.: 1974, 'Explanatory Models in Medicine: Facts, Theories, and Values', *Texas Reports on Biology and Medicine* **32**, 225–239.
17. Engelhardt, H.T., Jr.: 1976, 'Ideology and Etiology', *The Journal of Medicine and Philosophy* **1**, 256–268.
18. Foster, W.D.: 1965, *History of Parasitology*, Heinemann, London and Edinburgh.
19. Foucault, M.: 1963, *Naissance de la clinique*, P.U.F., Paris.
20. Gilbert, R.B.: 1965, 'Health and Politics: The British Physical Deterioration Report of 1904', *Bulletin of the History of Medicine* **39**, 143–153.
21. Gracia Guillén, D.: 1973, 'El estatuto de la Medicina en el "Corpus Aristotelicum"', *Asclepio* **25**, 31–63.
22. Gracia Guillén, D. and Peset, J.L.: 1972, 'La medicina en la baja edad media latina', in P. Laín Entralgo (ed.), *Historia universal de la medicina*, 3 vols., Salvat, Barcelona, pp. 338–351.
23. Hatzfeld, H.: 1971, *Du pauperisme a la sécurité sociale*, A. Colin, Paris.
24. Holmes, C.: 1966, 'Benjamin Rush and the Yellow Fever', *Bulletin of the History of Medicine* **40**, 246–263.
25. Jarcho, S.: 1970, 'Galen's Six Non-naturals', *Bulletin of the History of Medicine* **44**, 372.
26. Jones, M.O.: 1967, 'Climate and Disease: The Traveler Describes America', *Bulletin of the History of Medicine* **41**, 254–266.
27. King, L.S.: 1952, 'Dr. Koch's Postulates', *Journal of the History of Medicine* **7**, 350–361.
28. King, L.S.: 1963, 'Some Problems of Causality in Eighteenth Century Medicine', *Bulletin of the History of Medicine* **37**, 15–24.
29. King, L.S.: 1964, 'Stahl and Hoffmann: Study in Eighteenth Century Animism', *Journal of the History of Medicine* **19**, 118–130.
30. King, L.S.: 1969, 'Medicine in 1695: Friederich Hoffmann's *Fundamenta Medicinae*', *Bulletin of the History of Medicine* **43**, 17–29.

31. Kudlien, F.: 1968, 'Der Arzt des Körpers und der Arzt der Seele', *Clio Medica* **3**, 1–20.
32. Laín Entralgo, P.: 1946, *Bichat*, C.S.I.C., Madrid.
33. Laín Entralgo, P.: 1947, *Claude Bernard*, C.S.I.C., Madrid.
34. Laín Entralgo, P.: 1961, *La historia clínica. Historia y teoría del relato patográfico*, Salvat, Barcelona.
35. Laín Entralgo, P.: 1970, *La medicina hipocrática*, Revista de Occidente, Madrid.
36. Laín Entralgo, P. and Albarracin Teulón, A.: 1961, *Sydenham*, C.S.I.C., Madrid.
37. Lieber, E.: 1970, 'Galen on Contaminated Cereals as a Cause of Epidemics', *Bulletin of the History of Medicine* **44**, 332–345.
38. López Férez, J.A.: 1971, *Las ideas médicas de Demócrito y su influencia en el Corpus hippocraticum* (thesis unpublished), Madrid.
39. López Férez, J.A.: 1974–1975, 'Las ideas de Demócrito sobre salud y enfermedad y su posible influjo en el Corpus hippocraticum', *Asclepio* **26–27**, 157–165.
40. Lorch, J.: 1965, 'Latham on the Etiology of Plague, 1900', *Bulletin of the History of Medicine* **39**, 79–80.
41. Luria, S.E.: 1976, 'Biological Aspects of Ethical Principles', *The Journal of Medicine and Philosophy* **1**, 332–336.
42. Middleton, W.S.: 1964, 'Felix Pascalis-Ouvière and the Yellow Fever Epidemic of 1797', *Bulletin of the History of Medicine* **38**, 497–515.
43. Moravsik, J.: 1976, 'Ancient and Modern Conceptions of Health and Medicine', *The Journal of Medicine and Philosophy* **1**, 337–348.
44. Niebyl, P.H.: 1971, 'The Non-naturals', *Bulletin of the History of Medicine* **45**, 486–492.
45. Pagel, W. and Winder, M.: 1968, 'Harvey and the "Modern" Concept of Disease', *Bulletin of the History of Medicine* **42**, 496–509.
46. Peset, J.L. and Peset, M.: 1975, *Lombroso y la escuela positivista italiana*, C.S.I.C., Madrid.
47. Peset, M. and Peset, J.L.: 1972, *Muerte en España*, Seminarios y Ediciones, Madrid.
48. Rather, L.J.: 1968, 'The "Six Things Non-natural": A Note on the Origins and Fate of a Doctrine and a Phrase', *Clio Medica* **3**, 337–347.
49. Rogers, F.B.: 1965, 'Shadrach Ricketson (1768–1839): Quaker Hygienist', *Journal of the History of Medicine* **20**, 140–150.
50. Rosenberg, C.: 1960, 'The Cause of Cholera: Aspects of Etiological Thought in Nineteenth Century America', *Bulletin of the History of Medicine* **34**, 331–354.
51. Rosenkrantz, B.G.: 1976, 'Causal Thinking in Erewhon and Elsewhere', *The Journal of Medicine and Philosophy* **1**, 372–384.
52. Schiller, J.: 1967, *Claude Bernard et les problèmes scientifiques de son temps*, Les éditions du Cèdre, Paris.
53. Schmitz-Cliever, E.: 1968, 'Bevölkerungsstatistische Beobachtungen de Golbery's im rheinischen Roerdepartment (1804–1809)', *Clio Medica* **3**, 361–370.
54. Schipperges, H.: 1962, *Lebendige Heilkunde*, Walter, Olten.
55. Schipperges, H.: 1970, *Moderne Medizin im Spiegel der Geschichte*, G. Thieme, Stuttgart.
56. Shrewsbury, J.F.D.: 1970, *A History of Bubonic Plague in the British Isles*, University Press, Cambridge.

57. Shryock, R.H.: 1972, 'Germ Theories in Medicine Prior to 1870: Further Comments on Continuity in Science', *Clio Medica* **7**, 81–109.
58. Siegfried, A.: 1960, *Itineraires de Contagions: Epidemies et Idéologies*, A. Colin, Paris.
59. Stevenson, L.G.: 1965, 'Putting Disease on the Maps. The Early Use of Maps in the Study of Yellow Fever', *Journal of the History of Medicine* **20**, 226–261.
60. Temkin, O.: 1978, *Galenism. Rise and Decline of Medical Philosophy*, Cornell University Press, Ithaca and London.
61. Théodorides, J.: 1966, 'Les grandes étapes de la parasitologie', *Clio Medica* **1**, 129–145, 185–208.
62. Théodorides, J.: 1972, 'L' influence de la parasitologie sur le développement de la médecine clinique', *Clio Medica* **7**, 259–269.
63. Wartofsky, M. (ed.): 1976, 'Causality in Medicine', *The Journal of Medicine and Philosophy* **1**, 289–300.
64. Withington, E.T.: 1964, *Medical History from the Earliest Times*, The Holland Press, London.

DIETRICH VON ENGELHARDT

CAUSALITY AND CONDITIONALITY IN MEDICINE AROUND 1900

I. INTRODUCTION

At the turn of this century in Germany, there was an intense discussion focused on the topics of "causalism" and "conditionalism". This essay looks at the interplay between competing views of the concept of "cause" as it was understood in the early 20th century.[1] Although the following analysis is anchored in a particular historical time period in philosophy and medicine, one is encouraged to appreciate the more conceptual issues at play, namely, those which emerge in the interplay between competing views of the concept of "cause".

The themes found in discussions of the concept of "cause" as it was construed in the early 1900s reflect, in part, a general concern about the competitive relationship between the natural sciences and medicine. Assuming that medicine is primarily an art, the conflict engendered by this competitive relationship was taken quite seriously by philosophers and medical professionals alike at the turn of the century.[2] The way in which this dualism is interpreted determines the concept of medicine as theory and practice, as a discipline capable of rendering verifiable data as well as of fashioning the means to health from a diseased or ill state.

The discussions in the field of medicine have a historical relationship to those in philosophy and the theory of science.[3] Bacon, Spinoza, Locke, Hume, the Idealists, Schopenhauer, Comte, Mill, Kirchhoff, Mach, and Vaihinger might be mentioned as key influences. At the same time the nineteenth century tension between the humanities and the sciences becomes significant. Interest in theoretical problems decreases among medical scholars and philosophy is frequently labeled useless and damaging.[4] Thus, the ideological as well as methodological reflections of the naturalists and physicians are usually not up to the level of philosophical discourse. Within medicine, theoretical studies appear independently in the physical, biological, and medical disciplines, and even at that time the theory of science is a heterogeneous concept.

An analysis of medical problems at the turn of the century using categories and perspectives provided by positivist philosophy has yet to be carried out.

Corinna Delkeskamp-Hayes and Mary Ann Gardell Cutter (eds.), Science, Technology, and the Art of Medicine, 75–104.

Such a study might compare the relationship between understanding reality and predictability (as construed by Schlick, Bacon, Hume, or Comte) and the relationship between diagnosis and prognosis in medicine. Yet there are objective reasons for the considerable independence of these different strains of thought and their development. The problem of causality was reconsidered in positivism under the influence of the theory of relativity and quantum mechanics. In medicine, however, immanent changes led to new theoretical assessments.

Medical discussions of causality and conditionality were carried out against the background of the controversy between Virchow's cellular pathology and bacteriology, also referred to as the controversy between etiology and pathology. The controversy about the importance of environmental conditions in epidemiology and hygiene also contributed to the intellectual milieu. The categories had been historically established. Sydenham had pointed out the importance of environmental conditions in epidemics. The Brownian theory of sensitivity had dealt with an organism's sensitivity to irritation. According to Lotze and Henle, the causes of disease and disease events cannot be clearly separated from one another; there is a continuum between disease and health. The intensity of the irritation supposedly determines whether healthy life is maintained or destroyed. Next to irritation, disposition and constitution are further factors responsible for disease. Around the middle of the century, Henle wrote that infectious diseases were dependent on "parasitic organisms". According to Uhle and Wagner, the origin of disease cannot be understood by reference to a necessary cause-effect relationship. With auxiliary concepts such as "external" and "internal" causes, or "essential" and "incidental" causes, the theory of causes supposedly remained the weakest aspect of pathology. These conceptual and linguistic roots will not be pursued here any further. Rather, the deliberations around 1900 will be presented in light of their relationship to the controversy between cellular pathology and bacteriology and the development of the pathology of constitution/resistance.

II

In his view of cellular pathology, Virchow explained disease as an essentially cellular phenomenon. Disease causes are expressly distinguished from disease events. External stimuli can destroy or cripple cells, but a given stimulus does not always produce the same effect. The way in which the cell reacts is the decisive factor. This factor is the internal cause of disease in the cell or its predisposition. For Virchow, disease is, in addition to the well-known formula "life under changing circumstances", the inability of the organism to counteract harmful influences:

> Disease begins at that moment when the regulatory system of the body is not sufficient to overcome a disturbance. It is not life under abnormal circumstances, nor the disturbance as such which produces a disease, rather the disease begins with the insufficiency of regulatory mechanism ([191], p. 93).

This regulatory ability is, according to Virchow, different from individual to individual. When Virchow speaks of a disease entity (*ens morbi*), he refers to the cell as the central physical entity to which that disease is bound. Ontology is empiricized by Virchow. The cell is the essential locus of life, of healthy as well as diseased life. It is recognized as being third in importance next to blood and nerves, and even these developed from cells. Virchow's cellular pathology rules out the possibility of a humoral or solid pathology. For Virchow, disease is a normal cellular phenomenon of life at the wrong place (heterotropy), the wrong time (heterochrony), in the wrong amount (heterometry). It is characterized by danger and damage to various areas of the body, to a bodily function, or to the life of the organism. The problem of cause interests Virchow more than the course of the disease itself.

In the field of bacteriology, on the other hand, important objections to cellular pathology had existed from the end of the seventies. The old theory of an externally living cause of disease now acquired empirical verification through the discoveries of Pasteur and Koch, their colleagues and followers. In direct disagreement with Virchow, Edwin Klebs presented the bacteriologists' objection in 1877. At the 50th Congress of German Naturalists and Physicians, Klebs, a student of Virchow, proposed to the theoreticians and originators of cellular pathology as well as to the physicians of his time, that the principle of etiology had been given too little consideration in the study of disease. Physiology, histology, and anatomy had made astounding progress while the origin of disease processes "had been touched on only rarely and then usually in a totally superficial and theoretical way" ([86], p. 47).[5] This situation impeded the development of causal therapy and left medicine with symptomatic and exclusively expectative therapy. Only in most recent times under the influence of practiced antisepsis was a change visible:

> Our task will be, however, to show that in truth the experience we have had requires us to seek the cause of numerous and important diseases outside of the body and requires that these causes of disease be of a parasitic nature ([86], p. 45).

Mechanical and chemical effects are, according to Klebs, of less importance than those of parasitic infections. Mechanistic disturbances are to be regarded with the same attitude as congenital or acquired anomalies.

Virchow reacted to Klebs at the same congress of naturalists and physicians in his famous speech on "The Freedom of Science in Modern Society" in which he encouraged reservation on the part of scientists in order not to forfeit the independence and respect of science, and also to avoid dogmatism. One could, after all, not maintain that: "all contagious or infectious diseases (were) dependent on living causes" ([189], p. 70). Undoubtedly, there were infections which arose from organic poisons.

In the next year, again before the German natural scientists and doctors, Klebs maintained his conviction in his speech "On Cellular Pathology and Infectious Diseases". He agreed, however, with Virchow's scientific-political reflections and suggestions as well as with his criticism of Haeckel. His objections were now directed against cellular pathology itself. The varying reactions of the body to the same stimulus can only be explained, according to Klebs, with reference to additional external conditions, to a contagion, or infection. There is no actual cell power in and of itself: the cells suffered changes in a passive state. They are secondary "to specific transformations of the whole organ". Klebs denied "the autonomy of the cell as a principle of disease" ([87], p. 132), not only in parasitic diseases, but indeed in all other diseases as well. In his opinion, cellular pathology could not be recognized as a universal theory of disease. It could not be denied:

> that the theory of cellular pathology valid until now is not sufficient to explain the most important phenomena in this area and therefore must be considered a formidable hindrance to progress in this area which is of such great practical import as well ([87], p. 179).

Klebs' conviction that cellular pathology guided therapy in the wrong direction engendered his criticism. Therapy has to be directed at the true cause, and this cause lies outside the organism. Cellular pathology has not contributed to therapeutics because of its neglect of the preconditions of cellular change and of the causes of diseases in the absence of cellular changes.

> The great therapeutic results of which the present time can boast are outside the realm of cellular-pathology. *We have no cellular therapy, and if we did have one, we would not reach our goal through its use anyway* ([87], p. 133).

In 1880, Virchow attempted to refute Klebs' objections. As he explained, cellular pathology stands between a radical infectionistic perspective and a radical embryological, genetic perspective ([190], p. 196). Klebs had failed to formulate a basic principle. Cells can still be considered "carriers of life as well as of disease" ([190], p. 5). The pathological process is described by Virchow in the following manner:

Something external affects a living cell and changes it in a mechanical or chemical way. The external factor is the *Causa externa*, or, as one usually says, the *cause of disease*. The changed condition is called in contrast *Passio, suffering*. If an action (*actio-reactio*) occurs in the living cell after the change which it has experienced, then the change is called irritation (*irritamentum*) and the cause of disease is called an irritant. If instead no reaction occurs, and the situation is limited to the change which the cell had "suffered," then we are dealing with a mere *disturbance* (*laesia*) or with a *debilitating* (paralysis). Yet since the same cause can have an irritating effect on one cell, a disturbing effect on another, and a crippling effect on a third, one assumes that a certain difference in the internal conditions is the reason for this varied behavior. Thus, one comes to the internal cause or the *praedispositio* ([190], p. 84).

In Virchow's opinion, external agents cannot be the sole cause of disease. Even the physiological reaction of tissue demonstrates a specific disposition. It can only be argued as to whether external causes or internal arrangement is more important in the origin of disease. He had always emphasized the importance of the *causae externae*, but they do not constitute the essence of disease, the course of which cellular pathology sought to explore. For this reason it could deal only cursorily with the causes. One must be on guard against "*confusing the essence of disease with the cause of disease*" ([190], p. 10). Pathology cannot be restricted to etiology. Virchow admits that a cellular therapy was not yet available. At the same time, he emphasizes that therapy and prophylaxis were both goals of cellular pathology, and that pathology has the right to investigate disease, without always having to consider therapy at every turn. While recognizing the theoretical and practical achievements of bacteriology, Virchow held fast to this basic interpretation during the following years.

Later Klebs, also taking up Schopenhauer's theory of causes, adopted a more reserved point of view. Infectious diseases can be explained neither solely from the viewpoints of the bacteriologist, nor from that of the cellular pathologist. Infectious diseases are, to him, a battle between bacteria and cells ([88], 1, p. vi, 10ff; [89], p. 14ff).

In this conflict between pathologists and bacteriologists about the cause of disease, quite different viewpoints were espoused over these years. Many bacteriologists criticized cellular pathology and defended the monocausal exogenic viewpoint. The parasite was felt to be the only cause of disease, and disposition was explained away as unimportant. The tuberculosis bacillus is for Cornet the "sole cause of tuberculosis" ([24], p. 192). The "unfortunate assumption of the importance of disposition as a second cause for tuberculosis along with or above the bacillus" ([24], p. 300) is to be discounted. The concept "etiology", general in itself, was at this time understood in terms of exogenous (parasitic) monocausality. This is also what was meant by the expression "etiological period". Cellular pathologists of course defended one-sided convictions against the bacteriologists as well. However, influential bacteriologist, Ferdinand Hueppe, a student of Robert Koch, attempted a reconciliation between the two sides.

III

The view that bacteria are the sole pathogens was continually opposed by Hueppe who wanted to "defend Koch against Koch" ([71], p. 110; see also, [70], [72], [73], [74], [75], [76]).[6] His presentation on "On the Causes of Fermentation and Infectious Diseases and the Relation Thereof to the Causal Problem and to Energetics" at the 1893 Congress of German Naturalists and Physicians is an exemplary attempt to combine bacteriology with cellular pathology, as well as biology with energetics, and to devise a new causality in medicine. In this he was influenced by Johannes Muller, Robert Mayer, Helmholtz, Kirchhoff, and Mach. Muller's law of the specific energy of the sensory nerves had already shaped the principle of disposition in Virchow's cellular pathology. Robert Mayer and Helmholtz, according to Hueppe, provided the basis for a comprehensive theory of the origin of disease in terms of a mechanical theory of equilibrium. In Mach he found the causal relationship adequately replaced with a functional interrelationship. In contrast to the prevailing view in both bacteriology and cellular pathology, Hueppe opposed the importance of the infectious stimulus as well as of the diseased cell; simultaneously he recognized the fundamental importance of both Koch's and Virchow's views. He believes Virchow to be the greatest pathologist "of all time" and found it difficult to understand why he did not immediately recognize and support the significance of bacteria research for cellular pathology ([71], p. 107). Predisposition, stimulus, and external conditions are, in Hueppe's view, equally important determinants of disease; they form the causal scheme of disease. They each contribute to the disease process in a dynamic or energetic way. Disease is "a function of variable predisposition, variable stimulus, and variable external conditions" ([45], p. 217). In the disease process, latent energy is converted to kinetic energy. "Cause" as the substrate of the latent energy is assigned to disposition; the "external conditions" take account of the localist temporalist viewpoint of Pettenkofer, Simon, Farr, and Parkes. Causal therapy, like hygiene, according to Hueppe's causal scheme can set in at three places. Thus, the "hygiene of constitution" can well be rendered compatible with a "hygiene of cause" (Pasteur, Koch) and with " conditional hygiene" (Pettenkofer) ([71], p. 123).

For Hueppe, regard for the individuality of the sick person is crucial. For him it is a matter of placing "the person himself with his congenital and acquired predispositions at the center of medical thought instead of external conditions and bacteria" ([71], p. 123). Next to negative reactions from bacteriologists, as from Koch, Gottstein ([47], [48]) and Buchner [16] agreed with Hueppe's interpretation. In his diphtheria research, Gottstein reached the conclusion:

> The acceptance of a purely contagionistic viewpoint for the origin and spread of endemic diphtheria, as well as for the existence of a general disposition, is inadmissible in view of findings in bacteriology as well as observations at the bedside ([49], p. 595).

Science will overrule the one-sidedness of bacteriology and, even in the context of the history of science, affirm the Pflugerist law of teleological mechanics, "according to which the cause of a movement is also the cause of its ultimate neutralization" ([48], p. 595).

Next to Hueppe, the position of the clinician Friedrich Martius was particularly influential and worthy of note.[7] His critique of an absolutized "parasitic pathology" and his work in further development of the cellular pathological ideas of Virchow led Martius to constitutional pathology, or "constitutional thought", as he preferred to call it, in order to express the impossibility of a pathological system based on merely one principle. Before Martius, Rosenbach [150] had opposed bacteriology. He had relativized the pathogen-effect of bacteria and emphasized the importance of disposition, within the context of a Darwinian and energistic theory of disease. Bacteria are the instigation and stimulus of disease but not its cause; the cause lies in the weakness of the organism.

In the lecture "Causes of Disease and Disease Predisposition" (presented in 1898 to the German naturalists and physicians and considered by the author to be his "confession of faith in pathogenesis", [119], p. 21), Martius explicitly joined the Virchow-Klebs controversy and agreed with Hueppe's criticism of a one-sided 'bacteriologism' as well as with his interpretation of the concept of causality. No one can deny the tremendous positive contribution of bacteriology, yet at the same time, he wished to orient his general pathogenetic perspective more around Virchow, with the central reservation that, while indeed the cell is "the substrate, in which disease processes take their course" ([122], p. 91), the essence of disease remains incomprehensible. In addition, the law of the relativity of disease causes could not be denied ([122], p. 96). Martius, who was incorrectly considered an absolute opponent of bacteriology, supported, as did Hueppe, the energism of Mayer and

V. Helmholtz and the descriptivist and functionalist positions of Kirchhoff and Mach. Constitution pathology cannot depend on "cause" as a sole precondition; man has to free himself from that one-sided, naive etiological way of thinking ([122], p. 93). Multi-facetedness and descriptivist approaches are closely related:

> The more aspects of the exceedingly complicated process that we recognize, the better we can describe that process most thoroughly, most completely, and thereby most simply. Our human potential for knowing is capable of no more ([122], p. 92).

To be sure, for Martius the disease tendency is decisive. The cell is considered the location of resisting power and the location of the tendency toward disease. "Constitution" as the specific sensitivity to irritation is founded in it. Indeed, the cause of disease lies within the organism; the "triggering mechanisms", in previous terminology, "external causes of disease", are stimuli, poisons, irritants.

Martius interpreted disposition and stimulus, as did Gottstein [47], as a variable value in the mathematical sense. He formulated the mathematical symbol for the etiology of disease, where 'w' represents the resistance ability and 'p' represents the disease-effecting stimulus. Under 'p', Martius includes "not only bacilli, but also, need and social distress, worry and trouble, undernourishment and catching a chill, in short all weakening and debilitating stress in life" ([122], p. 105). This formula has been adopted and modulated in various ways. It was incorporated into medicine as the "Gottsteinian, Martiusian, and Strumpellian formula". However, the mathematical concept of function was not considered binding by all medical professionals of the period. The fact that etiological factors could not be given particular values set limits on the mathematization of medicine.

The indifference of bacteriology towards individual and social factors was also avoided by Martius who adopted a multi-factorial viewpoint. Constitutional pathology was valid for psychic disease as well. Hueppe[8] and Martius contributed significantly to the spread of the new interpretation inside and outside of Germany during the years following 1898. Constitution and immunity were investigated in many diseases and empirically tested. In the field of infectious disease a broad spectrum from an almost absolute effectivity of bacteria to nosoparasitism (Liebreich [98]) was revealed. Eugenics and eugenic research were given a new foundation. Even with regard to basic principles, however, the views during the following period were not at all unanimous ([165], p. 21).

The controversy between cellular pathology and bacteriology directed attention to the concept of cause in a fundamental way. A monocausal etiology had proven unacceptable. The importance of empirical findings of diagnosis and therapy has been emphasized again and again in the discussions of the theory of medicine which followed. At that time, the theory of medicine was not studied in isolation from the practice of medicine. With "conditionalism", the current understanding of causality in medicine was to have been overcome. Notable proponents of conditionalism were Verworn and von Hansemann. They called the opposing position "Causalism".

IV

The physiologist Max Verworn formulated a widely respected criticism of the causal concept in his monograph "Kausale und konditionale Weltanschauung" in 1912.[9] Verworn saw his concern with the concept of cause as a contribution to the philosophy and ideology of the period and as a discussion of Haeckel's monism and Ostwald's energeticism. Conditionalism or, as Verworn called it, "conditionism" serves as a foundation for his psychomonism ([180–187]). This view was to help overcome the problems of mysticism and illusion which had attended causalism. For employing the causal concept, natural science had allowed, "even in its most exacting areas, a remnant of the old mysticism" ([184], p. 8) to endure. Verworn's views are also influenced by Muller's laws about the specific energy of the sensory nerves and by the Kirchhoff-Mach description. The limits which Du Bois-Reymond had set for human knowledge had made it clear to him that even talking about strength as a cause was inadequate ([184], p. 8). Indeed, conditionalism was to render the two limits about which Du Bois-Reymond had spoken (the connection between energy and matter, the relationship between mind and matter) obsolete [183]. According to Verworn, conditionalism consists of five principles:

1. *There are no isolated or absolute things.* All things, that is, all processes or states are conditioned by other processes or states. (*Law of the conditioned character of all being and events.*)
2. *There is no process or state that is dependent on only one factor.* All processes and states are conditioned by numerous factors. (*Law of the plurality of conditions.*)
3. *Each process or state is univocally determined by the sum of its conditions.* Only under the same conditions do the same processes or states arise, and conversely, different states are engendered by different conditions. (*Law of univocal conformity.*)

4. *Each process or state is identical to the sum of its conditions.* The totality of the conditions represents the process or condition. (*Law of identity.*)
5. *All of the conditions of a process or state are equally important for its coming to existence, to the extent that they are necessary.* (*Law of the effective equivalence of the conditioning factors.*) ([184], p. 45)

In Verworn's opinion, the causal view fails in the face of the complexity of reality. "Reality consists of a continuous continuity of things." For this reason, he said, the concept "condition" has been introduced in natural science along with the concept "cause". Yet this position has not been consistently upheld, because from among the determining factors one would always be singled out as the cause. Absurdly, it is not common for the final contributing factor to be given the status of cause. Conditions, however, occur within dependent relationships characterized by necessity, and here a progression in decisiveness is unthinkable. For Verworn, finality is just as untenable as causality. Teleological considerations, a vitalistic dualism, have no place in natural science. Along with monocausality and any theoretical hierarchization of conditions, the distinction between a thing and its nature is eliminated. The natures of things are, according to Verworn, again only things which exist in a regular relationship with other things, and from this lawlike interrelation no thing can escape. Questions concerning the "thing in itself" are senseless. The conditions are the "natures" of things. Only processes and states in their regular relationship with a complex of conditions, understood in the sense of a Kirchhoff description and theoretically open to a mathematical mode of representation, could provide the goal of science.

In order to disprove the existence of any supposed limits to knowledge, Verworn applied conditionalism to concrete phenomena as well. He applied it to the body-soul problem where psychophysical parallelism was replaced by a psychomonism, to the principles of the organism where teleology dissolves into conditionalism, to the doctrine of heredity where the concept of inherited substance and tendency to inherit is rejected, to the question of immortality which he believed to be an impossibility, and, finally, to pathology where Verworn refers to the book of von Hansemann which appeared in 1912. The current etiological concept of disease is, according to Verworn, as untenable as the general interpretation of cause. In medicine, one should do without the causal concept. A supposed selection process among interpretations will eventually eliminate this concept. Pathology, with Virchow's cellular pathology as the basis of medicine and the key to the future, should not be oriented solely around the analysis of the diseased organ. Rather, it had to investigate "all of the changes in the whole organism in their dependency on one another" ([184], p. 32).

The pathologist and student of Virchow, David von Hansemann, who in 1912 applied the conditional principle to medicine in greater detail than Verworn, also blamed the mistakes and failures in medicine which had occurred for about 25 years during the etiological period of medicine on the causal principle. He argued that the causal principle could not satisfy the "rigorous standards of logic and science" ([58], p. 5).[10] He was not concerned with the elimination of causal thought, but rather with its competition. Etiology must not be equated with the doctrine of causes; rather, one should understand it as the "doctrine of determining conditions" ([58], p. 2). His theoretical concern with medicine and the problem of cause linked von Hansemann, like Verworn, with the more general scientific and ideological interests ([53], [56]).

The inadequacy of the generally accepted concepts of disease arises, according to von Hansemann, from the fact that "disease or, in a broader sense, pathological processes and phenomena are not absolute, but relative, and are in no way distinctly separable from what we call normal or physiological" ([58], p. 11). Disease is always dependent on the living entity and can be understood in the spirit of Virchow's definition as "an occurrence at the wrong time and at the wrong place and with altered energy" ([58], p. 11).

Within the descriptivist framework, von Hansemann also opposed a monocausal concept of medical causality. Cause is distinguished from condition in that a cause must always have an effect, whereas a determiner could produce a phenomenon only in conjunction with other conditions. In von Hansemann's opinion, it has been a merely superficial interest in therapy and the equation of the goals of medicine with goals of therapy which have promoted the monocausal viewpoint:

> Causal thought has actually become a dogma, particularly because it is exceptionally convenient. It is undeniable that, one can say right away that a disease is produced by a cause, and if one eliminates the cause then one also eliminates the disease, such a simple procedure sounds quite plausible and extremely convenient from a theoretical point of view ([58], p. 22).

In contrast to causal therapy, where causal means "monocausal", prevention requires that several conditions are considered and thus agrees more closely with the conditionalist theory ([58], p. 27). Besides, a therapy developed on the basis of the etiological knowledge of bacteriology had just revealed the faults of bacteriological theory. Conditions must be divided into "necessary" and "substitutable conditions". This distinction is important for von Hansemann with respect to both therapy and prophylaxis. The diagnosis must take multi-factoriality into account.

> The diagnosis should identify the disease as such, that is, as the sum of its phenomena, but it should not arbitrarily isolate individual phenomena and put these in the foreground. A diagnosis should not be arrived at bacteriologically, or anatomically, or biologically. Rather, it should evaluate all of these methods simultaneously and then to the same extent ([56], p. 7).

Von Hansemann based the conditionalistic theory specifically on empirical data.[11] Traumatic conditions and poisonings, tuberculosis and other infectious diseases, non-infectious diseases, tumors, prognoses, and epidemics were interpreted in a conditionalistic perspective. As one of the early opponents of absolutized bacteriology, von Hansemann could not see bacteria as the "cause of disease". Tuberculosis is not infection by tuberculosis bacilla, but rather the reaction of the body to the infection ([55], p. 9). Individual disposition must not be discounted. In contrast to the beginning of the bacteriological era, this factor is indeed rarely neglected by medicine of the early twentieth century. Disposition can be congenital or acquired, and is anatomically, chemically, or biologically demonstrable. Often, however, such attempts at proof do not succeed ([56], p. 29). Thus, "disposition" is used by von Hansemann in a specific sense and is not understood merely as disturbed immunity.

V

The conditionalism of Verworn and von Hansemann had great resonance not only in medicine but in other sciences and in foreign countries as well. The laws of plurality, equivalence, and identity received special attention. Conditionalism stimulated reflection on the concept of cause and spread the idea of the multiplicity of etiological factors. It thereby contributed to a change in the practical application of the causal concept and fell into place with developments in medicine which had already led to a correction of the contemporary medical understanding of cause. Hueppe and Martius, who had contributed to this development along with other medical professionals, could not accept conditionalism. But their interpretation also met with criticism which, however, did not always entail absolute agreement with conditionalism. Thus, the situation with regard to the theoretical account of medicine's views of the problem of causality was quite complicated at the beginning of the twentieth century.

Discussion of conditionalism concerned specific aspects of the problem and proceeded in various degrees of depth. One can criticize Verworn's theory on the grounds that his basic principles could be reduced to one principle. According to Hueppe and Martius who discussed the problem of causality many times in the years following their lectures of 1893 and 1989 ([70], [71], [72], [75], [116–123]), the alternatives of conditionalism and causalism should not be based on ideological viewpoints, but rather be decided only on the level of empirical scientific theory and practical science.

The majority of physicians and naturalists was not prepared to give up the principle of causality. For them, Kant (1724–1804) had correctly envisioned an *a priori* category in causality which made experience possible in the first place. Furthermore, causality was considered to be an essential component, a basic requirement for ordinary consciousness.[12] The self-concept of the patient and his relationship to the doctor depends on causality as well; the patient wishes to learn from the physician what is or was the cause of his disease. No consensus of opinion existed among physicians regarding the subjective or objective status of causality. This is a fact that bears witness to their limited interest in philosophical problems.

The law of the plurality of conditions was generally accepted. Due to the condemnation of monocausality, conditionalism found repeated support. Sahli believed that conditionalism was "one of the greatest advances which we were capable of making in medicine and that it would radically reform our etiological views" ([160], p. 39). Vorkastner agreed and said that conditionalism had supplanted causalism because it linked fewer prejudiced determiners with resulting phenomena than causalism did ([194], p. 360). For Bleuler, etiology is a perfect example of the autistic-undisciplined thought in medicine which he censured. Although physicians in particular, he said, were naturally led to the assumption of a multiplicity and variety of causes, they nevertheless forever fell back on monocausal deductions.

> In a thousand specific cases, they again and again consider only a single cause, instead of many ("conditions" in Verworn's sense) ([11], p. 73).

Kraus could only support the conditionalist intent to avoid emphasizing cause vis-à-vis a system of determiners. In addition, for him, "the unnecessary contrast between causal views and purpose views can thereby be avoided, and the difference between the psychic and the physical, rather than being related to different scientific subjects, can be explained in view of the direction of the investigation" ([93], p. 56). Any talk of an "irreplaceable" condition, however, in his opinion tends to re-introduce the general causal principle.

However, the critique of monocausality is not a special characteristic of conditionalism. Hueppe and Martius proposed a multi-factorial approach and still retained the causal concept. Conditionalism in their view can claim no priority in this respect. A century before Verworn and von Hansemann, Schopenhauer had already called attention to the difference between a cause and a collection of conditions or stimuli ([71], p. 125; [123], p. 481). They had re-emphasized this collection again and again. It was rather a one-sided etiology which at that time was considered contrary to the theory of constitution ([129], p. 4; [173], p. 12). Monocausality was rejected by the causalists Hueppe and Martius as well as by the conditionalists Verworn and von Hansemann. Since there are points of agreement even in this technical respect between conditionalists and causalists (as, for instance, in their emphasis on constitution and their acceptance of the descriptivist approach), it was not uncommon for their shared ideas to be emphasized and contrasted to a formerly existing monocausal viewpoint, and for their differences to be considered less important. That conditionalism in view of the law of plurality had in fact presented merely an apparent alternative to causalism, was often held against Verworn and von Hansemann, and was proven by reference to their definitions as well as to applications in philosophy, inorganic sciences, and organic sciences.[13] Roux called their opposition to causalism "a fiction formulated and consequently upheld" by Verworn ([157], p. 59). It was admitted that in colloquial language usage cause was often discussed in the singular even though one was thinking of the total effect of a number of conditions. Often, however, only one cause is chosen as the determining factor without anyone intending thereby to underestimate the importance of other conditions. Hering introduced the term "coefficient" into medicine so that the medical professional would always be reminded that the problem is never brought about by a single cause ([66], p. 1433), thus endorsing the conditionalistic criticism of the spread of monocausal thought.

Verworn's law of effective equivalence of conditions received particularly harsh criticism. It was rejected by Hueppe and Martius for energistic and constitutional pathological reasons. Hueppe called this law untenable ([71], p. 175). Martius also argued against considering the object which was capable of being sick a condition that carried weight equal to that of the stimulus. It cannot be denied that:

> the role which the substrate of disease plays, namely, the organism reacting to disease according to its particular constitution (specific for this disease) at the onset of tuberculosis is a factor of a different kind and value from the stimulus ([120], p. 103).

Martius had designated the sickened organism as "Ur-Sache" in order to distinguish it linguistically as a determining factor from the triggering stimuli and from external determiners. This was supposedly in agreement with Hueppe's energistic disease theory. Such usage was considered unconvincing by medical professionals who also held, while agreeing with Martius, that it could not be adopted by physicians or by lay people. In this suggestion from Martius, Bauer, who wished to continue to use the expression "cause" when an etiological factor was uniformly followed by a disease, saw "a rearranging of words and concepts which was as impractical as (it was] absurd" ([7], p. 2; [51], p. 27; [165], p. 4). However, for Bauer too the different degrees of importance among the etiological factors were as important as their multiplicity since they were partially substitutable. Bernhard Fisher counts "the evaluation of the various determiners of a natural event" ([36], p. 380) among the most important tasks of natural research, as equally necessary for the understanding of the process as for the procedure. Roux provided a similar criticism of conditionalism. Differentiation, not equivalence, is the goal of natural science.

Belief in the effective equivalence of conditions necessarily has, according to the causalists, an obstructive effect on practice. Medical assessments — e.g., in court, in insurance decisions concerning disease or death — are not expedited by conditional thought. "The existence of a family", asserts Martius, "that has been robbed of its provider, depends on the expert's fundamental interpretation and the decisions reached on the basis thereof" ([120], p. 119). Here a conditionalistic view would fail, because it would not ascribe relative importance to the determiners. This relative importance can indeed be the decisive factor in a judgment about accidental homicide.

If one recognizes the law of plurality, the conditionalist law of equivalence can be discounted without thereby committing oneself to the energistic interpretation of the causalists Hueppe and Martius. According to Winterstein, the specific conditions of tissue cannot be equated with potential energy, and infection cannot be equated with the triggering of its change into kinetic energy:

> there is no 'potential energy of being sick' which is converted into kinetic energy by the disease 'stimulus/trigger', rather, disease is the result of a reciprocal effect between the metabolic processes of the organism and those of the parasite ([198], p. 638).

Criticism of Verworn's "effective equivalence of conditions" often resulted from misunderstandings. In Roux's opinion, the word *equinecessitas* should be used instead of "equivalence", because Verworn had intended that they were equally necessary, rather than that they were of equal value ([157], p. 30). Verworn had held that theoretically all conditions were necessary, but that in practice, the neglect of certain conditions was legitimate and appropriate. This distinction was ignored by many opponents of conditionalism. The various natures of determiners were not denied by Verworn. The replaceability of conditions concerns, in his opinion, merely particular elements. Other particular elements, indeed those which are decisive for the law of equivalence, are retained during the replacement. In these, the exchanged conditions are identical. Roux, otherwise an opponent of conditionalism, supported the law of equivalence by arguing that if one takes into consideration all of the particular aspects of the total event, all conditions are actually equally necessary ([157], p. 24; [63], p. 16ff; [67], p. 354). Causal thought like practical thought assumes, however, an abstract selective perspective. This difference between the theoretical and practical perspective was also emphasized by von Hansemann, who himself spoke about necessary and substitutable determiners. He formulated the essential task as

> finding a useable middle path which avoids the faults of causal thinking, but also dodges the practical inadequacy of purely philosophical, conditional thought ([57], p. 250).

The causal concept of theoretical pathology need not be identical with that of therapy-oriented pathology. The evaluation of the various factors which participate in a disease process was considered the most important task during this time. Roux demands analysis of the configuration of the effective determinants of their direction, relative position, quality, and size. The central factor of a given complex or system of factors may, in medicine, be considered to be the cause. The choice of this factor, according to medical professionals like Fisher and Lubarsch, who wanted therapy to mediate between conditionalism and causalism, depends on the theoretical and practical interests at the given time ([35], [36], [38], [103], [104]; [58], p. 686ff; [69], p. 69). Economy of thought and the requirements of practice should determine medicine's concept of cause; here too Mach was influential.

A three-fold scheme of etiological factors won general acceptance. In it, the views of cellular pathology, bacteriology, and epidemiology are reflected. Multi-factoriality, interdependence, and specific evaluation and comparison are incorporated in this scheme. The terms for the three etiological categories are quite varied. Hueppe preferred the formula, "disposition, trigger, and condition." He found the great number of current terms confusing.

For an energistic description of causal connections the cause lies in the disposition or constitution which may also be referred to as the predisposition, diathesis, possibility of effect and possibility of function or readiness, bodily condition, reactive norm (K.H. Bauer) or factor of determination (W. Roux). For trigger, one finds stimulus (Haller), excitation (Liebig), occasion/motive (Helmholtz, later Ribbert as well), inducement, effector or realization factor. For conditions: conditional moment, accessory factor, or auxiliary or secondary cause. In other words, a rich assortment of terms which, without clear definition and insight into the causal context, become worthless words ([72], p. 125.)[14]

Roux, whose causal scheme originated in the area of developmental mechanics, was understood by Martius as an additional corroboration of his own interpretation, although the terminology varied from his:

Whether physicians should be trained to say regarding pathogenesis 'determinating factor' instead of 'constitutional moment', 'realizing factor' instead of 'trigger', 'exciter' or 'accessory factors' instead of 'conditions', can be argued. As for myself, I have used Roux's terminology repeatedly in my recent works ([119], p. 129; [156], p. 184).

This three-part scheme also appeared in the particular medical disciplines. In psychiatry, one speaks of the cause of disease, the conditions of disease, and the life constellation. According to the dominance of one of these categories, three groups of psychic disease are distinguished ([169], p. 1389; [41], [69]). The causal discussion in psychiatry is bound up with general questions about explanation and understanding ([69], [78]).

Verworn's law of the identity of the sum of the conditions with the resulting event or process found acceptance because of the implied refusal to speak of an essence behind phenomena. Causes were considered by medical professionals, in line with Vaihinger, to be fictional. "All ideas of causality, regardless of what name they go by, work with fictions," asserted Grote ([51], p. 30; [1], [22], [90], [96], [147]). Such identification of condition and resulting phenomena, however, was not accepted by the causalists. As Herzberg established in his discerning observations about the controversy, their scientific interest followed a different perspective:

The causal principle, in contrast, does not speak of the relationship between the sum of partial moments and the total process at all, but rather of the relationship between individual partial moments, the relationship, e.g., between impetus and movement, between bacterial production of poison and the destruction of tissue, between cortical process and sensation ([67], p. 356; [8], p. 135).

In medicine, as in other sciences, it is essentially the interrelationship of partial moments which are examined for theoretical as well as for practical reasons. The law of identity does not hold in these relationships, but neither does it preclude an investigation of them.

The discussions about the concept of cause just presented were not without consequence for therapy, practice, and social politics. Neither conditionalism nor causalism was able to change anything about the dual nature of medicine, being both science and an art. During this period, just as today, etiological therapy was a goal of medicine, especially a therapy geared towards external causes or conditions. At the same time, the impossibility of ever actually reaching this goal was evident. Experience in conjunction with a knowledge of the causes or conditions would continue to guide therapeutic intervention.

The issue of diagnosis is touched upon by these discussions insofar as diagnosis can be either oriented towards patho-phenomenological or prognostic knowledge or based on a causal system of determiners ([91], [113], [114], [170]). Transformation of diagnosis into etiology might offer itself more readily from the monocausal approach than from the functionalistic-multi-conditional perspective. Bleuler counts this transition as evidence for autistic-undisciplined thought ([11], p. 75). However, the expectations of the patient can also be a factor contributing to this tendency. Since the patient wishes above all to learn from the doctor the cause of his malady, indeed the single cause, and since disease usually has a series of determiners, it is, as Schmidt observes, too often "impossible for the physician to give a precise answer to his patient's question" ([166], p. 54). The patients are satisfied on the level of monocausality. Etiological thought would of course be influenced by this.

The arguments in the theory of medicine between conditionalism and causalism were not brought to any conclusion by the parties involved.[15] This has been characteristic of many controversies in the history of science. Discussions about the concept of cause greatly influenced the concept of disease in those years. They pervade the various attempts to distinguish and mediate between cellular pathology, bacteriology, and constitution pathology. They influence new, energistic and functionalistic approaches. Finally, they are to be found in Ricker's [143] "relational pathology" (*Relationspathologie*), in Grote's [52] "medicine of reality" (*Wirklichkeitsmedizin*), and in medical anthropology where respect for individuality has roots in constitution pathology.

VI

The debate about the medical concept of cause was carried on in the following decades up to the present time for its own ends, in connection with pathophenomenology, teleology, diagnosis, and therapy, and in other areas such as law. We cannot go into this further here.[16] This debate may be summarized by the following main features.

Discussions in the theory of medicine about causes and conditions around 1900 developed out of nineteenth century scientific medicine, specifically out of the controversy between Virchow's cellular pathology and bacteriology, and they were closely bound up with the development of constitution pathology, including considerations of epidemiology and hygiene. At the same time, they have a basic, theoretical meaning for medicine that is independent of time. Conceptual issues in philosophy and the theory of science can be traced in them. Disease entities and their agents, description and causal derivation, monocausality and multi-factoriality, energetism and functionalism, situation and environment were discussed in these debates, and they had practical consequences.

The great number of perspectives, which sometimes intersected or at least did not exclude each other, was concentrated above all in the two positions of conditionalism and causalism. Conditionalism can be understood as a contribution to the ideological battles of the period; the causalists were instead concerned with a theoretical clarification of the principles and concepts of medicine. Both positions discounted an ontological conception of disease, like the causal conception, and espoused a multi-factorial position. The conditionalists, e.g., Verworn and von Hansemann, incorrectly attributed to the causalists of their time a merely monocausal way of thinking. The causalists, e.g., Hueppe and Martius, overcame the one-sidedness of cellular pathology and bacteriology precisely because they allowed for the multi-factoriality of disease processes. Certainly, theory as well as practice proved again and again that the criticism leveled at monocausalism by conditionalism was justified.

The causalists insisted on a basic qualitative distinction among the preconditions of disease, while in conditionalism, the valuation of the theoretically equivalent determiners appeared to be only practically necessary. The conditionalist law of the effective equivalence of determiners was considered to be a hindrance to practice by the causalists who also considered the law of the identity of conditions and disease process irrelevant for a real understanding of disease. The energistic functionalistic interpretation of the causalist views of Hueppe and Martius was not supported by all those who accepted a pluralistic causal view. The causalistic scheme of 'disposition', 'trigger', 'condition' became representative of the medical thought of the period. For this scheme, which was also formulated in a different terminology, mathematical formulae were developed, though limits to the mathematizability of medicine did not remain unnoticed.

Discussions about the concept of cause have still not ceased; they are pertinent today. Multi-factorial, interdependent, and comparative thought continually encounters impediments. This is true not only for medicine, but is observed in other sciences as well. Specialization, therapeutic commitment, and ideological prejudice all encourage monocausal statements based on genetics, milieu theory, or the labeling approach. The necessity and justification for distinguishing between theoretical and practical dimensions are only too often not recognized or insufficiently considered. From these deliberations it has become clear that the idea of cause not only has consequences for the theory of disease origin, disease process, and disease condition, but also affects practice, diagnosis, prevention, and therapy, all of which in turn shape the individual and social situation of the patient.

Medizinische Universität Lübeck
Lübeck, Germany

NOTES

[1] On the notion of causality as understood in the history of medicine in general, and the turn of the century in particular, see Schmiz [167]; Diepgen [25]; Aschoff [4]; Riese ([145], [146]); Hasche-Klünder [61]; Stachowiak [171]; Anning [3]; Jandolo [77]; Diepgen, Gruber and Schadewald [26]; Regöly-Merei [137]; King [84]; Evans [30]; see also the autobiographies of Hueppe [71]; Martius [120]; Roux [157]; and Gottstein [46].

[2] See, for example, the works by Fedeli [32]; Donders [27]; Mendelsohn [126]; Fröhlich [43]; Pye-Smith [136]; Romberg [148]; Hemmeter [64]; Schweninger [168]; Chauffard [21]; Müller [127]; Sahli [160]; Koch [91]; and Grote [51].

[3] See, for example, Riehl [144]; Wundt [200]; Weber [195]; Mach [109], [110]; von Aster [5]; Erdmann [24]; Geyser [44]; Frank [39], [49]; Frankl [42]; Exner [31]; Meinong [125]; Hartmann [60]; Schlick [163], [164]; Nernst [128]; Brunschwieg [15]; Reichenbach [138]; Hessen [68]; Bergmann [9]; Planck [134]; for the history, see Lang [97]; Volkmann [193]; Schlechtweg [161]; Wentscher [196]; Enriques [28]; Lukowsky [105]; Bunge [19]; Korch [92]; and Kuhn [95].

[4] Lipp's 1906 speech to a scientific and medical convention on "Natural Science and World View" was, according to Verworn, a complete failure: "The language of the philosophers was different from that of the scientists. They did not understand one another" ([183], p. 9). Löhlein said in 1917: "Every proper scientist — and I want to be counted as one — approaches questions of scientific logic only reluctantly. The division of labor derived from the historical development of the sciences causes a necessary one-sidedness of knowledge. In particular, the literary and the current state of knowledge in each sub-specialty of scientific research leaves only a few of us the time and strength to pursue philosophical study in addition" ([101], p. 1314). Roux reports in his autobiography that the statement "he is a philosopher" was the worst judgment that an anatomist could pass on a colleague. The search for causal knowl-

edge of living beings was already subjected to this rebuke ([156], p. 153). Bleuler wrote in 1919: "There is no truth to the opinion that the sharp conceptual distinctions of the philosophers promote clarity in other sciences. On the contrary they encourage faulty thinking, since they create artificial borders where reality knows none" ([11], p. 160).

[5] For discussions of Klebs, see [85], [132], [148].

[6] For discussions concerning Hueppe, see [49], [71], [83], [118].

[7] For a discussion of Martius, see [119].

[8] Hueppe develops the formula: K = (PRA), where K stands for disease, P for predisposition, R for stimulus, and A for external conditions. Should the external conditions be constant or unimportant, the formula becomes $K = F (p + p^1, RA)$, ([72], p. 217). Graul developed a combination of the formulas of Hueppe and Strumpell/Martius ([50], p. 1567; [14], p. 13ff).

[9] Already referred to in [185], n.17, more completely in [187], p. 453, and also repeated later in physiological works; on Verworn see [6], [43], [82], [99], [124], [131], [135], [152], [161], [177], [197], [199].

[10] Von Hansemann is in favor of the replacement of cause by conditions even before 1912, in about 1908: [54], p. 796; on von Hansemann, see [23], [59], [130].

[11] More specifically, von Hansemann formulates his conditionalistic standpoint in 14 theses ([58], p. 1834).

[12] For further discussions, see [23], p. 242.; [63], p. 51; [71], p. 50f; [101], p. 1317; [119], p. 24; [166], p. 47; [176], p. 5ff.

[13] For additional related discussions, see [123], p. 210; [34], p. 985f; [62], p. 2f; [67], p. 353; [94], p. 1144f; [157], p. 20.

[14] Von Hansemann does not consider it very important whether one speaks of "conditions" with Verworn, or "functional concepts" with Mach, or of "relational concepts" with Thole. He, however, finds "condition" the most attractive term ([58], p. 28). To Bauer, moreover, it is irrelevant whether "one calls the etiological factors 'conditions' and distinguishes with von Hansemann 'chief' or 'necessary conditions' from 'replaceable conditions' or one labels them 'coefficients' with H.E. Hering" ([7], p. 2).

[15] Grote states in 1921: "The medical literature of recent years includes several thorough treatments of the problem of causality. This is a sign of the unsettling effect of uncertainty about these questions, but agreement on a solution has not been reached" ([51], p. 28). In 1950, Riese concludes: "After the first World War and with the ascension of the constitutionalist view of disease, conditionalism fell swiftly into obscurity" ([145], p. 23). Nevertheless, Verworn's *Kausale und Konditionale Weltanschauung*, after a second edition in 1918, also appeared in a third edition in 1928.

[16] See, for example, Tendeloo [175]; Bosma [12]; Ziehen [201]; Tischner [178]; Fischer [35], [37]; Burckhardt [18]; Jellinghaus [79]; Lukowsky [105], [106]; Gottschick [45]; Rothschuh ([153], [154]; Vallejo [179]; Jenny [80]; Susser [172]; Vlasiuk [192]; King [84]; Evans [30]; and Agassi [2].

BIBLIOGRAPHY

1. Aebly, I.: 1925, 'Fiktionen in der Medizin', *Allgemeine homö-opathische Zeitung* **172**, 19–24.

2. Agassi, J.: 1976, 'Causality and Medicine', *Journal of Medicine and Philosophy* **1**, 301–317.
3. Anning, S.T.: 1958, 'An Aspect of Medical History: Evolution of Medical Thought Concerning the Aetiology of Disease', *University Leeds Medical Journal* **7**, 50–58.
4. Aschoff, L.: 1938, 'Virchows Cellularpathologie', in L. Aschoff, E. Kuster, and W.J. Schmidt (eds.), *Hundert Jahre Zellforschung*, Berntraegen, Berlin, pp. 169–269.
5. Aster, E. von: 1905, 'Untersuchungen über den logischen Gehalt des Kausalgesetzes', in Th. Lipps (ed.), *Psychologische Untersuchungen*, Vol. 1, Engelmann, Leipzig, pp. 289–314.
6. Baglioni, S.: 1922, 'Max Verworn', *Rivista di Biologia* **4**, 126–133.
7. Bauer, J.: 1917, *Die konstitutionelle Disposition zu inneren Krankheiten*, 2nd ed., 1921, Springer, Berlin.
8. Becher, E.: 1914, *Naturphilosophie*, Teubner, Leipzig.
9. Bergmann, H.: 1929, *Der Kampf um das Kausalgesetz in der jüngsten Physik*, Vieweg, Braunschweig.
10. Bier, A.: 1922, 'Über medizinische Betrachtungsweisen, insbesondere über die mechanistische und über die teleologische', *Münchener Medizinische Wochenschrift* **69**, 845–849.
11. Bleuler, E.: 1919, *Das autistisch-undisziplinierte Denken in der Medizin und seine Überwindung*, 4th ed. 1927, repr. 1962, Springer, Berlin.
12. Bosma, H.A.: 1922, 'Cause, Condition, and Constellation in Medicine', *The Lancet* **202**, 410–412.
13. Boutroux, E.: 1874, *De la contingence des lois de la nature*, Bailliere, Paris.
14. Brugsch, Th.: 1918, *Allgemeine Prognostik*, Urban and Schwarzenberg, Berlin.
15. Brunschwieg, L.: 1922, *La expérience humaine et la causalité physique*, Alcan, Paris.
16. Buchner, H.: 1896, 'Biologie und Gesundheitslehre', *Verhandlungen der Gesellschaft Deutscher Naturforscher und Ärzte* **68** (1), 39–56.
17. Buchner, H.: 1953–1954, 'Von den Ursachen der Krankheiten', *Stimmen der Zeit* **154**, 188–187.
18. Burckhardt, H.: 1933, 'Über den Begriff der "Ursache"', *Frankfurter Zeitschrift fur Pathologie* **44**, 508–522.
19. Bunge, M.A.: 1963, *Causality: The Place of the Causal Principle in Modern Science*, World Publishing Company, Cleveland.
20. Bunge, M.A.: 1971, *Les théories de la causalité*, Presses Universitaires de France, Paris.
21. Chauffard, A.: 1911, 'Du degré de certitude de la médecine', *Revue scientifique* **49,** 321–330.
22. Cörper, C.: 1919, 'Bedeutung des fiktionalenen Denkens für die Medizin', *Annalen der Philosophie* **1**, 191–202.
23. Cohn, M.: 1912, 'Über Dr. v. Hansemanns Forderung des konditionalen Denkens in der Medizin', *Deutsche Ärzte-Zeitung* **185**, 209–211, 242–244, 259 –262, 274–276.
24. Cornet, G.: 1889, 'Die Verbreitung der Tuberkelbacillen ausserhalb des

Körpers', *Zeitschrift für Hygiene* **5**, 191–331.
25. Diepgen, P.: 1926, 'Krankheitswesen und Krankheitsursache in der spekulativen Pathologie des 19. Jahrhunderts', *Sudhoffs Archiv.* **18**, 302–327; also in W. Artelt, E. Heischkel and J. Schuster (eds.), *Medizin und Kultur. Gesammelte Aufsätze*, Enke, Stuttgart, 1938, pp. 261–282.
26. Diepgen, P., Gruber, G.B. and Schadewaldt, H.: 1969, 'Der Krankheitsbegriff, seine Geschichte und Problematik', in H.W. Altmann and F. Buchner (eds.), *Handbuch der Allgemeinen Pathologie*, Vol. 1, Springer, Berlin and New York, pp. 1–50.
27. Donders, F.C.: 1879, 'De la science et de l'art', *Revue Internationale des Sciences* **4**, 417–436.
28. Enriques, F.: 1941, *Causalité et déterminisme dans la philosophie et l'histoire des sciences*, Hermann, Paris.
29. Erdmann, B.: 1905, *Über Inhalt und Geltung des Kausalgesetzes*, Niemeyer, Halle.
30. Evans, A.S.: 1976, 'Causation and Disease: The Henle-Koch Postulates Revisited', *Yale Journal of Biology and Medicine* **49**, 175–195.
31. Exner, F.: 1919, *Vorlesungen über die physikalischen Grundlagen der Naturwissen-schaften*, Deuticke, Wien.
32. Fedeli, F.: 1850, *La medicina, scienza e arte*, Pamphlet, Pisa.
33. Feinstein, A.R.: 1967, *Clinical Judgement*, Robert E. Krieger Publishing Co., New York.
34. Fischer, B.: 1919, 'Der Begriff der Krankheitsursache', *Münchener Medizinische Wochen-schrift* **66**, 985–987.
35. Fischer, B.: 1931, 'Der Begriff der Krankheitsursache und seine Bedeutung für das ärztliche Gutachten', in H. Lininger, R. Weichbrodt, and A.W. Fischer (eds.), *Handbuch der ärztlichen Begutachtung*, Vol. 1, Barth, Leipzig, pp. 325–332.
36. Fischer, B.: 1913, 'Grundprobleme der Geschwulstlehre', *Frankfurter Zeitschrift für Pathologie* **12**, 369–385.
37. Fischer, B.: 1933, 'Der Ursachenbegriff in der Biologie', *Frankfurter Zeitschrift für Pathologie* **44**, 523–526.
38. Fischer, B.: 1920, 'Zum Ursachenbegriff', *Münchener Medizinische Wochenschrift* **67**, 74–75.
39. Frank, Ph.: 1907, 'Kausalgesetz und Erfahrung', *Annalen der Naturphilosophie* **6**, 443–451.
40. Frank, Ph.: 1932, *Das Kausalgesetz und seine Grenzen*, Springer, Wien.
41. Frankhauser, K.: 1915, 'Über Kausalität im allgemeinen sowie "Psychische Kausalität" im besonderen', *Zeitschrift für die gesamte Neurologie und Psychiatrie* **29**, 201–215.
42. Frankl, W.M.: 1917, 'Studien zur Kausalitätstheorie', *Archiv für systematische Philosophie* **23**, 3–22.
43. Fröhlich, Fr.W.: 1923, 'Max Verworn', *Zeitschrift für allgemeine Physiologie* **20**, 185–192.
44. Geyser, J.: 1906, *Naturerkenntnis und Kausalgesetz*, Schoningh, Münster.
45. Gottschick, J.: 1959, 'Das medizinische und juristische Kausaldenken', *Der*

Medizinische Sachverständige **55**, 137–148.
46. Gottstein, A.: 1925, 'Adolf Gottstein', in L.R. Grote (ed.), *Die Medizin der Gegenwart in Selbstdarstellungen*, Vol. 4, Meiner, Leipzig, pp. 53–91.
47. Gottstein, A.: 1897, *Allgemeine Epidemiologie*, Spohr, Leipzig.
48. Gottstein, A.: 1893, 'Die Contagiosität der Diphtherie', *Berliner klinische Wochenschrift* **30**, 594–598.
49. Gottstein, Al.: 1922, 'Ferdinand Hueppe. Zur Feier seines siebzigsten Geburtstages am 24. August 1922', *Klinische Wochenschrift* **1**, 1767–1768.
50. Graul, G.: 1917, 'Über die Erkenntnis des Krankheitsgeschehens', *Deutsche Medizinische Wochenschrift* **43**, 1541–1542, 1567–1568.
51. Grote, L.R.: 1921, *Grundlagen ärztlicher Betrachtung*, Springer, Berlin.
52. Grote, L.R.: 1936, 'Wirklichkeitsmedizin', *Die Medizinische Welt* **10**, 1351–1355.
53. Hansemann, D. von: 1905, 'Der Aberglaube in der Medizin und seine Gefahr für Gesundheit und Leben', 2nd ed. 1914, Tuebner, Leipzig and Berlin.
54. Hansemann, D. von: 1908, 'Ätiologie und Pathogenese der Epityphlitis', *Deutsche Medizinische Wochenschrift* **34**, 769–772.
55. Hansemann, D. von: 1916, 'Allgemeine ätiologische Betrachtungen mit besonderer Berücksichtigung des Lungenemphysems', *Virchows Archiv* **221**, 94–106.
56. Hansemann, D. von: 1910, 'Die Freiheit der Wissenschaft', *Die Grenzboten* **69**, 345–355.
57. Hansemann, D. von: 1912, 'Konditionales Denken', *Berliner klinische Wochenschrift* **49**, 2503–2504.
58. Hansemann, D. von: 1912, *Über das konditionale Denken in der Medizin und seine Bedeutung fur die Praxis*, Hirschwald, Berlin.
59. Hart, C.: 1921, 'D. von Hansemann', *Centralblatt für Allgemeine Pathologie und pathologische Anatomie* **31**, 113–114.
60. Hartmann, N.: 1920, 'Die Frage der Beweisbarkeit des Kausalgesetzes', *Kant Studien* **24**, 261–290.
61. Hasche-Klunder, I.: 1952 'Infektion und Infektionskrankheit, Bakteriologie, und Pathologie', *Centaurus* **2**, 205–250.
62. Heim, G.: 1914, 'Scheidung der Ursache von den Bedingungen pathologischer Vorgänge', *Virchows Archiv* **216**, 1–10.
63. Heim, G.: 1913, *Ursache und Bedingung. Widerlegung des Konditionalismus und Aufbau der Kausalitätslehre auf der Mechanik*, Barth, Leipzig.
64. Hemmeter, J.C.: 1908, 'Science and Art in Medicine: Their Influence on the Development of Medical Thinking', *Journal of the American Medical Association* **46**, 243–248.
65. Hering, H.E.: 1919, 'Über die Bedeutung der Begriffe Ursache, Bedingung und Funktion für den Mediziner', *Münchener Medizinische Wochenschrift* **66**, 499–501.
66. Hering, H.E.: 1912, 'Ueber die Koeffizienten für das Auftreten postmortaler Herzcontractionen', *Medizinische Klinik* **8**, 1733–1735.
67. Herzberg, A.: 1921, 'Konditionismus oder Kausalprinzip?', *Berliner klinische Wochenschrift* **58**, 352–357.
68. Hessen, J.: 1928, *Das Kausalprinzip*, 2nd ed. 1958, Filser, Augsburg.

69. Hoppe, A.: 1921–1922, 'Begriff der Kausalität in der Psychiatrie', *Monatsschrift für Kriminalpsychiatrie und Strafrechtsreform* **12**, 65–77.
70. Hueppe, F.: 1898, 'The Causes of Infectious Disease', *The Monist* **8**, 384–414.
71. Hueppe, F.: 1923, 'Ferdinand Hueppe', in L.R. Grote (ed.), *Die Medizin der Gegenwart in Selbstdarstellungen*, Vol. 2, Meiner, Leipzig, pp. 77–138.
72. Hueppe, F.: 1903, 'General Views on the Aetiology of Infectious Disease', *The Journal of State Medicine* **11**, 11–19; also London, 1904, *Archiv für Rassen- und Gesellschafts-biologie* **1**, 210–218.
73. Hueppe, F.: 1891, 'Über Erforschung der Krankheitsursachen und sich daraus ergebende Gesichtspunkte für Behandlung und Heilung von Infections-Krankheiten', *Berliner klinische Wochenschrift* **28**, 279–283, 305–310, 332–336.
74. Hueppe, F.: 1889, 'Ueber den Kampf gegen die Infectionskrankheiten', *Berliner klinische Wochenschrift* **26**, 989–994, 1014–1020.
75. Hueppe, F.: 1901, 'Ueber Krankheitsursachen vom Standpunkte der naturwissenschaft-lichen Medizin', *Wiener Medicinische Wochenschrift* **51**, 305–311, 371–376.
76. Hueppe, F.: 1893, 'Ueber die Ursachen der Gährungen und Infectionskrankheiten und deren Beziehungen zum Causalproblem und zur Energetik', *Verhandlungen der Gesellschaft Deutscher Naturforscher und Ärzte* **65** (1), 134–158.
77. Jandolo, M.: 1964, 'L'evoluzione del pensiero etiopatogenetico dall'antichità ad oggi', *Annali di Medicina Navale* **69**, 417–429.
78. Jaspers, K.: 1913, 'Kausale und "verständliche" Zusammenhänge zwischen Schicksal und Psychose bei der Dementia praecox (Schizophrenia)', *Zeitschrift für die gesamte Neurologie und Psychiatrie* **14**, 158–263.
79. Jellinghaus, K. -Th.: 1954–1955, 'Zum Verhältnis von Kausalität und Finalität im organischen Geschehen', *Philosophia Naturalis* **3**, 194–210.
80. Jenny, E.: 1970, 'Ätiologie und Therapie', *Medizinische Welt* **15**, 643–649.
81. Jensen, P.: 1921, *Reiz, Bedingung und Ursachen in der Biologie*, Borntraeger, Berlin.
82. Jensen, P.: 1922, 'Max Verworn', *Nachrichten der Gesellschaft der Wissenschaften in Göttingen, Geschäftliche Mitteilungen*, 61–78.
83. Kaup, J.: 1922, 'Ferdinand Hueppe (Zu seiner Würdigung beim Eintritt ins biblische Alter)', *Münchener Medizinische Wochenschrift* **69**, 1547–1549.
84. King, L.S.: 1975, 'Causation: A Problem in Medical Philosophy', *Clio Medica* **10**, 95–109.
85. Klebs, E.: 1913–1914, 'Edwin Klebs', *Verhandlungen der Deutschen Pathologischen Gesellschaft* **17**, 588–597.
86. Klebs, E.: 1877, 'Über die Umgestaltung der medicinischen Anschauungen in den letzten drei Jahrzehnten', *Amtlicher Bericht der 50. Versammlung Deutscher Naturforscher und Ärzte*, 41–55.
87. Klebs, E.: 1878, 'Über Cellularpathologie und Infectionskrankheiten', *Tageblatt der 51. Versammlung Deutscher Naturforscher und Ärzte*, 127–134.
88. Klebs, E.: 1987–1989, *Die Allgemeine Pathologie oder die Lehre von den Ursachen und dem Wesen der Krankheitsprozesse*, vols. 1–2, Fischer, Jena.
89. Klebs, E.: 1894, *Die causale Behandlung der Tuberkulose. Experimentelle und*

klinische Studien, Voss, Hamburg.
90. Koch, R.: 1924, *Das Als-ob im ärztliche Denken*, Gebr. Paetel, München.
91. Koch, R.: 1917, *Die ärztliche Diagnose. Beitrag zur Kenntnis des ärztlichen Denkens*, 2nd ed. 1920, Bergmann, Wiesbaden.
92. Korch, H.: 1965, *Das Problem der Kausalität*, Dt. Verlag der Wissenschaften, Berlin.
93. Kraus, F.: 1919–1920, 'Die allgemeine und spezielle Pathologie der Person', *Klinische Syzygiologie*, Vols. 1 and 2, Thieme, Leipzig.
94. Kronenberg, M.: 1913, 'Kausale und konditionale Weltanschauung', *Die Naturwissenschaften* **1**, 1413–1417.
95. Kuhn, T.S.: 1966, '*Les notions de la causalité* dans le développement de la physique', in M.A. Bunge (ed.), *Les théories de la causalité*, Presses Universitaires de France, Paris, 1971, pp. 7–18.
96. Kulenkampff, D.: 1924, 'Wert und Bedeutung der Als-Ob-Betrachtung im medizinischen Denken', *Virchows Archiv* **255**, 332–360.
97. Lang, A.: 1904, *Das Kausalproblem*, Bachem, Kölm.
98. Liebreich, O.: 1895, 'Über Lupusheilung durch Cantharidin und über Tuberculose', *Berliner klinische Wochenschrift* **32**, 293–296, 323–327.
99. Lippelt, O.: 1911, 'Verworns psychosomatische Weltanschauung', *Pädagogische Studien* **32**, 305–317.
101. Löhlein, M.: 1917, 'Ursachenbegriff und kausales Denken', *Medizinische Klinik* **13**, 1314–1317.
102. Löhlein, M.: 1918, 'Über das kausale Denken in der Medizin und Biologie', *Sitzungsberichte der Gesellschaft zur Beförderung der gesamten Naturwissenschaften zu Marburg*, 53–63.
103. Lubarsch, O.: 1919, 'Ursachenforschung, Ursachenbegriff und Bedingungslehre', *Deutsche Medizinische Wochenschrift* **45**, 1–4, 33–36.
104. Lubarsch, O.: 1919, 'Zur Frage des Ursachenbegriffs', *Münchener Medizinische Wochenschrift* **66**, 1169.
105. Lukowsky, A.: 1958, 'Kausale und finale Betrachtungsweise in Naturwissenschaft und Medizin', *Wiener Medizinische Wochenschrift* **108**, 293 –296.
106. Lukowsky, A.: 1966, *Philosophie des Arzttums. Ein Versuch*, Dt. Ärzte-Verlag, Köln.
107. Lukowsky, A.: 1955–1956, 'Über die Entwicklung des Kausalbegriffes', *Kant Studien* **47**, 359–366.
108. Mach, W.: 1906, 'Beschreibung und Erklärung', in E. Mach, *Populärwissenschaftliche Vorlesungen*, Barth, Leipzig, 1910, pp. 411–427.
109. Mach, E.: 1896, 'Causalität und Erklärung', in E. Mach, *Prinzipien der Wärmelehre*, Barth, Leipzig, pp. 430–437.
110. Mach, E.: 1905, *Erkenntnis und Irrtum, Skizzen zur Psychologie der Forschung*, Barth, Leipzig.
111. Mach, E.: 1882, 'Die ökonomische Natur der physikalischen Forschung', in E. Mach, *Populär-wissenschaftliche Vorlesungen*, Barth, Leipzig, 1910, pp. 217–244.
112. Mach, E.: 1894, 'Über das Prinzip der Vergleichung in der Physik', in E. Mach, *Populär-wissenschaftliche Vorlesungen*, Barth, Leipzig, 1920, pp. 266–289.

113. Mainzer, Fr.: 1925, *Über die logischen Prinzipien der ärztlichen Diagnose*, Borntraeger, Berlin.
114. Marchand, F.: 1920, 'Klinische, anatomische und ätiologische Krankheitsbegriffe und Krankheitsnamen', *Münchener Medizinische Wochenschrift* **67**, 681–686.
115. Martini, P.: 1947–1948, 'Kausalität und Medizin', *Studium Generale* **1**, 342–350.
116. Martius, F.: 1900, 'Constitutionsanomalien und die constitutionellen Krankheiten', in F. Martius, *Pathogenese innerer Krankheiten*, Vol. 2, Deuticke, Wien and Leipzig, pp. 158–260.
117. Martius, F.: 1922, 'Einige Bemerkungen über die Grundlagen des ärztlichen Denkens heute', *Klinische Wochenschrift*, **1**, 49–53.
118. Martius, F.: 1922, 'Ferdinand Hueppe', *Medizinische Klinik* **18**, 1106–1108.
119. Martius, F.: 1923, 'Friedrich Martius', in L.R. Grote (ed.), *Die Medizin der Gegenwart in Selbstdarstellungen*, Vol. 1, Meiner, Leipzig, pp. 105–140.
120. Martius, F.: 1914, 'Das Kausalproblem in der Medizin (Kritik des Konditionalismus)', *Beihefte zur Medizinischen Klinik* **10**, 101–128.
121. Martius, F.: 1914, *Konstitution und Vererbung in ihren Beziehungen zur Pathologie,* Springer, Berlin.
122. Martius, F.: 1898, 'Krankheitsursachen und Krankheitsanlage', *Verhandlungen der Gesellschaft Deutscher Naturforscher und Ärzte* **70**, (1), 90–110.
123. Martius, F.: 1918, 'Die Lehre von den Ursachen in der Konstitutionpathologie', *Deutsche Medizinische Wochenschrift* **44**, 449–452, 481–484.
124. Matthaei, R.: 1922, 'Max Verworn', *Deutsche Medizinische Wochenschrift* **48**, 102–103.
125. Meinong, A.: 1919, 'Zun Erweise des allgemeinen Kausalgesetzes', *Sitzungsberichte der Akademie der Wissenschaften in Wien, Philosophisch-historische Klasse* 189, No. 4.
126. Mendelsohn, M.: 1893, *Ärztliche Kunst und medizinische Wissenschaft. Eine Unter-suchung über die Ursachen der "ärztlichen Misere"*, 2nd ed. 1894, Bergmann, Wiesbaden.
127. Müller, O.: 1913, 'Medizinische Wissenschaft und ärztliche Kunst', *Medizinisches Correspondenzblatt des Württembergischen ärztlichen Landesvereins* **83**, 25–41.
128. Nernst, W.: 1922, 'Zum Gültigkeitsbeweis der Naturgesetze', *Die Naturwissenschaften* **10**, 489–494.
129. Neuburger, M.: 1913–1914, 'Zur Geschichte der Konstitutionslehre', *Zeitschrift für angewandte Anatomie und Konstitutionslehre* **1**, 4–10.
130. Ostertag, B.: 1936, 'David von Hansemann. 5. XI. 1858–28. VIII. 1920', *Verhandlungen der Deutschen Pathologischen Gesellschaft* **20**, 370–378.
131. Pathe, O.: 1914, 'Der Psychomonismus Verworns', *Praxis Volksschule* **24**, 290–296.
132. Paul, E.: 1914, 'Edwin Klebs', *Münchener Medizinische Wochenschrift* **61**, 193–196, 251–257.
133. Pflüger, E.Fr.W.: 1877, *Die teleologische Mechanik der lebendigen Natur*, Cohen, Bonn.
134. Planck, M.: 1932, *Der Kausalbegriff in der Physik*, Barth, Leipzig.

135. Pütter, A.: 1921, 'Max Verworn', *Münchener Medizinische Wochenschrift* **68**, 1655–1656.
136. Pye-Smith, P.H.: 1900, 'Medicine as Science and Medicine as Art', *Nature* **62**, 356–357.
137. Regoly-Merei, G.: 1970, ['Robert Koch and His School in the Conceptualization of Etiology', hung.], *Orvosi Hetilap* **111**, 1414–1417.
138. Reichenbach, H.: 1926, 'Die Kausalstruktur der Welt und der Unterschied von Vergangenheit und Zukunft', *Sitzungsberichte der Bayerischen Akademie der Wissenschaften, Mathem.-Naturwissenschaftliche Abteilung*, 133–175.
139. Ribbert, H.: 1918, 'Über den Begriff der Krankheit', *Deutsche Zeitschrift für Nervenheilkunde* **60**, 169–178.
140. Ribbert, H.: 1892, *Über Wesen, Ursachen und Heilung der Krankheiten*, Meyer and Zeller, Zürich.
141. Ribbert, H.: 1913, 'Über den Ursachenbegriff in der Medizin', *Deutsche Medizinische Wochenschrift* **30**, 1106–1109.
142. Ricker, G.: 1905, *Entwurf einer Relationspathologie*, Fischer, Jena.
143. Ricker, G.: 1924, *Pathologie als Naturwissenschaft. 'Relationspathologie'*, Springer, Berlin.
144. Riehl, A.: 1877, 'Kausalität und Identität', in A. Riehl, *Philosophische Studien aus 4 Jahrzehnten*, Quelle and Meyer, Leipzig, pp. 202–218.
145. Riese, W.: 1950, *La pensée causale en médecine*, Presses Universitaires de France, Paris.
146. Riese, W.: 1968, 'The Principle of Individual Causality from Aristotle to Claude Bernard', *Episteme* **2**, 111–120.
147. Rietti, F.: 1924, 'Das Als-Ob in der Medizin', *Annalen der Philosophie* **4**, 385–416.
148. Röthein , O.H.: 1962, *Edwin Keebs* (1834–1913). *Einfrüher Vor-Kämpfer der Bakteriologie und seine Irrfahrte*, Diss. med., Zürich.
149. Romberg, E.: 1905, 'Erfahrung und Wissenschaft in der inneren Medizin', *Therapie der Gegenwart* **46**, 145–153.
150. Rosenbach, O.: 1909, *Ausgewählte Abhandlungen*, W. Gutmann (ed.), Vols. 1 and 2, Barth, Leipzig.
151. Rosenbach, O.: 1891, *Grundlagen, Aufgaben und Grenzen der Therapie nebst einen Anhange: Kritik des Kochschen Verfahrens*, Urban and Schwarzenberg, Wien.
152. Rothschuh, K.E.: 1976, 'Max Verworn', in C.C. Gillispie (ed.), *Dictionary of Scientific Biography*, Vol. 14, Scribner, New York, pp. 2–3.
153. Rothschuh, K.E.: 1965, *Prinzipien der Medizin*, Urban and Schwarzenberg, München.
154. Rothschuh, K.E.: 1973, 'Zu einer Einheitstheorie der Verursachung und Ausbildung von sematischen, psychosomatischen und psychischen Krankheiten', *Hippokrates* **44**, 3–17.
155. Roux, W.: 1922, 'Prinzipielles der Entwickelungsmechanik', *Annalen der Philosophie* **3**, 454–474.
156. Roux, W.: 1923, 'Wilhelm Roux', in L.R. Grote (ed.), *Die Medizin der Gegenwart in Selbstdarstellungen*, Vol. 1, Meiner, Leipzig, pp. 141–206.

157. Roux, W.: 1913, *Über kausale und konditionale Weltanschauung und deren Stellung zur Entwicklungmechanik*, Engelmann, Leipzig.
158. Roux, W.: 1922, 'Über Ursache und Bedingung, Naturgesetz und Regel', *Deutsche Medizinische Wochenschrift* **48**, 1232–1233.
159. Sahli, H.: 1913, 'Über das Wesen des Morbus Basedowi', *Correspondenz-Blatt für Schweizer-Ärzte* **43**, 269–276.
160. Sahli, H.: 1914, 'Über den Einfluss der Naturwissenschaften auf die moderne Medizin', *Verhandlungen der Schweizerischen Naturforschenden Gesellschaft* **2**, 29–66.
161. Schlechtweg, W.: 1919, 'Der Ursachbegriff bei Hume, Kant und Verworn', *Pädagogische Warte* **2b**, 69–75.
162. Schlechtweg, E.: 1925, 'Fiktionen in der Medizin (Die Philosophie des Als Ob)', *Deutsche Zeitschrift fur Homöopathie* **4**, 2–10; also see, 209–212.
163. Schlick, M.: 1931, 'Die Kausalität in der gegenwärtigen Physik', *Die Naturwissenschaften* **19**, 145–162.
164. Schlick, M.: 1920, 'Naturphilosophische Betrachtungen über das Kausalprinzip', *Die Naturwissenschaften* **8**, 461–474.
165. Schmidt, A.: 1916, *Konstitution und ihre Beeinflussung*, Niemeyer, Halle.
166. Schmidt, A.: 1913, 'Wahre und vermeintliche Krankheitsursachen', *Deutsche Revue* **38**, 47–54.
167. Schmiz, K.: 1921, 'Werden und Wege der Pathologie', *Die Naturwissenschaften* **9**, 803–812.
168. Schweninger, E.: 1906, *Der Arzt*, Rutten and Loening, Frankfurt/Main.
169. Seelert, H.: 1923, 'Krankheitsursachen in der Psychiatrie', *Klinische Wochenschrift, N.F.* **2**, 1389–1391.
170. Slotopolsky, B.: 1919, 'Über den logischen Charakter der Diagnose', *Deutsche Medizinische Wochenschrift* **45**, 997–998.
171. Stachowiak, H.: 1957, 'Über kausale, konditionale und strukturelle Erklärungsmodelle', *Philosophia Naturalis* **4**, 403–433.
172. Susser, M.: 1973, *Causal Thinking in the Health Sciences. Concepts and Strategies of Epidemiology*, Oxford University Press, New York.
173. Tandler, J.: 1913–1914, 'Konstitution und Rassenhygiene', *Zeitschrift für angewandte Anatomie und Konstitutionslehre* **1**, 11–26.
174. Tendeloo, N.Ph.: 1913, 'Die Bestimmung von Ursache und Bedingung: Ihre Bedeutung besonders für die Biologie', *Die Naturwissenschaften* **1**, 153–156.
175. Tendeloo, N.Ph.: 1921, *Konstellationspathologie und Erblichkeit*, Springer, Berlin.
176. Thöle, F.: 1909, *Das vitalistisch teleologische Denken in der heutigen Medizin, mit besonderer Berücksichtigung von Bier's wissenschaftlichen Erklärungen*, Enke, Stuttgart.
177. Thörner, W.: 1922, 'Max Verworn', *Medizinische Klinik* **18**, 130–139.
178. Tischner, R.: 1925, 'Kausales und konditionales Denken in der Medizin', *Blätter fur biologische Medizin* **7**, 69–75.
179. Vallejo, E.A.: 1969, 'El médico ante la etiolo´gía de la enfermedad', *Revista española de las enfermedades del aparato digestivo*, **28**, 407–418.

180. Verworn, M.: 1922, *Aphorismen*, Fischer, Jena.
181. Verworn, M.: 1907, *Die Erforschung des Lebens*, 2nd ed. 1911, Fischer, Jena.
182. Verworn, M.: 1914, *Erregung und Lähmung. Eine allgemeine Physiologie der Reizwirkungen*, Fischer, Jena.
183. Verworn, M.: 1908, *Die Fragen nach den Grenzen der Erkenntnis*, 2nd ed. 1917, Fischer, Jena.
184. Verworn, M.: 1912, *Kausale und konditionale Weltanschauung*, 3rd ed. 1928, Fischer, Jena.
185. Verworn, M.: 1901, *Naturwissenschaft und Weltanschauung*, Barth, Leipzig.
186. Verworn, M.: 1907, 'Die vitalistischen Strömungen der Gegenwart', *Die Deutsche Klinik* **11**, 251–168.
187. Verworn, M.: 1905, 'Prinzipienfragen in der Naturwissenschaft', *Naturwissenschaftliche Wochenschrift* **20**, 449–456; also enlarged, 1905, 2nd ed. 1917.
188. Virchow, R.: 1858, *Cellularpathologie*, 4th ed. 1871, Hirschwald, Berlin.
189. Virchow, R.: 1877, 'Die Freiheit der Wissenschaft im modernen Staatsleben', *Amtl. Bericht der 50. Versammlung Deutscher Naturforscher und Ärzte*, 65–77.
190. Virchow, R.: 1880, 'Krankheitswesen und Krankheitsursachen', *Archiv für pathologische Anatomie, Physiologie und klinische Medizin* **79**, 1–19.
191. Virchow, R.: 1869, 'Über die heutige Stellung der Pathologie', in K. Sudhoff (ed.), *Rudolf Virchow und die Deutschen Naturforscher-Versammlungen*, Akademische Verlagsgesellschaft, Leipzig, 1922, pp. 77–97.
192. Vlasiuk, V.V.: 1973, ['Problems of the Causality of Diseases', russ.] *Vestnik Akademii Meditsinskikh Nauk S.S.S.R.* **28**, 32–39.
193. Volkmann, P.: 1912, 'Historisch-kritische Studie zum Kausalitätsbegriff', in *Festschrift fur H. Weber zu seinem siebzigsten Geburtstag am 5. März*, Teubner, Leipzig, pp. 428–442.
194. Vorkastner, W.: 1914, 'Ueber hereditäre Ataxie', *Medizinische Klinik* **10**, 360–362, 404–407, 448–450, 495–497.
195. Weber, H.: 1881, *Über Causalität in den Naturwissenschaften*, Engelmann, Leipzig.
196. Wentscher, E.: 1921 *Geschichte des Kausalproblems in der neueren Philosophie*, Meiner, Leipzig.
197. Wichert, R.: 1943, *Die philosophischen Grundlagen und Folgerungen der Psychologie von Max Verworn*, Diss. med., Königsberg.
198. Winterstein, H.: 1919, 'Causalität und Vitalismus vom Standpunkt der Denkökonomie', *Anatomische Hefte* **57**, 679–724.
199. Wüllenweber, R.: 1968, *Der Physiologe Max Verworn*, Diss. med., Bonn.
200. Wundt, W.: 1880–1883, *Logik*, 2 Vols., Enke, Stuttgart.
201. Ziehen, Th.: 1922, 'Ueber kausale und teleologische Denkweise in der Medizin', *Deutsche Medizinische Wochenschrift* **48**, 1233–1234.

ANNE M. FAGOT-LARGEAULT

ON MEDICINE'S SCIENTIFICITY—DID MEDICINE'S ACCESSION TO SCIENTIFIC 'POSITIVITY' IN THE COURSE OF THE NINETEENTH CENTURY REQUIRE GIVING UP CAUSAL (ETIOLOGICAL) EXPLANATION?*

I. THE NINETEENTH CENTURY 'TURN': MEDICINE, EMPIRICAL ('POSITIVE') SCIENCE

Both José Luis Peset [25] and Dietrich von Engelhardt [11] raise a question about the nature of the conceptual 'turn' through which medicine is seen as a 'positive' science in the nineteenth century. I shall concentrate on that question.

Peset's thesis is that towards 1800 the paradigm of etiological explanation changes under the influence of 1) a Humean-Kantian philosophical axis; and 2) a socio-economical axis (struggle against infectious diseases, cost of illness). In this, one goes from an *a priori*, dogmatic, essentialist form of causal thinking to an *a posteriori*, empiricist, phenomenalist form of causal thinking. This shift results in radically different theories of illness. For example, the ancients talked of morbid entities (nosography), while modern scientists look for anatomical or functional alterations whose development can be understood from the perspective of normal physiological laws.

In what follows, I will be challenging the thesis offered by Peset by raising the following points.

1. If nineteenth century Medicine rejects a causal paradigm, it is the paradigm of classical science whose model is to be found in Newtonian mechanics and Laplace's cosmological determinism. Its philosophical claims are best expressed by the positivistic school — from Comte, through Mach, to the Vienna circle. Positivists hold that for any phenomena necessary and sufficient conditions can be established (strict determinism). They tend to do away with causal vocabulary — a remnant of the 'metaphysical era'. They take "the fundamental dogma of invariability of natural laws" as a "logical" principle on which scientific investigation is grounded ([10], I, 1, 3, § 16).

2. Medicine seems to turn 'positive' with a reluctance or incapacity to turn 'positivistic', as though causal vocabulary was forced in by the very nature of the thing studied. It thinks causally in a naive, traditional, Aristotelian way —

Corinna Delkeskamp-Hayes and Mary Ann Gardell Cutter (eds.), Science, Technology, and the Art of Medicine, 105–126.

or else furtively, with apologies. At a time when the task of science is conceived to be inductively to infer *general laws* which allow *strict prediction* of phenomena given certain sets of determined conditions (and not to explain why such regularities can be exhibited or to know the 'deep causes' of things), there come the clinicians affirming the primacy of *individuality* over general laws, and Louis Pasteur exclaiming (about fermentations): "Small microscopic beings are the *hidden agents* of the phenomenon" ([24], Vol. 7. p. 37). Pierre-Simon Laplace said: "The surest method in the search for truth, is to inductively work one's way from phenomena up to laws, and from laws up to forces" ([20], p. 63). He alluded to a gravitational force that the eyes are unable to observe — sort of a 'Ding-an-sich', an 'occult cause', on the nature of which Newton in 1867 had refused to make hypotheses, and the concept of which is entirely unfolded in the equations describing its 'effects'. As G.R. Kirchoff (1874) and E. Mach (1883) will attempt to show, there is no need to assume that such forces correspond to any 'reality'. Cells and bacteria are *seen,* however, through the microscope, and even though viruses are hypothesized before being observed, they are meant to be real beings, not a mere way of speaking.[1] Rom Harré ([16], Ch. 6) stresses the difference between the *descriptive* character of the laws of mechanics, and the *explanatory* character of the virus theory. They exemplify, in his view, "opposing paradigms" of the character of scientific knowledge. The biological paradigm is closer to the Aristotelian paradigm than to the classical positivistic one.

3. Positivistic stands are indeed not entirely absent from scientific medicine. Claude Bernard and the French school were influenced by Comte. The German school around 1900 — as von Engelhardt shows in his essay — naturalizes some of the concepts of Mach and others. The outcome, however, is neither an elimination of etiological thinking from medicine nor an exclusive domination of the deductive-nomological model of explanation: this will be the force of my point.

My position might suggest that nineteenth century medicine has not yet reached 'positivity' and, as a result, is conceptually *backwards.* However, one might equally well be tempted to think — as does, e.g., M. Foucault ([13], p. 203) — that medical positivity is elsewhere and of a more secret nature. In conserving a pattern of scientific explanation more ingenuously Aristotelian and diversified than the pattern of classical mechanics, medicine may have contributed in setting up the circumstance that a scientificity established on other grounds authorizes a rational discourse on individuality.

II. PERSISTENCE OF THE ARISTOTELIAN PARADIGM

The doctor knows, says Aristotle, that circular wounds heal more slowly than others; the geometer knows why (*Post. Anal.* I, 13, 79a, 14–16). Both kinds of knowledge (of the fact, of its 'formal' ground) are scientific.

Peset rightly emphasizes the fact that for the ancients 'scientific knowledge' and 'causal knowledge' are synonymous (cf. the well-known passage of Aristotle's *Metaphysics* A, 3, 983a–b). This amounts to saying that the variety of meanings ascribed to the word 'cause' covers the variety of scientific questions which may be asked about an object (cf., *Metaphysics* Δ 1013a 24–1014a 25): questions of fact (τὸ ὅτι), of reason (τὸ διότι), of existence (εἰ ἔστι), of essence (τί ἐστιν) (*Post. Anal.* II 89b 20–35). That is, once it has been established that the thing is, one asks: 1) what is it made of (its *matter*)?; 2) what are its distinctive features (its 'quiddity' or *form*)?; 3) how was it produced or brought to its present state (its *efficient* cause, in medieval terminology)?; and 4) for what purpose does it exist (its *end*, or final cause, which is "always some good")? The important point here is this: not only are there for Aristotle several senses in which a thing can be said to have causes, but there may be a variety of different causes to produce the same effect; only in the case when "the effect is proved to *essentially* belong to the thing", should it have a unique cause. In contrast, if the link is *accidental*, then it may have several different causes (*Post. Anal.* II, 99a 1–6). Causal *necessity*, then, is encountered only in the realm of formal (*definitional*) causality.

Thomas Aquinas in his *Commentary* on Aristotle's *Posterior Analytics* remarks that it was generally believed by the ancients that whereas in the celestial (supralunar) world motions are regular and connections are necessary, the terrestrial (sublunar) world includes irregularities in the causal tissue of events (e.g., monsters, coincidences). Some causal *contingency* has to be admitted in the realm of *efficient* causes (the causes of "generation and corruption"); so that our (imperfect) knowledge is not always responsible for our not finding a strictly deterministic link between things: there may not be one . . . .[2]

There may be *chance* events: an idea not so foreign to contemporary biomedical sciences, in so far as they part from the deterministic ideology of positivism, to admit of 'chance' as well as of 'necessity' (cf. J. Monod [21]).

So the Aristotelian paradigm perdures in nineteenth century medicine, although apparently it splits up into a pattern of explanation fit to clinical medicine (hospital work) and a pattern of explanation fit to experimental medicine (laboratory work).

A. *Clinical Medicine: Formal* vs. *Material Causes, and a Somewhat Essentialistic Trend*

In the work referred to by Peset, Michel Foucault holds that medical positivity, originating at the dawn of the nineteenth century, is rooted in a way of "saying" that which is "seen", "in a sometimes so really *naive* discourse, that it appears to be situated at a more archaic level of rationality"([13], pp. viii–x). The anatomo-clinical method, initiated by Bichat, opens the way to a practice of letting the configuration of 'signs' be uttered at the patient's bedside. In the anatomy amphitheater, those signs are then implanted into the concrete quality of a matter.

It seems to me that clinical medicine does not break up with the Aristotelian paradigm of etiological inquiry, but rather resumes it above and over the linear paradigm of Humean causality. The Scottish philosopher Thomas Reid bluntly says that Aristotelian distinctions must be tied to an oddity of the Greek language: "We do not indeed call the matter or the form of a thing its cause" ([27], p. 394). In Reid's opinion, the only admissible use of the word 'cause', in either Latin or English, is that of 'efficient cause': "the real agents or causes which produce the phenomena of nature" ([27], p. 393). Moreover, the post-Newtonian thinker naturally knows that science does not, in that sense, unveil the causes of phenomena, but only "the law or rule according to which the unknown cause operates":

> ...natural philosophers, who think accurately, have a precise meaning to the terms they use in the science; and when they pretend to show the cause of any phenomenon of nature, they mean by the cause, a law of nature of which that phenomenon is a necessary consequence ([27], p. 396).

This was a clear statement of the modern paradigm of scientific explanation, and some forty years before Comte, quite a profession of positivistic faith.

In saying that Bichat and his successors are free from causal thinking, Foucault ([13], p. 142) means they are free from Humean causal thinking. He contrasts J.B. Morgagni's analysis [22] which identifies the *seat* of the illness with the first link of a *causal chain,* with the clinicians' notion of *localization* or 'infectious focus'. However, when in his 1861 paper read to the Société d'Anthropologie Broca sets up a correlation between a language disorder — named by him 'aphemia' — and a brain lesion localized in the posterior part of the third frontal circonvolution of the left hemisphere ([8], pp. 337–393), he may well be thought of as pursuing another type of causal analysis. Isn't identifying a syndrome (*clinical diagnosis*), and referring it to a tissular lesion, i.e., to the ultimate elements within reach of observation (*histological diagnosis*), isn't that referring a *form* to a *matter*, that is to say, pointing to both the distinctive pattern 'defining' the thing (e.g. as a motor — or 'Broca type' — aphasia, as opposed to a sensory — or 'Wernicke type' — aphasia), and to "what the thing is made of", i.e. what its elemental substratum is (a lesion of F_3, as opposed to a lesion of T_1–T_2), laid open by the *post mortem* examination?[3]

Two of Aristotle's senses of the word 'cause' are involved here. One may indeed object that the anatomo-clinical diagnosis is not etiological in the *strict* sense, insofar as the meaning of the term 'etiological' in the Humean tradition has been cut down to only part of its etymological resources (reduced to efficient causes, and possibly final ones, if one thinks of psychosomatic interpretations). However, it does not appear unreasonable to consider anatomo-clinical diagnosis as etiological in a *wide* sense, congenial to ordinary medical language: physicians commonly 'attribute' observed signs to the syndrome as being 'due to' the anatomical lesion. There is even in these causal imputations the *necessity* of what Aristotle calls the "demonstrations of essence". The patient exhibiting a 'Broca type' aphasia will *by definition* bear a lesion of F_3 in his left brain, and understand what he is told without being able to articulate a verbal answer.

Thus, if clinical medicine transgresses a causal paradigm, it is less the ancient paradigm than the modern one, which reduces causal inquiries to the detection of immediate antecedents in the superficial sequence of observed events. Not only is the anatomo-clinical method pluralist on the surface, but it looks into the depth.

Its parting from Aristotle is elsewhere: clinical medicine holds possible a rational discourse about individuals (cf. Foucault ([13] p. x)). The epistemological reach of such a position is considerable: it gives medicine the status of the first of the anthropological disciplines, rather than that of the last-born of the natural sciences. By granting to the mortal individual the ultimate objectivity (not to the eternal, general law, as classical science does), it reinstates finitude into knowledge's positivity ([13], p. 201). Only should it be admitted that scientific knowledge is possible, even if one isn't sure that there are any universal laws: clinicians were undoubtedly more inclined to admit it than experimentalists.

B. *Experimental Medicine: Efficient* vs. *Final Causes, Mechanism*

Claude Bernard typifies the medical experimentalist. He consciously borrows his explanatory model from the newborn physical-chemical sciences (his teacher Magendie had rendered him aware of the works of Lavoisier and Laplace ([18–20]); he accordingly holds that physiology should look for the *general laws* governing normal and pathological phenomena (i.e., regularities allowing prediction). He, however, following Bichat after all, acknowledges the need to recognize *special causes* accounting for those irregularities which, in biomedical sciences, make *prognosis* fall short of *prediction*.[4]

Very much in the spirit of Comte's positivism, Bernard takes the principle of *absolute determinism* to be indispensable as a methodological postulate. He first deplores Bichat's indulgence in the belief that 'vital properties' may at some point deceive a rational attempt to seize them; quoting Bichat:

> "vital properties being essentially unstable, all vital functions being susceptible to a great many variations, nothing can be predicted or calculated among their phenomena" (Bichat ([7], I, p. 7), quoted and summarized by Bernard ([3], p. 58).

Note that Bernard has no reluctance admitting the notion of 'vital properties' and saying with Bichat that, in the last analysis, "phenomena have their cause in properties inherent in living matter" ([3]), p. 57). Bernard, however, operated on that which we would nowadays call 'dispositional properties' of the living thing, a positivistic reduction in the Kantian style, placing the 'thing in itself' out of the reach of human knowledge, without denying its actuality: "Nowhere can one reach the primary causes; physical forces are as obscure a principle as the vital force... One cannot act on those entities, but only on the physical and chemical conditions which bring about the phenomena" ([3], p. 55). Ultimate or deep causes being unknown to us, we have to limit ourselves to looking for "immediate and accessible causes" ([3], p. 64); and what guarantees the scientific character of such research is the belief that phenomena "always appear in the same way following certain laws, not arbitrarily or capriciously at the whim of a lawless spontaneity" ([3], p. 61). Bernard remembers having been rebuked in his youth's enthusiasm by a professor of surgery who forcefully reminded him that it was an *error* to hold that vital phenomena be constantly identical in identical conditions ([3], p. 59). Nonetheless, Bernard held on to his view in the works published during his life-time, he consistently maintained the presupposition of a strict and linear determinism — identical causes cannot have dissimilar effects, any phenomenon is absolutely and invariably determined by its 'conditions of existence' which *necessarily entail* it ([3], p. 56). In brief: 'the cause' of E is its *necessary and sufficient condition*, so that the experimenter will not be satisfied until he has established 1) that the cause never occurs without its effect obtaining (experimental *proof*: $C \rightarrow E$); and 2) that the removal of the cause brings the effect to disappear (*counter-proof*: not C $\rightarrow$ not E, i.e., E $\rightarrow$ C). So that finally, once we know the *law* of the phenomenon (C if and only if E), we always and with absolute certainty can infer either the cause from the effect, or the effect from the cause ([2], 1, 2, § "De la preuve et de la contre-épreuve"). That is determinism in the Laplacean sense, rigorously authorizing both prediction and retrodiction.

However, in his *Principes de médecine expérimentale* (a posthumous work), Bernard reluctantly moves away from the classical paradigm. Patients, he says, want to be cured, and general laws are not sufficient to help us meet that need; we have to further look into special, "etiological determinisms" ([4], p. 7). From 'observation' we have to move on to actual 'experimentation'. Clinicians from Hippocrates to Laënnec have been content to *describe* morbid entities, classify them, draw their lines of evolution, set up rules for diagnosis and prognosis, and *wait* for nature to cure the patient; to this descriptive tradition belong Sydenham, Pinel, Bichat, and *even Virchow* ("a pathologizing Bichat, not an experimenting physician") ([4], Ch. X. p. 98). The experimenter goes beyond such essentialism: he wants to *explain* morbid events, that is, "analyse, artificially reproduce" them, and uncover their *mechanisms* ([4], Ch. XIV, p. 137). Indeed, Bernard does not deny that the understanding of mechanisms also involves some *deep* insight into *elemental* properties — whether it be his own 'milieu interieur' or Virchow's cells. He explicitly parallels his theory of the 'internal milieu' to Galen's theory of 'humors' ([4], Ch. XIV, p. 131). He grants to Virchow's cellular pathology an essential significance, though he deplores the fact that by focusing on elements, Virchow was led to neglect functional organization, and in particular to neglect the role of the nervous system ([4], Introd. p. 12). Nevertheless, on Bernard's view, properties are one thing, mechanisms another. He consequently devises a "program of experimental medicine" in three chapters (physiology, pathology, therapeutics). The objective of this program is systematically to replace empirical treatments based on generalities drawn from observation (e.g., "quinquina cures fever") with *etiological* treatments based on a knowledge of the specific determinisms involved (e.g., the treatment of scabies by sulphur is rational, and its success guaranteed without exception if correctly applied, because we know the *acarus scabiei* responsible for the disorder) ([4], Ch. VII, p. 73).[5]

At this juncture the experimentalist's reflection stumbles, for he has to account for the fact that similar external agents (e.g. a poison, a parasite) may have diverse effects according to cases (e.g., according to whether the subject had been mithridatized, or vaccinated) ([4], Ch. XIV, p. 157), and that a wide variety of stimulations (e.g. section, heat, cold...) applied to the same organic element (e.g., a nerve) may have very similar effects. Should we then admit that in the detail of particular determinisms, identical causes induce diverse effects, and different causes, similar effects? *No*: *if* the principle of *strict* scientific determinism is to be saved, external causes leading to physiological disturbances have to be considered *occasional* and inadequate to their effects ([4], Ch. XIV, p. 162), and *the true cause* which specifically determines the illness is the *property* of the injured tissue (or: cell, he later says with Virchow). Thus, illness does not fall upon the organism from outside; the determining, actual and proximate cause is the organism's *predisposition*.

A paradoxical consequence of this conclusion is that a strictly *etiological* diagnosis implies an *expectant* therapeutic attitude, whereas effective therapeutics is tied to an approximate knowledge of "indefinitely varied" causal links — the very consequence the author had meant to avoid. Namely, the task of 'preventive' medicine is to take individuals away from "morbigenous" influences which may modify their dispositions: its action is effective in spite of hazardous knowledge of the link between predisposing conditions and their effects. Once the individual has fallen ill, however, 'curative' medicine watches the development of a process which is essentially a function of the organism's reactive capability — it may intervene and modify the internal 'milieu' with another 'poison' (a drug), but it does not create the healing process which is spontaneous ("Hippocrates's *natura medicatrix*") ([4], Ch. XIV, pp. 160–164).

In order to save the classical principle of determinism (one cause, one effect), Bernard has therefore reached the surprising conclusion that illness is determined with *necessity* by the nature of the thing ill, but from such an etiological diagnosis no curative therapeutics can be derived: causal necessity goes along with expectative.

Reckoning that such is exactly the attitude he had reproached Virchow with, Bernard comes to then question his first theoretical position: "the task of science is undoubtedly to bring particulars within general cases and comprehend all varieties into the unity of a type... *but*... is there anything else than individuals?" ([4], Ch. XIV, p. 142). Reality always departs from the type; now, physicians have to deal with individuals, even the laboratory physician, who shall beware of generalizations, because laws have to be formulated in terms of: "in identical conditions, all..." and living beings, including experimental animals, never are identical, they have their "idiosyncrasies" ("constitution" + individual history, i.e., "conditions") resulting in an "individual predisposition". The physician therefore has to pay less attention to the type than to the "relation between the individual and the type", in order to understand, e.g., why different subjects react differently to the same "contagions" ([4]. Ch. XIV, pp. 142–145, 157).

Confronted with the facts of medical research, the experimenter first abandoned efficient causality (because it is a hazardous multicausality) in order to save determinism. Determinism (which authorized scientific prediction) finally dissolves, however, into idiosyncrasies of individual dispositions. Yet, the initial project intuitively looked sound (thus, for example, find out the mechanism by which curare, i.e., poison, an external cause, induces death in order to be able eventually to cure a curare intoxication). But the linear causal sequence it presupposed (phenomenally binding a succession of events), suddenly split up into a regular but 'profound' causality (dispositional property of the substrate) and an irregular, multi-factorial, 'phenomenal' causality: an irritating—or meaningful—revival of a complex paradigm of explanation which Bernard himself had originally rejected as prescientific and metaphysical.

C. *'Field' Medicine (Medical Officers, Hygienists, Statisticians): A 'Wild'—and Efficient—Causal Thinking*

Peset rightly underscores the enormous amount of work done in the course of the nineteenth century in the field of public health. Its initiators generally concede that industrialization and urban overpopulation had been responsible for a dramatic deterioration of sanitary conditions, calling for an equally dramatic counteraction. The most striking feature of the movement is, however, that it is backed up by an apparently very crude form of causal thinking: any climatic, or environmental, or 'zymotic' (etc.) factor is considered a possible cause of illness and death *prima facie*, and all the resources of popular traditions and medical empiricism are drawn upon. Of the collected heterogeneous data, a mathematical (statistical) treatment then permits a critical appreciation, and discrimination of 'genuine' and 'spurious' causal factors, and a measurement of causal 'influences'.[6]

The theoretical background for the method is set up by Bayes [1], and by Laplace: "Mémoire sur la probabilité des causes par les événements" [18]. The method is used extensively, and refined, by hygienists and statisticians, throughout the nineteenth century. They themselves collect numerous data from local practitioners, of whom all contemporaries agree to deplore for lack of education and (sometimes) praise for devotion and humanity.

Nevertheless, the considerable informative contribution of these practitioners should not be forgotten.

Meanwhile, medicine proper ('grand' medicine) is largely unaware of the newly delineated concepts, or at least does not grant them a scientific 'dignity' until the acceptance of the notion of 'risk factor' in the second half of the twentieth century. The controversy between causalists and conditionalists keenly analyzed by von Engelhardt shows what clinicians and fundamentalists missed for ignoring it.

Peset mentions the initiators of the hygienists' movement; the precursor statisticians, namely, John Graunt [14] and Johann Peter Süssmilch [30] (who laid the foundations of population statistics), may also be recalled. I shall here concentrate on one example from Great Britain.

The occasion for keeping an account of burials in London arose from the plague (1592 epidemics), according to Graunt ([14], Ch. I, Art. 2), and as early as 1629 a mention of the disease or casualty causing death was added on the register ([14], I, 5). From his examination of those 'bills of mortality', Graunt delineates a number of 'general laws' (e.g., a third of all children die under five years of age, women die less than men), as well as investigates some of the causes (e.g., only 7% of the people die "of age", other (premature) deaths are due to either acute illnesses, or chronic illnesses, or accidents). Further, he "venture(s) to make a standard of the healthfulness of the *Air* from the proportion of *Acute* and *Epidemical* diseases, and of the wholesomeness of the *Food* from that of the *Chronical*" ([14], II, 18). The novelty of this approach is not the belief that air can cause infectious diseases, or bad food physical impairment: the Hippocratic tradition had put strong emphasis on the pernicious influence of 'miasmas', and insisted on dietetics. The novelty lies in his attempt at grounding a causal inference on statistical data.[7] A century later (1774–1794), the Société Royale de Médecine de Paris conducts an inquiry involving 150 physicians with the purpose of assessing meteorological influences on epidemics. There again, the idea is no news; the method is. Its being put into common practice is noteworthy.

Public offices for vital statistics were created in most European countries towards the beginning of the nineteenth century.[8] The outstanding contribution of William Farr dominated the British scene. Appointed in 1838 as "Compiler of Abstracts" to the newly-established Registrar General Office, Farr soon drew up a 'statistical nosology', in fact a classified list of causes of death appropriate for registration of data entered on death certificates and for statistical treatment. As he put it: "the nomenclature is of as much importance in this department of inquiry as weights and measures in the physical sciences" (*First annual report*, p. 95, in [12], p. 231). Farr's nomenclature was considered together with the nomenclature proposed by d'Espine (of Geneva) at the Second International Congress of Statisticians in Paris (1855) at which a compromise was adopted. Then, the Chicago meeting of the International Institute of Statistics (1893) followed the 'Bertillon proposition'. The latter, adhered to by the American Public Health Association (1898), became the "International List of Causes of Sickness and Death" and the official reference for the compilation of cause of death returns. Revised every ten years, it now is published under the authority of the World Health Organization under the name of "International Classification of Diseases, Injuries and Causes of Death" and serves as a worldwide base for vital statistics. In its ninth revision [33], the classification's *eclecticism* is remarkable. Besides over a thousand various illnesses, intoxications, and traumas (medical causes), and their occasions (accident, poisoning, etc.: external causes), it also enumerates a number of socio-economic factors (circumstances influencing health) to be taken into consideration in analyzing a population's standards of health.

The purpose of statistical analysis, as repeatedly stressed by Farr in his *Annual Reports*, is to identify the chief causes "which are injurious and fatal to men" and "contribute to the removal of evils which shorten human life" (*16th Annual Report, Appendix*, p. 79, in [12], p. 255). Now statistical studies did not only point to phthisis, typhus, or cholera as to the causes of many premature deaths, they also showed overcrowding and squalid urban conditions to be significantly correlated with high death rates. The masterwork here is Chadwick's *Report on the Sanitary Condition of the Labouring Population of Great Britain* (1842) [9]. Chadwick's enquiry, amply relying on local reports and information supplied by medical officers, inexorably establishes a correlation between the incidence of major diseases and insanitary living conditions (e.g., overcrowded slums, inadequate water supply, defective sewerage). It assesses the economic and social cost of squalor and ill-health, and advocates legislative action to impose ventilation in work places, effective drainage, and the removal of sewage from the towns by suspension in water, etc. The Report's ultimate outcome was the Public Health Act of 1848. Great Britain is only an example. It is a well-known fact that in the course of the nineteenth century similar preventive measures aimed at safeguarding the health of the populations were taken by governments in most industrialized countries.

Noteworthy enough, Chadwick (as did many of his contemporaries in the medical profession) adhered to the *miasmatic* theory of disease's propagation

—a theory which was to be rigorously proved *false* thirty years later, when Pasteur and Koch established the bacteriological causation of contagious diseases. Dr. W.P. Alison, an influential professor at the Edinburgh school of medicine, disagreed with Chadwick, arguing that putrid smells released by decaying organic matter would not generate fever if the bodies were not enfeebled by poor nourishment, overwork, and destitution. On the other hand, as soon as 1848, J. Snow showed that a water distribution conforming to certain sanitary norms stopped the spreading of cholera.[9]

Now, the last significant cholera epidemic in Europe took place in 1867, while the choleric *vibrio* was not isolated until 1884; the incidence of tuberculosis was at its heights roughly between 1780 and 1840, and had long started decreasing when in 1882 Koch discovered the bacillus. Peset and von Engelhardt both like to say that the pertinence of causal thinking in medicine is best gauged by the efficiency of therapeutics; along their line one should be tempted to claim (a point made by Dubos (1953)) that sewerage, legislation on labor, and quarantine (1851: first European conference on the quarantine against plague, cholera and yellow fever) were a more causal step towards the eradication of cholera and 'consumption' than 'etiological therapeutics' which became available after the battle was over.

Indeed, Pasteur both determined the causal agents of infectious diseases and an immunization procedure. Long before Pasteur, however, mathematicians attempted to compute the comparative gains in life expectancy attributable to either vaccination or other sanitary measures (cf. Bernoulli [5], Trembley [31], Laplace [20]). In the statisticians' and hygienists' perspective, immunization was *a priori* neither more nor less 'etiological' than was the amelioration of the standards of living—as a matter of fact, better be both immunized and nourished than get immunization with malnutrition. In brief, statisticians seem liberally to admit of a possible *overdetermination* (there may be over-sufficient causes) as well as underdetermination (non-sufficient conditions, non-necessary conditions, are not always negligible).

It thus appears that field enquiries were first conducted in a naive and bold causal spirit, huddling together physical, economical, social, and medical items into the category of etiological factors without the orderly systematics of Galenic medicine which assigned to each factor a proper place. Secondarily, however, a statistical screening could reassess causal claims and lead to a virtually ever-refinable appreciation of 'significant' and 'negligible', 'direct' and 'indirect' causal dependencies, through a quantitative evaluation of the respective 'influences' of multiple factors and their evolution over time. A good example of such refinement is to be found in the development of national and international statistics of 'causes of death': detection of 'simple' causes, then the treatment of 'joint' causes, and finally the question of 'complex' causes.

The probabilistic approach brought up a profound, though slowly and unheededly pervasive intellectual revolution. It conveyed the idea that there is no obvious ground for granting *a priori* a privilege to 'external' or 'internal', 'proximate' or 'remote', 'immediate' or 'underlying', 'predisposing' or 'occasional', 'essential' or 'accidental', 'formal' or 'material', causal factors (cf., Nysten [23]). This is the bushy landscape in which medical ratiocinations got lost. All presumed causes were to be treated as equivalent *prima facie*, and equally submitted to *statistical testing*, through which 'spurious' causal hypotheses could be detected and discarded, and the actual importance of 'genuine' causes *measured*.

The multicausal scheme underlying the probabilistic approach reduces the case in which 'the cause'. C, is a necessary and sufficient condition of its effect, E, to the particular — any possibly never occurring — circumstance in which the probability of E, given C, is one. In the general case, 'a cause' is merely that which increases the probability of an effect (cf., Suppes [29]); and probabilistic prediction falls short of deterministic prediction. The resort to probabilistic predictions, though very congenial to common medical experience, certainly embarrassed many: Bernard considered statistics to be "the highest expression of empiricism"([4], Ch. VI, p. 59) and wanted medicine to raise its scientific standards by going from empiricism up to the knowledge of 'mechanisms'. He could not admit (it hardly started being admitted in the recent years) that statistics offered a means of finally departing from *mere empirical generalizations* and precisely evaluating causal dependencies in a universe in which, at first sight, causal relations are neither linear nor simple. The non-recognition of the significance of a mathematical tool such as that heavily bears on the dispute between causalists and conditionalists told by von Engelhardt.

III. THE ANCIENT PARADIGM REUNITED AROUND 1900 IN A GALENIC SPIRIT

In his essay, von Engelhardt analyzes the controversy between German 'causalists' and 'conditionalists' around 1900; two striking facts are evident in his analysis: 1) dissatisfaction generated by a monocausal perspective, whether everything be charged to the internal disposition (cellular pathology), or everything charged to the external infecting agent (bacteriology); and

2) dissatisfaction with the concept of causal 'necessity', and attempt at 'relativizing' the causal relation.

There is some ambiguity, however, about what such a 'relativization' amounts to, and a brief digression using simple logical terminology might help clarify it. One may want the 'causal law' to be 'absolutely' true, i.e., true in all possible worlds, or one might insist on the fact that causal laws are true in a particular model **M** of a particular world which we happen to have observed (or rather deducible in a particular axiomatized system (call $A_1, \ldots, A_n$ the system's axioms) meant to characterize such a model). 'Relativizing' the causal relation then amounts to saying that it is not the case in general that $\models C \leftrightarrow E$, but it is the case that relative to $\mathbf{M} \models C \leftrightarrow E$; or, it is not the case that $\vdash C \leftrightarrow E$, but it is the case that $A_1, \ldots, A_n \vdash C \leftrightarrow E$.

The effect was so far supposed to obtain *if and only if* the causal event itself occurred, that is, the causal link was supposed to be reducible to a logical equivalence. There is, however, another form of 'relativization' consisting in dropping the requirement that the causal link be logically analyzable as a biconditional. Causal 'necessity' then merely reduces to the 'factuality' of the causal sequence: we cannot pervert facts, or reverse time.

The latter 'relativistic' point of view (dissociable from the first) is quite familiar to medical thinking and any form of thinking concerned with particulars. It already was explicitly stated by the Roman physician A.C. Celsius in his *De re medica* (first century A.D.), where after rejecting both 'empiricism' and 'dogmaticism', and objecting to the sort of positivistic attitude called 'methodicism', he concluded

> that nothing can be attributed to a single cause, and we take as cause anything likely to have much contributed to produce an effect; that such a thing which has no action by itself may induce the worst disorders when joined to others; [that a general physiological theory] cannot explain why of two equally plethoric individuals, one shall fall sick and the other remain in health; [that finally] there almost is no principle in medicine from which one should not deviate in certain cases (*De re medica* I, 26–27).

Celsius's scheme of medical diagnosis requires to look for 1) obvious external causes (e.g., cold, heat); 2) the patient's constitution (temperament); and 3) his mode of life, resulting in acquired dispositions. As for "hidden causes" (the nature of the patient's 'vitality') Celsius holds that they always are conjectural: "inductions based on them" should therefore be banished from the 'Art', "though not from the Artist's mind" (*De re Medica* I, 33).

Celsius's reflections would by no means sound out of place in the midst of nineteenth century controversies, insofar as they lay stress upon the cogency of a flexible multicausal approach: this is precisely the perspective on which both causalists and conditionalists converge, as shown by von Engelhardt. If that were all the dispute came down to, one might just conclude that they all finally fall back onto a kind of 'relativism' which is more like the ancients' than that of Einstein's. There is, however, more to the controversy.

I shall not insist on the fact that Ferdinand Hüppe uses the term 'energy' when talking about disposition. His terminology might be considered imprudent, for amounting to an implicit decision about the 'essence' of vital phenomena — rather far from Mach's 'reduction' of the notion of 'force' in physics. But after all, the 'potential energy of disposition' is treated by Hüppe like one of the three 'variables' of which illness is a 'function' (phenomenalist vocabulary), as much as like a deep property ('Ur-sache') of the thing (essentialist vocabulary).

More crucial is the fact that the *causalists'* explanatory scheme offers a clear *reorganization of etiological factors* in a *qualitative* spirit highly reminiscent of the Galenic scheme.

In his book on hygiene — as mentioned by Peset — Galen defines 'health' as the dynamic equilibrium of a smoothly functioning system: "health is a sort of harmony", consisting in "never having a pain in any part" and "being unimpeded for any of the functions of life" (*De sanitate tuenda* I, 5 and III, 5). The system may be shoved into an unsettled state by two kinds of causes:

1. causes "inevitable and intrinsic, having at their root the sources of generation": this is the 'diathesis' or constitutional disposition, whose defectiveness is called "discrasia' (*De sanitate tuenda* I, 2); it also is a natural tendency to age, i.e., to "progressively dry up" (an idea that goes back to Aristotle, e.g. *De longitudine et brevitate vitae* V, 466 a21), an "innate destiny of destruction" and universal fate of living creatures. Those are the *res naturales*, on which nevertheless diet and exercise (i.e., *res non-naturales*: artifices) have a relative modifying effect (a 'dyscrasia' may be corrected by an appropriate diet, aging may be accelerated by worries or ill health) (*De sanitate tuenda* II, 9).

With respect to illness, the latter set of factors is specified as the 'primary' (proègoumenon), 'proximate' (prosekhes) or 'internal cause' (synektikon aition) (*De symptomatum causis* I, 5) — traditionally the factors predisposing to diseases: αἱ προηγούμεναι αἴτιαι τῶν νοσημάτων (*De morborum causis* 2).

2. causes "not inevitable nor arising from ourselves" (*De sanitate tuenda* I, 2): those are the *res contranaturales*, of which Galen distinguishes two types (*De sanitate tuenda* I, 4): a) first type: the violent aggressive agents

causing an obvious deterioration (such as a wound); and b) second type: the environment's sly and commonly undetected aggressions (which again may be prevented or corrected by a sensible mode of life: *res non-naturales*). With respect to illness, that set of factors is specified as the 'initial' (prokatarchonta) and 'primitive causes' (prokatarktika aitia) and hitting the body from outside: τὰ δ' ἔξωθεν προσπίπτοντα... τὸ σῶμα προκατάρχοντά τε καὶ προκαταρκτικὰ... αἴτια (*De morborum causis* 2).

It is not very difficult to decipher Hüppe's tricausal scheme with Galen's grid:

ta aitia proègoumena	ta aitia prokatarktika	ta aitia prokatarktika
Prädisposition	<of the first type>	<of the second type>
	Reiz	äußere Bedingung

Neither is it difficult to view as *res non-naturales* the hygenic or therapeutic interventions aimed at modifying or mending either the *res naturales* (Konstitutionshygiene), or the *res contra-naturales* of the first type (Auslösungshygiene), or the *res contra-naturales* of the second type (Konditionalhygiene).

The causalists' insistence on centering their approach on man is also very faithful to Galen, who likes to repeat that the norms of health are "different in each" (*De sanitate tuenda* I, 5). In contrast, Martius's pseudomathematical formula

$$K = \frac{W}{P}$$

which in Galen's terms can be written:

$$\text{illness} = \frac{\text{aitia proègoumena}}{\text{aitia prokatarktika}}$$

looks like a poor summary for a qualitatively complex causal analysis. Moreover, it cannot be put to any use, since Martius fails to offer a method for quantifying any of the qualified factors.

As to the *conditionalists*, they obviously mind the qualitative heterogeneity of the types of causes in the causalists' scheme. Their principle according to which "all the factors are equivalent" (Verworn's Proposition 5) expresses a sensible view rejoining that which hygienists and statisticians somehow had suggested: all causes are to be treated in the same way *prima facie*. The 'relativistic rationalism' of Verworn certainly could be interpreted as a less 'practical' than 'metaphysical' position (or rather, 'antimetaphysical': there is no in-itself of the thing, the thing *is* the set of its phenomena — positivistic radicalism of the times). It also may be understood as a reminder of the doctor's indispensable obligation to question his medical knowledge and convictions *a propos* of each concrete case: no element should be neglected. In more familiar terms: it doesn't suffice to ask for the bacteriological diagnosis and antibiogram, then to prescribe the appropriate antibiotic, to kill the germ and save the patient; rather the patient's age, nutritional state, clinical history, etc., must be taken into account to make a pertinent diagnostic and therapeutic judgment. That, however, is an endlessly repeated and endlessly forgotten truism, well known to all hospital residents, and its statement should not irritate the causalists.

What the causalists perceive behind the principle of the "effective equivalence of the conditions" is a declaration in favor of a logico-mathematical kind of determinism, by which the preeminence of certain major etiological factors is erased. Let us remember that Bertrand Russell at the time popularizes logical atomism (cf. [28], lecture 8). Not that the causalists would much scruple to adopt a pseudomathematical vocabulary ("any state or event is strictly determined by the *sum* of its conditions": Proposition 3) — a vocabulary which, again, is of very little use, as long as no indication is given on the way to *measure* the said conditions. But by suggesting the schema:

$$C_1 + C_2 + \ldots. + C_n = E$$

and adding that this equation comes to an *equivalence* or identity (Proposition 4), Verworn is in his turn a victim of the mirage according to which, in order to turn medicine into an (at last) respectable scientific discipline, one should align oneself with the logico-positivistic view of science, treating *events* as *propositions* deductible within an axiomatic system. In the present case, the system of the conditions is supposed equivalent to the effect ($C \leftrightarrow E$), a necessity which is conceivable only if it is purely definitional — as stated in Proposition 4. From the world of events (the diseased organism's history), one has been surreptitiously crept into the world of essences (formal causality), even though the thing's essence reduces to the set of its phenomena.

When the causalists retort that some conditions are *more negligible* than others in practice, they probably blame Verworn for a sin he hasn't really committed, but their intention is fully accountable. Namely, they resist the slip and talk from the other point of view, that of historical sequences of events in which common intuition perceives — if not determining influences — at least influences more decisive than others, and nevertheless contingent with respect to the final outcome (e.g., 'causes of death'). The logical 'reduction' of such causal sequences is chimerical. The presence of tubercle bacilli in a child's environment is not a sufficient condition for him to get a primary tuberculosis; it is a *risk factor*. The presence of asbestos in the atmosphere is not a necessary (neither is it a sufficient) condition to induce a pleural cancer (mesothelioma); it only augments the chances. The postulate that there be a set of necessary and sufficient conditions determining an event is by no means indispensable for the detection of significant causes.

There was another course to take. Rather than believing — as does von Hansemann — that medicine has to leave the 'etiological era' to enter the positive era, one could hold that there is a positive way of thinking etiologically — not only in medicine, as a matter of fact, but in all anthropological disciplines.

Wanting to turn medical diagnoses into pure phenomenological 'descriptions', is underestimating the spontaneous medical tendency to speak causally — and nothing is fundamentally changed when the word 'condition' is substituted for the word 'cause'. If, however, an etiological approach can hardly be eradicated from medical thinking, it does not reach positivity until, after considering *prima facie* any factor as a *possible cause* (even perhaps the whole universe — *à la* Einstein conceivably, and why not a la Paracelsus and van Helmont?), one is able to use an effective screen to discriminate between causally relevant and irrelevant factors. Indeed, that is what von Hansemann's secondary distinction between 'necessary' and 'subordinate' conditions ('notwendige Bedingungen' and 'Substitutionsbedingungen') is aimed at: a weak distinction, however, for it does not admit of degrees in the causal influence.

The absent is the mathematical tool: in order to get a positive grasp on reality, the conditionalists' intuitions should need to be backed with the resources of statistical analysis. One thing is to apprehend reality as a 'knot of relations'; another thing is to know how to untie the knot and disentangle the intricated factors in order to appreciate their respective influences and adopt a therapeutic attitude which can both be rational and fit on to the individual case.

IV. CONCLUDING REMARKS

Both the essays by Peset and von Engelhardt call for a reassessment of causal (etiological) thinking in medicine. Peset's thesis is that the accession of medicine to scientific positivity at the beginning of the nineteenth century is characterized by a renunciation to the Aristotelian paradigm of causal explanation (as developed by Galen) and an alignment on the disciplines whose paradigm of explanation is that of Newtonian mechanics. I argue that, in spite of the perceptible influence of the positivistic ideology, medicine could not so easily do away with etiological vocabulary; that, even though the Aristotelian paradigm seems first to split up, the anatomo-clinical method (hospital work) leaning on to the 'vertical' dimension of the causal explanation (formal vs. material causes), and the experimental method (laboratory work) orienting researchers towards the 'horizontal' problematics of establishing causal sequences (efficient vs. final causes), the two dimensions are found intermingled. I argue that at the end of the nineteenth century, in the controversy surrounding 'causalism' and 'conditionalism' — as analyzed by von Engelhardt — the paradigm is virtually reunited into a comprehensive structure highly reminiscent of Galen's scheme, with refinements which might even evoke the scholastic subtleties of medieval disputes.

However, *behind* the scene of an apparent continuity with tradition, a 'positive' medicine emerges, breaking off with the requirement that universal laws be the norm of scientific discourse. In a seemingly rather *crude* perspective, it takes as a cause anything that significantly raises the probability of an effect, no matter if the 'cause' be internal and dispositional (e.g., a blood or tissue group) or external (e.g., bacterial or viral agents, environmental factors). The degree of the causal 'influence' is measured through a correlation analysis. An exhaustive determination of causes is not required. Diagnosis and prognosis, taking into consideration a variety of causal influences, are probabilistic judgments. An effective therapeutics is that which shields one from significant 'risk factors'.

The method characterizes a form of operational rationality in those areas of experience which do not exhibit rigid regularities. For instance, Bayes's procedure (successive conditionalization of probabilities, given the successive states of available information) may be taken as sensibly describing the sensitivity of clinical judgment to individual cases and their evolution in time. At the level of populations, the possibility of isolating major risk factors and computing the degree to which they contribute to a given illness, together with the possibility of estimating the efficiency and cost of appropriate preventive or curative measures and of following the evolution of certain disease patterns throughout time, seems to have tardily brought about the acceptance of statistical methods.

Finally, it should not be forgotten that the very detection of risk factors is itself contingent on the objectives of medical research and sanitary policies, and those in turn are contingent on how a given civilization conceives of the standards of health: so that one might perhaps be tempted to think that the end of the twentieth century is actually taking a step more decisive towards the relegation of a certain medieval era, than the 'positive' step taken by clinical medicine a century ago, insofar as it bears the consciousness that a rational medical attitude less consists in scrutinizing the details of the causes of illnesses than in first choosing the objectives of health.

University of Paris-X, Nanterre, France

NOTES

* The passages quoted from the Greek and French texts in the above have been translated by the author of this paper.

[1] The virus of rabies was called by Pasteur and Roux an "être de raison" (cf. Pasteur [24], VI, pp. 603–607) because it eluded the best optical microscopes; the methodology of immunization, however (attenuation of the germ's 'virulence' by successive passages on rabbits' medullas and dessication), unambiguously shows that they realistically thought they were dealing with an actual being, which more powerful techniques later allowed to measure (Loeffler and Frosch, 1897), filtrate (Levaditi, 1898), isolate in crystalline form and chemically characterize (Stanley, 1935), then visualize and photograph (electron microscopy, 1940).

[2] On this and the history of causal thinking in general, see Wallace [32].

[3] Virchow opens the era of cellular localization in 1850, as Bichat did the era of tissular localization in 1800. Recall that between the two, achromatic microscopes started being constructed (around 1830).

[4] Here is how Bernard describes the situation in the first half of the nineteenth century: Magendie and himself were forcing medicine into the laboratory, England followed at a distance, and "Germany was dozing or dreaming in the clouds of the 'philosophy of nature', calling in question the legitimacy of empirical knowledge and roving in the abstractions of the *a priori* method" ([3], pp. 9–10). Germany, he acknowledges, caught up some years later, and led then the movement of experimental physiology; in his 1869 opening lecture at the Museum d'Histoire Naturelle, Bernard describes for his students the several laboratories newly created throughout European countries. He displays in front of them the plans of the University of Leipzig's laboratory "with its sophisticated equipment, of which we can't even conceive, here in France" ([3], p. 15).

[5] The dissatisfaction felt by the reader about this example is also felt by Bernard. As he adds in a footnote: "I can't consider that the problem is entirely solved, for we should need to know where the parasite comes from, and what causes its springing on some individuals and not on others" ([3], Ch. VII, p. 73). So, he swings from the parasitic etiology (external cause) to the dispositional etiology.

The agent responsible for scabies was first identified by Renucci (1834). The hypothesis of a parasitic etiology of all contagious diseases seems to have been formulated by F.V. Raspail (1794–1878) [26] at about the same time as by J. Henle (1809 –1885) [17]. Raspail, in fact, offers a comprehensive view of both cellular (internal) pathology and parasitic (external) pathology.

Retrospectively, the controversy between bacteriologists and cellular pathologists, evoked by von Engelhardt, mostly appears as a debate between those who are mainly aware of infectious and epidemic diseases, and those who are more aware of neoplasic and cardiovascular ('degenerative') diseases. As soon as one acknowledges the role of dispositional factors in the former and environmental factors in the latter, one falls back on the multicausal Galenic scheme. Bernard is an interesting case because he oscillates between the two points of view *for theoretical reasons*: he wants to save the principle of determinism, in order to warrant etiological diagnosis.

[6] That naive or spontaneous causal thinking is multi-factorial and not strictly deterministic, is a fact excellently stressed by Suppes [29].

[7] For an analysis of Graunt's methodology, see I. Hacking [15].

[8] a) France: first official 'Bureau de Statistique Générale': 1800 (Francois de Neufchateau, Lucien Bonaparte, Chaptal). Then, in 1833: 'Statistique générale de la France' (Thiers) and publication, from 1829 on, of the *Annales d'Hygiène* (Parent-Duchatelet, Villermé).

b) Germany: Statistical Offices of Prussia: 1810, of Bavaria: 1813. From 1872 on (United Germany): Statistische Reichsamt.

c) Austria: 1828.

d) Great Britain: Registrar General Office of England and Wales: 1837, etc.

[9] The main source of information for section 5 was the W.H.O. library in Geneva. See, in particular, Farr [12] and Chadwick [9].

BIBLIOGRAPHY

1. Bayes, T.: 1764, 'An Essay Toward Solving a Problem in the Doctrine of Chances', *Philosophical Transactions* **53**, 370–418.
2. Bernard, C.: 1865, *Introduction à l'étude de la médecine expérimentale*, repr. ed., Flammarion, Paris, 1952.
3. Bernard, C.: 1878, *Leçons sur les phénomènes de la vie communs aux animaux et aux végetaux*, repr., Vrin, Paris, 1966.
4. Bernard, C.: 1947, *Principes de médecine expérimentale*, P.U.F., Paris (posthumous).
5. Bernoulli, D.: 1760, 'Essai d'une nouvelle analyse de la mortalité causée par la petite vérole, et des avantages de l'inoculation pour la prévenir', *Histoire de l'Acad... Paris* **66**, 1–45.
6. Bichat, X.F.: 1801, *Anatomie générale*, repr. Lagrange and Lheureux, Paris, 1818, 2 vols.
7. Bichat, X.F.: 1800, *Recherches physiologiques sur la vie et la mort*, repr. Gérard, Verviers 1973.
8. Broca, P.P.: 1865, 'Sur le siège de la faculté du langage articulé', *Bulletin de la société d'anthropologie* **6**, 337–393.
9. Chadwick, E.: 1842, *Report on the Sanitary Condition of the Labouring Population of Great Britain*, repr. M.W. Flinn (ed.), Edinburgh University Press, 1965.
10. Comte, A.: 1844, *Discours sur l'esprit positif*, Société Positiviste Internationale, Paris, 1914.
11. Engelhardt, D. von: 1992, 'Causality and Conditionality in Medicine Around 1900', in this volume, pp. 75–104.

12. Farr, W.: 1885, *Vital Statistics*, Humphreys, London (posthumous).
13. Foucault, M.: 1963, *Naissance de la clinique, une archéologie du regard medical*, P.U.F., Paris; English translation by A.M. Sheridan Smith in 1973: *Birth of the Clinic*, Pantheon Books, New York.
14. Graunt, J.: 1662, *Natural and Political Observations... Made Upon The Bills of Mortality*, repr. W.F. Willcox (ed. and intro.), Johns Hopkins University Press, Baltimore, Maryland, 1939.
15. Hacking, I: 1975, *The Emergence of Probability*, Cambridge University Press, Cambridge.
16. Harré, R.: 1972, *The Philosophies of Science*, Oxford University Press, London.
17. Henle, F.: 1840, *Pathologische Untersuchungen*.
18. Laplace, F.S.: 1774, 'Mémoire sur la probabilité des causes par les événements', in *Oeuvres complètes*, publiées sous les auspices de l'Académie des Sciences, Gauthier-Villars, Paris, 1878–1912, Vol. 8, pp. 27–65.
19. Laplace, P.S.: 1778, 'Memoire sur les probabilités', in *Oeuvres complètes*, publiées sous les auspices de l'Académie des Sciences, Gauthier–Villars, Paris, 1878–1912, Vol. 9, pp. 383–485.
20. Laplace, P.S.: 1812–1820, *Théorie analytique des probabilités*, in *Oeuvres complètes*, publiées sous les auspices de l'Académie des Sciences, Gauthier-Villars, Paris, 1878–1912, Vol. 7.
21. Monod, J.: 1970, *Le hasard et la nécessité, essai sur la philosophie naturelle de la biologie moderne*, Seuil, Paris.
22. Morgagni, J.B.: 1761, *De Sedibus et causis Morborum per anatomen indagatis*, Ex Typographia Remondimiasma, Venetiis.
23. Nysten, P.H.: 1814, *Dictionnaire de médecine*, Brosson, Paris.
24. Pasteur, L.: 1922–1939, *Oeuvres complètes*, réunies par Pasteur Vallery-Radot, Masson, Paris, 7 vols.
25. Peset, J.L.: 1992, 'On the History of Medical Causality', in this volume, pp. 57–74.
26. Raspail, F.V.: 1843, *Histoire naturelle de la santé et de la maladie*, 3rd ed., Bruxelles, Paris, 1860, 3 vols.
27. Reid, T.: 1788, *Essays on the Active Powers of the Human Mind*, in *The Works of Thomas Reid of Edinburgh*, Etheridge, Charlestown, 1813, Vol. 3.
28. Russell, B.: 1914, *Our Knowledge of the External World*, 6th impr., G. Allen, London, 1972.
29. Suppes, P.: 1970, 'A Probabilistic Theory of Causality', *Acta Philosophica Fennica* **24**.
30. Süssmilch, J.P.: 1741, *Die göttliche Ordnung in den Veränderungen des menschlichen Geschlechts aus der Geburt, dem Tode und der Fortpflanzung desselber erwiesen*, Berlin (Enlarged edition, 1761).
31. Trembley, P.: 1796, 'Recherches sur la mortalité de la petite vérole', *Mémoires de l'Acad... Berlin* **99**, 17–38.
32. Wallace, W.A.: 1972–1974, *Causality and Scientific Explanation*, The University of Michigan Press, Ann Arbor, Michigan, 2 vols.
33. World Health Organization: 1977, *Manual of the International Classification of Diseases, Injuries, and Causes of Death*, 9th ed., W.H.O., Geneva, 2 vols.

ERIC T. JUENGST

CAUSATION AND THE CONCEPTUAL SCHEME OF MEDICAL KNOWLEDGE

I. INTRODUCTION

Philosophers often find themselves in debt. Imre Lakatos, for example, borrows a formula from Kant in order to show how he can repay his larger debt to historians of science. He suggests that while "philosophy of science without history of science is empty, history of science without philosophy of science is blind" ([22], p. 91).[1]

This paper also owes much to rich historical narratives, and, moreover, attempts to repay them in Lakatosian coin. Here, this means pointing out some of the interesting formal features of a "conceptual system for medical knowledge" that emerge from the work of Peset [29], von Engelhardt [11], and Fagot-Largeault [12] on the development of causal thinking in medicine.

The essay's focus is on the two intriguing conclusions that Fagot-Largeault draws from her response to Peset and von Engelhardt. She argues, first, that medicine continues to use "Aristotelian" as well as "positivistic" interpretations of causation into the modern era. This is an historical point that has a philosophical question tripping at its heels: *why* does the "Aristotelian paradigm" have such a hold on medical thought? The first two sections below seek to amplify Fagot-Largeault's point by offering a philosophical response to this question. Fagot-Largeault suggests it is "as though causal vocabulary was forced in by the very nature of the thing studied" ([12], p. 105). The suggestion here will be that a much more forceful source is the nature of the studies themselves. I will argue that one of the reasons these causal paradigms coexist is that the two kinds of explanations they invoke play interdependent roles in a basic medical explanatory strategy.

Fagot-Largeault's second conclusion expresses a thesis about medical progress that is palpably philosophical on its own. She suggests that the emerging probabilistic interpretation of causation might ground a more progressive conceptual scheme for medicine ([12], p. 106, 123). The question here is whether she is right. In the last sections of the paper some doubts are raised about how far Fagot-Largeault's advocacy of probablistic causal views can be taken, in light of lessons about medical explanation that can be drawn from her historical points.

Corinna Delkeskamp-Hayes and Mary Ann Gardell Cutter (eds.), Science, Technology, and the Art of Medicine, 127–152.

In capsule form, the conclusions I draw are the following. First, Fagot-Largeault's "Aristotelian" causal thinking tends to be realized in explanations that focus on the functional organization of biological systems. On the other hand, the "positivistic" model of causation is usually manifested by accounts explaining those systems as a consequence of their parts. Fagot-Largeault finds both forms of explanation active in modern "scientific" medicine because, in combination, they provide one important way in which medicine approaches its domain. By first asking for one type of explanation and then for the other, medicine parses the phenomena of its domain into hierarchies of compositional levels, organized and related through functionally defined processes.

This strategy is important for three reasons. The pluralistic ontology it produces allows medicine to bring order to the phenenomenal complexity that living organisms and their diseases display. Methodologically, the heuristic reciprocity between these explanatory perspectives provides impetus and guidance through complex medical problems. Most importantly, though, this explanatory strategy allows medicine to shape its inquiry with special practical goals, by using them to define the functions (or dysfunctions) of concern and the organic systems they, in turn, delineate.

II. "CAUSAL THINKING" AND MEDICAL EXPLANATIONS

Fagot-Largeault follows Peset and von Engelhardt in abstracting from medical thought the two philosophical models of causation that she labels "Aristotelian" and "positivistic". Her focus is, with Peset's, on the general historical interplay of these models, though she elegantly shows how the same patterns underlie the "causalist" and "conditionalist" positions von Engelhardt details in nineteenth century German medicine. In teasing out these models, of course, Fagot-Largeault already takes a first Lakatosian step back from medical history to see what it contains.

The Aristotelian Paradigm

Under her "Aristotelian paradigm" of causal thinking, Fagot-Largeault includes both Aristotle's own account of causation and Galen's application of it to medicine ([12], pp. 107, 117). She finds these causal schemes reappearing in several forms during nineteenth century medicine, long after Peset suggests they should have succumbed to positivistic views. First, she identifies Aristotelian elements in Broca, when he accounts for clinically observed syndromes in terms of underlying anatomical lesions. She asks:

Isn't identifying a syndrome... and referring it to a tissular lesion... referring a form to a matter, that is to say, pointing to both the distinctive pattern "defining" the thing... and to "what the thing is made of", i.e, what its elemental substratum is...? ([12], pp. 108–109).

She also sees the paradigm at work in Bernard's attempt to reconcile his commitment to determinism with the variety of physiological reactions he observes to similiar experimental stimuli. She surmises that "at this juncture the experimentalist's reflection stumbles: and he is forced back into an essentialist view: that the reactions represent individual deviations from a typical dispositional nature of a substance" ([12], pp. 111–112). Thirdly, applying Peset's explication of the Galenic account of causation in medicine, she also finds the model alive in the German "causalists" scheme. She matches Galen's "inevitable and intrinsic" constitutional causes, (in which Peset sees Aristotle's formal and material causes combined) with Hueppe's "potential energy of disposition", and two subsets of Galen's category of efficient causes with the Causalists' background "conditions" and external "stimuli" ([12], pp. 119 –120).

In each of these cases, again, Fagot-Largeault must provide a "rational reconstruction" of her texts to explicate the "Aristotelian" thinking within them: she translates Broca's language, reconstructs the course of Bernard's reflections, and "decipher[s] Hueppe's tricausal scheme with Galen's grid" ([12], p. 120)[2]. These interpretations are insightful: they do indicate, *contra* Peset, that there is something more than strictly positivistic causal thinking going on in these cases. But is it really a commitment to Aristotelian or Galenic accounts of the causal relation? What is the epistemic point of reinterpreting these explanations in the full historical dress of those specific theories of causation?

In fact, unlike Peset and von Engelhardt, Fagot-Largeault does not seem to be primarily interested in showing what specific causal doctrines medicine has affirmed. Rather, her concern seems to be to explicate some broader currents in medical thought, for which the doctrines of Aristotle and the positivists might be taken as emblems. Thus she describes what her "paradigms" exemplify as "modes of etiological inquiry" and "styles of causal thinking", rather than as specific concepts or theories of causation. Beyond these labels, Fagot is not precise about the features of medical inquiry she means to uncover through her "paradigms". Her use of that term is not strictly Kuhn's

[20], and she does not avail herself of the other constructs — such as Ludwik Fleck's notion of medical "thought-styles" [14] — that might help indicate her meaning. At best, one might say she is looking for patterns of explanation that are broader than the specific doctrines she uses to symbolize them, but in which they, as iconic metaphors, participate.

But if Fagot-Largeault is really after the epistemic bones of the "Aristotelian" style of causal thinking, fleshing out the already skeletal statements of Broca and Bernard in Aristotle's terms does not seem the most direct approach. In doing this, she risks confusing the suggestive value of her metaphor with its literal content. A more direct approach would be to search for the common ground between her metaphor and the historical cases it illuminates.

For example, it is interesting that in each of Fagot-Largeault's cases, there is an appeal to the causal efficacy of something's definitive nature — to "deep", "essential", or "formal" causes. Perhaps it is this causal appeal to *form* that is the most important conceptual cue Fagot-Largeault uses to identify examples of "Aristotelian" thinking. Certainly with Broca, Fagot-Largeault stresses his identification of "distinctive" pattern defining the "thing" and the "demonstrations of essence" by which it is linked to "what the thing is made of". Similarly, the "finalistic" thinking she sees in Bernard is consistently focused on the diseased object's dispositional nature, not on the *telos* that so disposes it. She says that Bernard falls into "Aristotelian" causal thinking when he concludes that

> the *true cause* which specifically determines the illness is the *property* of the injured tissue... Thus, illness does not fall upon the organism from the outside; the determining actual and proximate cause is the organism's *predisposition* ([12], p. 112).

She concludes from this that "Bernard has therefore reached the surprising conclusion that illness is determined with *necessity* by the nature of the thing ill" ([12], p. 112). Again, with the German causalists, it is their qualitative emphasis on internal or constitutional causes that marks their scheme as peculiarly Galenic; without it they would not differ from other adherents to efficient theories of causality. What does this appeal to formal causes suggest about the kind of explanation involved in these examples?

Formal Causes and Functional Explanation

To appeal to a formal cause roughly involves identifying the influences of something's design on its behavior. In Fagot-Largeault's examples, it is the influences that the "design" of a biological system has on the events that occur within it, or more precisely, the effects of the system's functional organization. The contribution of one's "constitution" to disease, then, might be seen as the effect of the way one functions as a system. Similarly, a "predisposition" to disease is a built-in functional tendency to manifest it, and the "essential property" of a tissue is its function in the body. Certainly, the "distinctive pattern" "defining" a syndrome clinically will be the pattern of functional processes that constitute its natural history. In each of the cases in which Fagot-Largeault identifies "Aristotelian causal thinking", in short, an explanatory appeal seems to be made to the way the body or one of its parts does the things it does: the way they function. The generalized form of this sort of explanation, explaining the dynamics of a biological system by attributing functions to it or its parts, is commonplace in biology. It is a variety of this functional explanation that Fagot-Largeault seems to have isolated by her "Aristotelian paradigm".

The philosophical literature on functional explanation in the life sciences is helpful in clarifying this connection between "functional" and "formal causal" explanation. One approach has been to describe biological systems teleologically, and explicate explanatory appeals to their functions in terms of the goals of their constitutive activities.[3] This tradition has had a defense by Christopher Boorse, who promotes a view built around the idea that "functions are, purely and simply (causal) contributions to goals" ([4], p. 77). For Boorse, functional explanations are attempts to show how some system works by showing how the activities that make it up contribute to its goals. Thus, to give the function of a bird's "insect-eating behavior" is to explain where it fits in to the organism's scheme for surviving. Boorse says that "to say 'X is performing function Z in the G-ng of (system) S' means that X is Z-ing, and the Z-ing of X makes a causal contribution to goal G of System" ([4], p. 80).

Boorse stresses, however, that because functional ascriptions depend simply on there being a goal-directed system with contributory parts, and not on any *particular* goals, "any goal pursued or intended by a goal-directed system may serve to generate a function statement" ([4], p. 77). Moreover, the selection of these goals, and the identification of biological systems in their terms may vary with the interests of the observer. Boorse argues that "since organisms contain no separate mechanisms that distinguish among the various goals that biological processes achieve, there is no way of finding a unique goal in relation to which traits of organisms have functions "([4], p. 79). For example, the function of avian insect eating might be seen, from an ecological perspective, to be its role in the preservation of the stability of some eco-system. Boorse must amend his analysis, then, to say

> "The function of X is Z" means that in some contextually definite goal-directed system S, ... the Z-ing of X falls within some contextually circumscribed class of ... causal contributions to G ([4], p. 82).

This point is helpful in clarifying the relation between Fagot-Largeault's "formal causal" interpretation of the medical explanations she cites and their interpretation as functional explanations. The difference seems to be largely a matter of contextual change. From one point of view, the "system" involved is the patient's body, or one of its parts, and its goal is the systemic state of "health". From this perspective, it is counter-intuitive to indicate the way this system *functions* to account for its being in an unhealthy, diseased state. Rather, it is the *dysfunctioning* of the system that explains its current state. Most functional explanation in biology does attempt to show how some system achieves its goal. Still, the pattern of explanation itself does not depend on whether the system is approaching or receding from this reference point. In medical contexts the point is often precisely to explain how a system is *not* working. To do so by indicating the dysfunctions it involves still is to show how the system's features contributes to its current goal-related activities.

However, one might also take a second point of view on these explanations. If the first perspective emphasized the diseased body, a second might focus on the "embodied disease". The system picked out, in other words, would be the set of features that ground the disease process. From this standpoint, it *is* the way features like one's constitution function in the disease process that explain the dynamics of the system in question. Thus, in explaining the role of constitutional predisposition in a disease, one is pointing out how one's constitution contributes to the course of events in the "natural history" of that disease. If the disease is abstracted even further within this context, as a third thing emerging from the combination of its "definitive pattern" and its bodily substratum, that pattern is most congenially seen as its formal cause, *à la* Fagot-Largeault. To the extent that these formal causal explanations explain the disease in terms of the goal-oriented processes that delineate its embodied natural history, they fit the pattern of functional explanation Boorse describes.

Boorse labels his goal-oriented analysis of functional explanation "operational explanation", and distinguishes it from other ways of understanding functions by pointing out that it is aimed at "an explanation of how some system currently achieves the goals it does, or how it works, rather than how it got that way, or why it has these mechanisms" ([4], p. 85). It is this kind of appeal to the organizational influences on biological processes that is the aim of Fagot-Largeault's "Aristotelian" explanations as well.

The Positivistic Paradigm

A similar analysis might be provided for Fagot-Largeault's "positivistic" paradigm of causal thinking. Here, the historical positivist doctrines themselves are more directly at work in the medical texts she examines. For example, Fagot-Largeault recounts Bernard's view that:

> In brief: 'the cause' of E is its *necessary and sufficient condition*, so that the experimenter will not be satisfied until he has established (1) that the cause never occurs without its effect obtaining ... and (2) that the removal of the cause brings the effect to disappear. ... So that finally, once we know the *law* of the phenomenon ... we always and with absolute certainty can infer either the cause from the effect, or the effect from the cause ([12], pp. 110–111).

Again, however, it is clear that Fagot-Largeault is more interested in the fate of the pattern of explanation this causal account involves than in the theory itself. She is after the role of "positivistic" ideology, or "the deductive-nomological model of explanation" in nineteenth century medicine ([12], pp. 105–106). Thus, in the section on Bernard, she asks how he reconciled his positivistic "explanatory model" ("that physiology should look for the general laws governing normal and pathological predication") with "the need to recognize *special causes* accounting for those irregularities which, in biomedical sciences, make *prognosis* fall short of *prediction*" ([12], p. 110).

Fortunately, getting at the pattern of explanation underlying this "paradigm" does not require the kind of analytic excavation that the "Aristotelian paradigm" did. According to the positivist tradition, explanation consists largely in appealing to a set of antecedent conditions and a universal generalization that links them with the phenomenon to be explained.[4] Bernard is seeking to use this approach to account for the "mechanisms" underlying pathological dysfunction. Similarly, the point of Verworn's laws of conditionality is to explain why a physiological system is in any given functional state. These explanations are, in other words, attempts to account for the same (dys)functional phenomena as the "Aristotelian" explanations. Philosophically, however, they fall in line with positivistic theories of functional explanation.

The positivists agree that the point of functional explanation is to understand the maintenance of the characteristic activities of living things. Rather than focus on the *purposes* of those activities, however, they concentrate on the conditions that must obtain for those activities to occur. Functions are seen from behind, as it were, as consequences of a biological system's constituent parts (cf. [26], pp. 106–109). The epistemic backbone of the explanation is composed of the entities and conditions that together produce the phenomenon to be explained. Thus, for example, in Larry Wright's view, functional explanations

> are in some sense etiological, i.e., concern the causal background of the phenomenon under consideration. And this is indeed what I wish to argue: functional explanations, although plainly not causal in the usual restricted sense, do concern how things *got there* ([38], p. 227).

Wright knows that his approach to functional explanation explicates functions from the bottom up. He says that "functional and telelogical explanations are usually contrasted with causal ones, and we should not abandon that contrast lightly" ([38], p. 227). Nevertheless, he takes the position that: "saying that the function of X is Z is saying at least that X is there because it does Z or that doing Z is the reason X is there" ([38], p. 228). Thus, to explain the function of the epiglottis in covering the wind-pipe is merely to give a shorthand account of the causal chain of events that produced the organ, first developmentally, then genetically, and finally in terms of the laws of evolution.

This analysis is very much the spirit of the explanations offered by the positivistic physicians that Fagot-Largeault and von Engelhardt discuss: they seek always to refer back to the conditions and laws that determine "why a given pathological state is here." For example, Bernard insists that

> Diseases are at bottom only physiological phenomena in new conditions still to be determined; toxic and medicinal action... We must therefore get used to the idea that science implies merely determining the conditions of phenomena; and we must always seek to exclude "life" entirely from our explanations of physiological phenomena as a whole. Life is nothing but a word which means ignorance, and when we characterize a phenomena as vital, it amounts to saying that we do not know its immediate cause or its conditions ([3], pp. 198, 201).

The two views of causation that Fagot-Largeault isolates in nineteenth century medicine, then, might be seen as emblems for two sorts of functional explanation, which I will label *Operational* and *Causal*. To discover some of the reasons why both forms appear so recurrently in modern medical thought, their relationship needs to be examined. It appears that they may not only be complementary forms of explanation, but actually interdependent ones within an explanatory strategy basic to the system of medical knowledge.

III. AN EXPLANATORY STRATEGY IN MEDICINE

The Correlation of Operational and Causal Explanations

Boorse and Wright present their analyses as rival accounts of functional explanation, but in the process give clues to their real complementarity. Boorse argues that Wright's account is not sufficient to cover all the sorts of functional explanation biologists employ. He says that

> When the question at issue is evolutionary explanation, the appropriateness of this Wrightian convention is apparent. But it is equally apparent that such a convention is inappropriate to those other biological contexts where the aim is an explanation of how some system currently achieves its goals ([4], p. 85).

In other words, while biologists are concerned to know why systems have the parts they do, which could be explained causally they also wonder how systems work. Answers to this second sort of query, Boorse maintains, are best understood as Operational forms of functional explanation.

Boorse goes on to argue that his Operational analysis of functional explanation can also preserve the successes of Wright's account in explaining "why this part is here". In seeking to subsume the Causal view under his Operational one, however, he neglects the possibility that the two analyses identify distinct kinds of functional explanations. Yet, he seems to recognize this himself in his criticisms of Wright:

> Function statements do often provide an answer to the question "Why is X there?", and it is this explanation pattern that Hempel called "functional explanation." There is, however, another sort of explanation using function statements that has an equal claim to the name. This sort answers the question "How does S work?" where S is the goal directed system in which X appears... It is operational explanation, not the evolutionary sort, with which physiology has traditionally been concerned ([4], p. 75).

This is important, because if these are distinctly different ways of explaining phenomena, they are likely to isolate different aspects of these phenomena as significant, and to offer complementary rather than rival accounts. As Morton Beckner argues, the stumbling-block in attempts to reinterpret teleological forms of functional explanation causally is that the two forms "involve different methods of classifying the elements of a system", so that each tells different things about it [2].

Beckner says that "functional explanations presuppose conceptual schemes of a certain logical character" and that "the teleological character of a sentence is not a matter of vocabulary but rather of the logical structure of the conceptual scheme involved" ([2], pp. 203, 209). In order to give an operational explanation of some phenomenon, one must first see it as the sort of thing that is susceptible to being explained operationally (i.e., as a functional system or contributory piece of a system). Thus, Beckner says, even the terms used to name the phenomenon in question will be defined by reference to its contributions to a teleological system ([2], pp. 206–207).

For example, "the heart's function is to pump blood" presupposes that by "heart" we mean a physiological system delimited by the activity of blood pumping, and that "blood pumping" is a contribution to the body's (functionally isolated) circulatory system. Without these assumptions about the nature of the heart and the context of what it does, we would not know what to make of the suggestion that "blood pumping is the heart's function". On the other hand, while the "heart" could also be identified causally (as the product of a certain ontogeny caused by genetic mechanisms resulting from selective evolutionary pressures), a description of it in terms of such a conceptual scheme would not tell us the same things as an Operational explanation, or even necessarily *about* the same things. In short, the two explanatory patterns apply different ontologies to the domain they examine.

That the ontological schemes presupposed by Operational and Causal explanations *differ*, however, does not mean that they may not be compared and correlated. For example, it is instructive to look at the examples that Wright and Boorse use to illustrate their rival accounts of functional explanation. While Wright's examples tend to be about the functions of anatomical pieces or parts of organisms, the "functional features" that Boorse cites are all activities, like "insect eating" or processes like "fermentation". This is not surprising, given the emphasis on explaining "how systems work" through which Boorse interprets functional explanation, and Wright's emphasis on "why systems have the parts they do". But it does suggest, following Beckner, that the conceptual scheme presupposed by Wright's Causal explanations makes it easier for them to explain the places or parts of structures in systems, while Operational explanations explain the dynamics of systematic processes or activities more easily.

However, there is more to this complementarity than a simple division of labor. In fact, these ontological schemes do not always isolate mutually exclusive parts of the biological domain. Thus, the heart can be taken as primarily an object, and explained in terms of its causal relations with other objects, or understood through its activity, and explained in terms of the process to which it contributes. It is just in this perspectival shift that the two ontological schemes complement each other most importantly. Each exhibits a certain transparency to the other, so that both their explanations are recognizably points of view on a common phenomenon. As we shall see, it is this "cognitive transparency" that grounds the interdependence of these forms of explanation in medical contexts, and their special role in shaping medical knowledge.

In brief, this special role is three-fold. Together these forms of explanation provide: (1) a usefully pluralistic ontology through which to structure medicine's domain; (2) a method for applying that ontology in medical inquiry; (3) a mechanism for informing that inquiry with medicine's special therapeutic goals. Each of these functions has been recognized before, to various extents and under different guises, by theoreticians of biomedicine. The sketch that follows rests heavily on these accounts, and attempts to show how their insights support Fagot-Largeault's observations from medical history, and how her history provides a locus of correlation for their philosophical views.

Medicine's Pluralistic Ontology

The hallmark of the phenomena that medicine includes in its domain is their complexity. They are complex enough, as Beckner suggests, to support interpretation by several basic conceptual schemes. Even under each of those, as Boorse argues, they are complex enough to be usefully explained from more than one context. Even worse, instead of being conceptually insulated from each other, systems picked out by one descriptive perspective often seem to exhibit relations with systems identifiable only under another. These "transparency" relations range from the simple congruence of the boundaries of the functionally and structurally delineated "hearts", to the more complex causal interactions William Wimsatt traces between biological levels of organization [36].

One of the ways in which medicine frames a coherent account of its domain in the face of this complexity is by combining and coordinating the conceptual screens that Causal and Operational explanations involve. Put (still too) simply, by alternating the type of explanation sought, medicine can organize a biological phenomena as a "layered" composite of parts and processes, in which the entities of each successive level are identified through the "transparency" of the last. Thus an operational vision of how the circulatory system works ("through a combination of activities like the pumping of the blood, etc.") suggests the set of functionally construed parts to look for (the heart-as-pump, etc.) and the structures these desiderata turn out to isolate (the heart-as-organ), influence further Operational interpretations of how they work. To explicate this further, points need to be made about: (a) the nature of these ontological "levels"; (b) how this scheme helps sort out descriptive complexity; (c) why both of this scheme's perspectives are required.

(a) First, the levels of organization posited here do not mutually exclude each other's contents, because they are not strictly structural levels. Conceptually interposed between each structural level (e.g., the heart-as-organ and the anatomical circulatory system) is a level of functional organization (the heart-as-pump in the process of blood circulation) that relates them. Thus, as in Stephen Toulmin's view of the "relations between [Operational] functions, [Causal] mechanisms, and causality in medicine"

> there is no clear division of natural processes in the real world into "functions" on the one hand and "mechanisms" on the other. Rather, we draw a distinction between the functional and mechanistic *aspects* of any natural process,... and whatever can be viewed as a mechanism from one point of view and in one context, can alternatively be seen as a function, from another point of view or in another context. Indeed, the very *organization* of organisms — the organization that is sometimes described as though it simply involved a "hierarchy" or progressively larger structures — can be better viewed as involving a ladder of progressively more complex systems. All these systems, whatever their levels of complexity, need to be analyzed and understood both in terms of the functions they serve and also of the mechanisms they call into play ([34], p. 53).

In other words, this strategy gives medicine the ontology to organize biological phenomena into a hierarchy of anatomical parts explained in terms of their causal background, but *identified* and related to one another through their roles in operationally construed biological processes pursuing some context-specific goal. As a result, in Toulmin's terms:

> when we shift the focus of our attention from one level of analysis to another — from one fineness of grain to another — even those very processes which began by presenting themselves to us under the guise of "mechanisms" will be transformed into "functions" ([34], p. 53).

(b) This explanatory pattern is especially helpful in organizing the biomedical domain, moreover, just because its hierarchy is not simply a series of physical reductions and a derivative set of functional interpretations. Rather, as even Bernard stressed, the descriptive reconstruction of the phenomena is guided through the variety of alternative structural "decompositions" (cf. [18]) by the functional processes of concern in the study. Bernard notes that in medical inquiry:

We must distinguish between two classes of things: (1) the passive mechanical arrangements of various organs and apparatus which, from this point of view, are really nothing but instruments of animal mechanics; (2) the activity of vital units which put in play this diverse apparatus. The anatomy of corpses can certainly take account of the mechanical arrangements of the animal organism; ... (but, about the "active or vital elements"), dead anatomy teaches nothing; it merely leans on what experimental physiology teaches; and a clear proof of this is that, where experimental physiology has learned nothing as yet, anatomists can interpret nothing by anatomy alone. ... But as soon as physiologists have discovered something about the functions of these organs, anatomists will put the physiological properties noted into relation with their anatomical observations ([3], pp. 107–108).[5]

This guidance is provided by what I loosely called the "transparency relations" between the points of view. After the identification of each successive structural system, the researcher is led to an operational analysis of how its (sub) parts work by referring back to the original explanatory goals of the study and the functions derived from them, and positing a system of functionally construed (sub) parts that could contribute to those overarching functions. This functional schema, then provides the orientation for identifying the parts in structural terms.

For example, once the heart-organ is identified, one refers back to the overarching 'vital' function of blood circulation to see that, to contribute to its function, the heart's parts would have to display a particular arrangement of pumps and valves. This schema, then, leads one to isolate ventricles and mitral membranes, etc., as the heart's significant parts. While the original anatomical description sets constraints on the operational analysis, it is this functional prefiguration of further structural decompositions that is the important "transparency relation" guiding the linkage of conceptual schemes. As Toulmin says, "The different levels of organization of different physiological systems are a reflection of the respective degrees of complexity of the *functions* they serve, rather than of their mere size and scale" ([34], p. 60).

(c) The point of correlating these schemes has also been argued by Marjorie Grene. Both perspectives are needed, she says, because it is only in the parts of a system that its functional organization exists, and only for the sake of that organization that the parts are combined as they are.

Organized systems cannot be understood in terms of the least parts alone, but only in terms of those parts *organized* in such systems. Organized systems are *doubly determinate*; they exist on at least two levels at once ... The higher level — form, organizing principle, code, fixed action pattern, or what you will, — exists only in its elements and depends on them for continuance, yet the laws of the elements in themselves, ... do not as such account for the principle which in this case happens to constrain them ([16], p. 20).

In other words, it is ontological interdependence of these conceptual schemes that produces this multi-leveled organization of the medical domain. Only through functional interpretations at one level can the features of the next be selected, and only in their terms in turn, can the processes that they manifest be causally explained. This is part of Fagot-Largeault's point about Broca's anatomo-clinical explanations. Fagot-Largeault says:

> it does not appear unreasonable to consider anatomo-clinical diagnosis as etiological in a wide sense, congenial to ordinary medical language: physicians commonly 'attribute' observed signs of the syndrome as being 'due to' the anatomical lesion. There is even in these causal imputations the necessity of what Aristotle calls the "demonstrations of essence" ([12], p. 109).

Grene also draws Fagot-Largeault's conclusion from this ontological interdependence: "Biological explanation, then, works in terms of form *or* matter, systems-theoretic wholes or parts-analysis of their constitutents, with the two kinds of explanations complementing each other" ([16], p. 21). Moreover, she goes on to agree with Toulmin (and Bernard)

> that complementarity, ... however, is not symmetrical. ... All natural things, organized parts of such things and processes exhibited by them, are inherently informed matter and can be studied on both levels; but form is prior. In non-Aristotelian language: although the upper level of a doubly determinate system depends on the lower level, ... for its existence, and is inseparable from it, it is the upper level that makes the system the kind of system it is. We have to refer to the upper level, we have seen, to generate a problem, to describe the system we are studying, and ... to explain the operation of the system as such ([16], p. 21).

Methodological Considerations

As Grene's last sentence suggests, this correlational strategy is not just important for organizing an ontological scheme for medicine. It also plays an important methodological role. In a discussion of why "typical explanations in biology exhibit the manner in which parts and processes articulate together to cause the system to do some particular thing" ([19], p. 247), Stuart Kaufman sketches a procedure that captures the role of this strategy in the "logic of search" in biology.

Kaufman agrees that "an organism may be seen as doing indefinitely many things, and may be decomposed into parts and processes in indefinitely many ways," depending on the interests of the researcher ([19], p. 246). The first step in explanation, then, is to derive from one's explanatory goals "some initial view about what the system is doing." This functional view "sets the explanandum and also supplies criteria by which to decide whether or not a proposed portion of the system with some of its causal consequences is to count as a part and process of the system" ([19], p. 248). These criteria are a "descriptive set of sufficient conditions"—an account of what would be needed for the initial description to be true—from which an operational "cybernetic" model of the phenomena is generated. Since the "cybernetic model exhibits the manner in which the processes of the parts of the system must articulate," Kaufman says that:

> With the cybernetic model in hand, and background knowledge ... we can search for the kind of causal processes which are likely to [produce the phenomenon]. ... With the suggestion of a specific set of causal mechanisms, the model has become a hypothesis requiring verification. If verified, it will specify which processes—that is, which of the many causal consequences of each of the portions of the system—are to count as processes of the system and which are irrelevant to this account ([19], p. 251).

This methodological procedure clearly rests on the same explanatory strategy that underlies the pluralistic biomedical conceptual scheme. It is again the selective "transparency" of operational explanations to particular causal explanations that provides its methodological momentum. Moreover, Kaufman goes on to complete the cycle, by showing how causal explanations can reflect back on operational ones to reconstrue whole inquiries. He says:

> it is clear that new information about a part picked out in an old decomposition of the system can lead to new views of what the system is doing. For example, Harvey's discovery that the blood circulates led to a new view of what the heart, specified on old anatomic grounds, does. ... The points to notice about such new syntheses are that, in general, the new synthesis may decompose the system in a new way, cutting across parts specified in an old decomposition ([19], p. 260).

Thus, explanations of one type lead one to frame research questions answerable through the second, which pose new questions for the first type, in a heuristic dialectic.

The Therapeutic Goals of Medical Knowledge

Finally, a third reason the approach Fagot-Largeault uncovers is important to medicine—a third reason one should not be surprised to find "Aristotelian" strains in the modern era—is that it allows medical inquiry to be fully informed by its own unique explanatory goals. Recall that both ontological and methodological contexts, the strategy's correlational movement between perspectives is guided by reference to the basic goals of the inquiry. Kaufman says that the conditions of the initial, necessarily selective, view of what an organism is "speak, if you will, in the imperative mood. They are an injunction to the scientist to direct his attention to those conditions, for around them it should be possible to build a cybernetic, and later a causal, model to explain the phenomena" ([19] p. 251). In medicine, these conditions incorporate another injunction as well, because the research interests they represent are responses to the medical imperative to heal.

As H.T. Engelhardt, Jr., points out, by identifying the processes and parts of interest in terms of the "telos of the medical enterprise", health, medicine can systematically inform its explanations with a "regulative ideal of autonomy, directing the physician to the patient as person, the sufferer of the illness, and the reason for all the concern and activity" ([9], p. 139). Pathological explanations built by this strategy will also incorporate this imperative, so that an account of a disease "acts not only to describe and explain, but also to enjoin to action. It indicates a state of affairs as undesirable and to be overcome. It is a normative concept; it says what ought not to be" ([9], p. 127).

Toulmin underscores this role for the correlational strategy in his article. By isolating phenomena in terms of evaluatively selected (dys)functions, he says, medical researchers build their practical goals into their understanding of the world:

> Having arrived at this point, they take it for granted that the integrity of the organs and the systems involved on each level is a "good thing"... There is, correspondingly, very little substance to the idea that a scientific physiology could, even in principle, be "value neutral." The chief vital functions of the human body are not merely "good in themselves." They are preconditions for almost any other imaginable human good... So it is no longer the case that the physician has nowadays to make any specifically ethical commitment to sustain health... his whole enterprise is nowadays intelligible and capable of being rationally expounded only in terms of a systemic picture of vital organization or functioning ... The *rational* significance of vital organization or functioning is inseparable from its *ethical* significance ([34], pp. 61–62).

In sum, then, this explanatory strategy — the "correlational strategy" — is why it is not surprising that Fagot-Largeault should find both "Aristotelian" and "positivistic" styles of causal thinking in modern medicine. Both styles are basic to the enterprise. Together they provide an important conceptual scheme for medical knowledge, a 'logic for medical discovery', and a way to keep both of those focused on medicine's therapeutic quest. The next section reexamines pathological explanation more closely, to draw out some of the implications of this strategy for how we understand diseases, and for the probabilistic etiological approach Fagot-Largeault cautiously advocates.

IV. PATHOLOGICAL APPLICATIONS

Interpreting patho-physiological accounts of disease in terms of this explanatory strategy has a number of interesting implications. From the ontological scheme the strategy involves, lessons might be drawn about the relation of medicine to the other natural sciences and the "reducibility" of its domain (cf. [24]). Similarly, one might examine its methodological features for insights into the dynamics of conceptual change in contemporary medicine [18]. Here, though, it seems most appropriate to end by showing how this account reflects back on two of the philosophical points already made about disease explanation in the preceding papers. The first, suggested by all three authors, is that there are important conceptual ties between different "forms of causal thinking" and particular views of what diseases are. Explicating this point here will prepare the way for commenting on the other: Fagot-Largeault's second conclusion that the multi-factorial interpretation of causation might provide a more progressive basis for pathological inquiry.

Causal Explanations and Disease Concepts

Both Fagot-Largeault and the authors she comments on agree that particular kinds of causal views and theories of disease are conceptually interdependent (cf. [12], p. 123; [29], p. 94; [11], p. 70). They associate essentialistic etiologies with "ontological" construals of disease, and phenomenalist causal thinking with "physiological" view.[6] They do not, however, detail these relations very closely. This neglect becomes noticeable, and notable, in the cases at the crux of the dispute between Fagot-Largeault and Peset, where more detail might shed significant light. How does a "constitutional" or "dispositional" etiology affect the way a disease is conceptualized, especially when it is combined with other etiological approaches? Does Fagot-Largeault view the essentialistic elements in the etiologies of the Causalists and Claude Bernard as dictating 'ontological' components for their theories of disease? If these components were identified and their links to "Aristotelian" forms of explanation clarified, Fagot-Largeault's case against Peset would be strengthened and the force of their common point clarified.

In fact, both the Causalists and Bernard were willing to integrate ontological conceptions into their interpretations of disease. In discussing his position, for example, Bernard recognizes "the taxonomical or nosological point of view," the "anatomical point of view and the physiological point of view" on disease ([3], p. 112). The first two of these construe diseases ontologically, either as Pinel's logical species or, among the patho-anatomists, as concrete things. Though his stress is clearly on the third, he insists that each is a necessary "partial point of view" to an understanding of disease. Thus he agrees with Broussais and Virchow, that "in the changes of tissues they found proper characteristics for defining disease" even while he attacks the sufficiency of their views as global accounts of disease ([3], p. 113).

Similarly, von Engelhardt's account of Hueppe and Martius stresses their efforts to synthesize the "constitution pathology" derived from Virchow with the reactive conceptions of disease embodied in etiologies positing "external causes of disease" (cf. [11], pp. 80–83). Thus, von Engelhardt quotes Martius' claim that:

> The more aspects of the expressed, composite process that we recognize, the better we can describe that process most thoroughly, most completely, and thereby most simply ([11], p. 82).

The logic of this integration becomes clear when these pluralistic pathologies are seen as prototypes of the 'correlational' explanatory strategy Fagot-Largeault uncovers. In short, Bernard and his nineteenth century colleagues are willing to interpret diseases pluralistically because the combination of Operational and Causal forms of explanation provides a conceptual scheme which can usefully integrate the different interpretations. Under this scheme, Bernard can "recognize and classify diseases" as either (a) reified (dys)functional systems (*disease entities*) proceeding through a "natural history"; or (b) the specific "changes of the tissues" that contribute to these processes, and still explain them in terms of (c) the "laws of physiology" (i.e., as a consequence of the causal relations between physico-chemical structures within the local environment). Each interpretation finds its place on a different level of organization in the scheme, and is related to the others through the correlations that connect the levels as "points of view" on the disease.

Seeing these pathologies in terms of the correlational strategy also clarifies *how* explanation is linked to disease conceptualization. Forms of explanation influence how a disease is conceptualized through the conceptual schemes the explanations presuppose. Operational explanations are likely to emphasize goal-directed processes and their contributory components, which can be abstracted as taxonomic disease 'types' or reified as essentialistic Virchowian disease entities. Causal explanations, similarly, will depict diseases as the consequence of the causal relations between structural parts, construing them as deviations from a physiological steady state due to some departure from "normal" relations among these parts. Thus, the identification of diseases in ontological terms, as entities or objects, will lead to their explanation at the next level in operational terms, and the identification of diseases as physiological reactions will suggest their analysis in terms of the causal relations of the contributory parts involved.

These points come through even more clearly when one examines the contemporary counterparts of "constitutional" etiological appeals. In biomedical explanations of *genetic* disease, and in their reconstruction by medical theorists, genetic diseases are also explained both 'ontologically' and 'physiologically'. Thus, diseases like sickle cell disease are construed typologically as genetic "traits", "transmitted vertically" between "carriers" [27], or as specific concrete entities: "molecular lesions" or "molecular morbi" ([28], [33], pp. 92–111). Similarly, they may be interpreted physiologically, from being a function of the "genetic load" of a population [1] to being a metabolic "error" within an individual system ([6], pp. 171–179).

Again, however, in the light of the account of pathological explanation that Fagot-Largeault suggests, this conceptual pluralism is not problematic. In fact, it seems clear that this diversity simply represents explanations concerned with systems at different levels of biological organization. Thus, genetic disease is construed 'ontologically' as a "lesion" or "defect" at the molecular level; as a metabolic "error" or "abnormality" within functionally defined physiological systems; as an ontological quality, a "trait" of the individual organism, and as a reactive "adaptation" of a population to environmental pressures.[7] Sickle cell disease, for example, can be properly construed both as a human hereditary trait, identifiable across and transferrable between individuals, and as an adaptive response of a population to environmental pressure, if it is understood that these interpretations pursue different, but related, levels of explanation.

The Probabilistic Causal Paradigm and Medical Progress

Another point on which all the authors agree, but on which Fagot-Largeault places special epistemic weight, is the emerging dominance in medicine of the multi-factorial analysis of causation. Fagot-Largeault draws from Peset's account of early epidemiology and von Engelhardt's depiction of the Conditionalists to trace the growth of the use of this analysis, "behind the scene of apparent continuity with tradition." She summarizes the view as follows:

> The multicausal scheme underlying the probabilistic approach reduces the case in which 'the cause', C, is a necessary and sufficient condition for its effect, E, to the

particular — and possibly never occurring — circumstances in which the probability of E, given C, is one. In the general case, 'a cause' is merely that which increases the probability of an effect ([12], p. 117).

Again, however, Fagot-Largeault's concern is for something broader than the history of this analysis of causation *per se*. Rather, as her summary suggests, she is after the "probabilistic approach" it supports; something, again, somewhere between a specific etiological theory and a basic style of medical thinking. She calls the approach "a 'positive' medicine, " in which

[the] degree of the causal 'influence' is measured through a correlational analysis. An exhaustive determination of causes is not required. Diagnosis and prognosis, taking into consideration a variety of causal influences, are probablistic judgments. An effective therapeutics is that which shields one from significant 'risk factors' ([12], p. 123).

In other words, like her other "paradigms", the "probabilistic approach" represents an explanatory strategy that is not restricted to etiological theory, but manifests itself in the full array of medical explanations. Moreover, it is a strategy which Fagot-Largeault explicitly contrasts with her two nineteenth century "paradigms" as the basis for a more progressive medicine. She says:

It should not be forgotten that the very detection of risk factors is itself contingent on the objectives of medical research ... and those in turn are contingent on how a given civilization conceives of the standards of health: so that one might perhaps be tempted to think that the end of the twentieth century is actually taking a step more decisive towards the relegation of a certain medieval era, than the 'positive' step taken by clinical medicine a century ago, insofar as it bears the consciousness that a rational medical attitude less consists in scrutinizing the details of the causes of illness than in first choosing the objectives of health ([12], p. 123–124).

As this passage suggests, what makes this approach particularly progressive for Fagot-Largeault is its openness to influence by the social goals and values of medicine. In fact, insofar as the identification and weighting of causal factors must be dependent on pragmatic concerns for Fagot-Largeault, her "probabilistic approach" might be better represented by a "manipulability theory" of causation like Collingwood's [8], than by Suppe's probabilistic views [32]. It is not just the probability of a factor's contribution that identifies it as a cause, but of a contribution recognized as such by its susceptibility to human control (cf. [9], [35], [31]). Thus, as H.T. Engelhardt writes:

Disease causality is equivocal ... The multiple factors in such well-established diseases as coronary artery disease suggests that the disease could be alternatively construed as a genetic, metabolic, anatomic, psychological or sociological disease, depending on whether one was a geneticist, an internist, a surgeon, a psychiatrist, or a public health official. The construal would depend upon the particular scientist's appraisal of which etiological variables were most amenable to his manipulations ([9], p. 133).

This interpretation of multi-factorial analyses in medicine is often advocated because it is taken to explicate actual medical practice ([35], p. 626). Fagot-Largeault advocates her "probabilistic approach" in this interpretation's terms because she sees the openly evaluative character of multi-factorial analyses as a way of maintaining the focus of medical inquiry on its unique goal of promoting health.

But, even if the (value-laden) multi-factorial view of disease causation has advantages for medical practice over either of the causal concepts derived from Operational and Causal forms of functional explanation, it is still not clear that its general adoption would mean moving beyond the basis explanatory strategy in which they participate to some "probabilistic approach." First, how would such an approach improve medical methods? Even if causes are identified statistically as "risk factors," the identification of sickled erythrocytes as a significant risk factor for an illness still raises complementary questions about what they do in the physiological system that embraces them, and what dysfunctions occurred within their own intracellular systems to produce them. Probabilistic etiologies merely provide a useful way of finding the starting places for the patho-physiological explanations that can ground the "effective (preventive) therapeutics" that Fagot-Largeault seeks.

Moreover, the 'progressive advantage' of the probabilistic view is no advantage, since the correlational strategy admits of just as much influence by social goals or values. The explanatory dialectic it involves is predicated on medicine's choice of the goals that will be used to identify functions and isolate systems, and these are the same "objectives of health" to which Fagot-Largeault refers.

In fact, the contemporary turn to multi-factorial analyses might be better seen as a manifestation of the "correlational strategy" than as the standard bearer of its successor. To use instrumental interests in the "objectives of health" to isolate a set of contributing 'factors' from which *the* cause might be statistically selected, is merely to pick out antecedent parts (or processes, if they are construed as "formal causes") in terms of their functional roles in a system delineated by particular medical explanatory goals. Thus, in the sickle cell case, it is the interests of the genetic counselor that focuses his analysis on the level of the individual and his trait, while it is the molecular geneticist's concern to "cut the heart of the matter" that constrains him to see the disease molecularly.[8] The multi-factorial view, in other words, is merely a way of expressing for etiology the (pragmatically determined) contextual pluralism of his strategy. It provides, in fact, additional evidence for Fagot-Largeault's claims about the forms of explanation operable in modern medicine.[9]

V. CONCLUSIONS

The framework for medical explanation that is erected in parts III and IV above is clearly just a scaffolding, and one that rests precariously on the narrow foundation of Fagot-Largeault's historical observations. At this point, it seems best to refrain from adding more weight to the structure, but to back away carefully, calling for more support. In uncovering both of the forms of explanation she calls "Aristotelian" and "positivistic" at work in modern medicine, Fagot-Largeault has hit upon a very important strategy within medical inquiry. The complementarity of explanations explaining the functions of parts and processes in pathological phenomena provides medicine with a conceptual system and a method that are both receptive to its unique non-epistemic goals of controlling illness and promoting health. This account seems to allow one to make sense of the persistent and interdependent appeal that both traditionally competing conceptualizations of disease have for modern medicine, and of the flourishing growth of multi-factorial analyses in etiology. But it is far from well illustrated here, and, for that reason, far from fully explicated.

I suggest three areas in medicine that might serve as further testing for this account. The first, introduced briefly above, is explanation in medical genetics. It is here where explanations are likely to range over the most levels of organization, from molecular to populational. How are the different perspectives related? Moreover, medical genetics shares much of its domain with genetics proper. Do genetic explanations in medicine differ from genetic explanations in other biological fields, reflecting different explanatory goals, as this account suggests it might? A second promising area is oncology. "Cancers" are often interpreted in strongly 'ontological' ways, but are also analyzed as abnormal physiological reactions. How are these interpretations integrated? Finally, one might try to apply the correlational account to psycho-pathology. The multi-leveled conceptual scheme seems well suited for somatic disorders; how might it apply to psychiatric ones? One might take as a starting point Engelhardt's work on John Hughlings Jackson and levels of structure and function in the nervous system [10]. The hypothesis to test in each of these areas would be that in modern medicine, to borrow third hand now, "Causal explanations without Operational explanations are blind, while Operational explanations without Causal accounts are empty."

National Institutes of Health
Bethesda, Maryland, U.S.A.

NOTES

[1] A revision of Kant's epistemological dictum that "thoughts without content are empty, intuitions without concepts are blind" (*Critique of Pure Reason*, B 75, A 51, tr. Smith, N.K.). Fortunately, Lakatos' slogan was a profitable one: as a methodological token, his aphorism has become the coin of the realm for many philosophers of science. Whether the value of this currency is inflated, of course, is a matter of ongoing debate (see, for example [5] and [15] for opposing views, and the moderating stance of [25]).

[2] "Rational reconstruction" is being used here in Lakatos' sense: he says that in studying science one should

> adopt the following procedure. (1) One gives a rational reconstruction (of the scientific concepts in question); (2) one tries to compare this rational reconstruction with the actual history and criticize both one's rational reconstruction for lack of historicity and the actual history for lack of rationality ([21], p. 138).

For some criticism of this approach, see [23].

[3] A good review of these discussions appears in [37].

[4] Cf. [17] for a comparative review of this tradition and its more recent rivals.

[5] Of course, while this passage is illustrative here, Bernard's case is actually more complex, since, as we have seen (p. 129, above), he hoped ultimately to be able to understand "the activity of vital units" in strictly physico-chemical terms. Bernard may still exemplify the point about the regulative role of operational perspectives, however, because he did hold that physico-chemical explanations of these phenomena were only possible in the context of an account of the *milieu intérieur* of an organism. Insofar as the constraints of this account would lead one to isolate physico-chemical relations underlying the uniquely operational biological phenomena — what he calls "the activity of the vital units" — then his approach fits the correlational strategy even more closely.

[6] The traditional philosophical distinction between 'ontological' and 'physiological' conceptions of disease is usefully explicated by H.T. Engelhardt, Jr., in his [9]. Also see the several historical treatments of this distinction collected in [7].

[7] For a more systematic treatment of the case of explanation in medical genetics, see [18].

[8] For example, molecular geneticist Richard Roblin writes:

> with all of the current research interest in molecular biology, it should not be surprising to find human genetic diseases being brought about in molecular terms. Conceptualization of diseases at the level of the gene

> strengthens the intellectual drive to "cure" the disease at the same level. There is something *aesthetically compelling* about cutting to the heart of the problem by treating the disease at the molecular level where it originates ([30], p. 111).

[9] If one explains the multi-factorial view in these terms, moreover, one can escape the problems of resting a general account of medical explanation on Baysian grounds, which Fagot-Largeault herself ably points out in her [13].

BIBLIOGRAPHY

1. Allison, A.C. : 1954, 'Protection Afforded by Sickle Cell Trait Against Subterain Malerial Infection', *British Medical Journal* **1**, 290–294.
2. Beckner, M.: 1976, 'Function and Teleology', in M. Grene and E. Mendelsohn (eds.), *Topics in the Philosophy of Biology*, D. Reidel, Boston, pp. 197–213.
3. Bernard, C.: 1957, *An Introduction to the Study of Experimental Medicine*, H.C. Greene (tr), Dover Publishers, Inc., New York.
4. Boorse, C.: 1976, 'Wright on Functions', *The Philosophical Review* **85**, 70–86.
5. Burian, R.: 1977, 'More than a Marriage of Convenience: On the Inextricability of History and Philosophy of Science', *Philosophy of Science* **44**, 1–42.
6. Canguilhem, G.: 1981, *On the Normal and the Pathological*, D. Reidel, Boston.
7. Caplan, A., Engelhardt, Jr., H.T. and McCartney, J.C. (eds.): 1981, *Concepts of Health and Disease: Interdisciplinary Perspectives*, Addison-Wesley Publishers, Reading, MA.
8. Collingwood, R.G.: 1974, 'Three Senses of the Word "Cause"', in T. Beauchamp (ed.), *Philosophical Problems of Causation*, Dickenson Publishing, Co, Encino, CA, pp. 118–126.
9. Engelhardt, H.T., Jr.: 1975, 'The Concepts of Health and Disease', in H.T. Engelhardt, Jr. and S. Spicker (eds.) *Evaluation and Explanation in the Biomedical Sciences*, D. Reidel, Boston, pp. 125–141.
10. Engelhardt, H.T., Jr.: 1971, 'John Hughlings Jackson and the Mind-Body Relation', *Bulletin of the History of Medicine* **49**, 137–151.
11. Engelhardt, D. von: 1992, 'Causality and Conditionality in Medicine Around 1900', in this volume, pp. 75–104.
12. Fagot-Largeault, A.: 1992, 'On Medicine's Scientificity—Did Medicine's Accession to Scientific 'Positivity' in the Course of the Nineteenth Century Require Giving Up Causal (Etiological) Explanation?', in this volume, pp. 105–126.
13. Fagot-Largeault, A.: 1984, 'About Causation in Medicine: Some Shortcomings of a Probabilistic Account of Causal Explanation', in L. Nordenfelt and B. Lindahl (eds.), *Health, Disease, and Causal Explanation in Medicine*, D. Reidel, Boston, pp. 101–129.
14. Fleck, L.: 1979, *On the Genesis and Development of a Scientific Fact*, F. Bradley and T. Trenn (trs.), University of Chicago Press, Chicago.
15. Geire, R.: 1973, 'History and Philosophy of Science: Intimate Relationship or Marriage of Convenience?', *British Journal of the Philosophy of Science* **24**, 282–297.

16. Grene, M.: 1976, 'Aristotle and Modern Biology', in M. Grene and E. Mendelsohn (eds.), *Topics in the Philosophy of Biology*, D. Reidel, Boston, pp. 3–37.
17. Hanna, J.: 1979, 'An Interpretative Survey of Recent Research on Scientific Explanation', in P. Asquith and H. Kyburg (eds.), *Current Research in Philosophy of Science*, Philosophy of Science Association, East Lansing, MI, pp. 291–317.
18. Juengst, E.: 1985, *The Concept of Genetic Disease and Theories of Medical Progress*, Ph.D. Dissertation, Georgetown University, Department of Philosophy.
19. Kaufman, S.: 1976, 'Articulation of Parts Explanation in Biology and the Rational Search for Them', in M. Grene and E. Mendelsohn (eds.), *Topics in the Philosophy of Biology*, D. Reidel, Boston, pp. 245–267.
20. Kuhn, T.: 1970, *The Structure of Scientific Revolutions*, 2nd ed., the University of Chicago Press, Chicago.
21. Lakatos, I.: 1970, 'Falsification and the Methodology of Research Programmes', in I. Lakatos and A. Musgrave (eds.), *Criticism and the Growth of Knowledge*, Cambridge University Press, Cambridge, pp. 91–195.
22. Lakatos, I.: 1971, 'History of Science and Its Rational Reconstruction", in R. Buck and R. Cohen (eds.), *Boston Studies in the Philosophy of Science* **8**, pp. 91–136.
23. Maull, N.: 1976, 'Reconstrued Science as Philosophical Evidence', *PSA 1976* **1**, 119–130.
24. Maull, N.: 1981, 'The Practical Science of Medicine', *Journal of Philosophy and Medicine* **6**, 165–183.
25. McMullin, E.: 'The Ambiguity of "Historicism"', in P. Asquith and H. Kyburg (eds.), *Current Research in the Philosophy of Science*, Philosophy of Science Association, East Lansing, MI, pp. 55–84.
26. Nagel, E.: 1970, 'Teleological Explanations and Teleological Systems', in B. Brody (ed.), *Readings in the Philosophy of Science*, Prentice-Hall, Inc., Englewood Cliffs, NJ, pp. 106–121.
27. Neel, J.: 1947, 'The Clinical Detection of Genetic Carriers of Inherited Disease', *Medicine* **26**, 115.
28. Pauling, L. *et al.*: 1949, 'Sickle Cell Anemia: A Molecular Disease', *Science* **110**, 543–548.
29. Peset, L.: 1992 'On the History of Medical Causality', in this volume, pp. 57–74.
30. Roblin, R.: 1979, 'Human Genetic Therapy: Outlook and Apprehensions', in G. Chacko (ed.), *Health Handbook*, North Holland Publishers, Amsterdam, pp. 104–114.
31. Sadegh-zadeh, K.: 1984, 'A Pragmatic Concept of Causal Explanation', in L. Nordenfelt and B. Lindahl (eds.), *Health, Disease, and Causal Explanation in Medicine*, D. Reidel, Boston, pp. 201–209.
32. Suppe, P.: 1970, 'A Probabilistic Theory of Causality', *Acta Philosophica Fennica*, Fasc. 24, North Holland Publishers, Amsterdam.
33. Taylor, F.: 1979, *The Concepts of Illness, Disease and Morbus*, Cambridge University Press, Cambridge.
34. Toulmin, S.: 1975, 'Concepts of Function and Mechanism in Medicine and

Medical Science', in H.T. Engelhardt, Jr. and S. Spicker (eds.), *Evaluation and Explanation in the Biomedical Science*, D. Reidel, Boston, pp. 51–66.
35. Whitbeck, C.: 1977, 'Causation in Medicine: The Disease Entity Model', *Philosophy of Science* **44**, 619–637.
36. Wimsatt, W.: 1976, 'Complexity and Organization', in M. Grene and E. Mendelsohn (eds.), *Topics in the Philosophy of Biology*, D. Reidel, Boston, pp. 174–197.
37. Wimsatt, W.: 1972, 'Teleology and the Logical Structure of Function Statements', *Studies in History and Philosophy of Science* **3**, 1–80.
38. Wright, L.: 1976, 'Functions', in M. Grene and E. Mendelsohn (eds.), *Topics in the Philosophy of Biology*, D. Reidel, Boston, pp. 174–197.

ANNE MARIE MOULIN

THE DILEMMA OF MEDICAL CAUSALITY AND THE ISSUE OF BIOLOGICAL INDIVIDUALITY*

Anne Fagot-Largeault has claimed that medicine still adheres to the Aristotelian paradigm of four causes while at the same time it tries to illuminate physico-chemical mechanisms as cause and effect sequences. This contradiction has been obvious since the so-called revolution of medicine at the end of the nineteenth century, and she makes her point with abundant quotations from Claude Bernard who found himself unable to stick to his own reductionist methodology. Fagot-Largeault rightly shows that this dilemma stems from the exceptional character of classical causal inferences in medicine of the Laplacian type which means when the cause A of B is the necessary and sufficient condition of B ([8], p. 9). The multi-factorial nature of causation in medicine is the main obstacle preventing a rigorously deterministic approach: but overall difficulties are added, due to the individual character of the organism, source of "major irregularities" ([8], p. 15).

While I agree with Fagot-Largeault on the persistent importance of this multi-factorial causality, I think that the connection with the issue of the "uniqueness of the individual" demands further analysis [27]. Moreover, I do not think that the consideration of multi-factorial causality necessarily leads physicians to adopt a probabilist approach, even if computer facilities are increasingly used for several medical purposes. Alternative strategies in contemporary medicine can be documented. Fagot-Largeault is, however, herself very cautious in her prediction of the future of medicine.

I will be very brief on that point. The statistical approach has been repeatedly explored since the middle of the last century ([20], [23]) for diagnosis and later for comparative evaluation of treatments. The methodology for randomized experimentation has been available since the 1930s and has raised more issues than it solved [24]. Up to the present, the debate raging over the interpretation of the results from randomized clinical trials has not come to an end. The absence of agreement suggests that the medical resistance does not come only from professionals defending their privileges and expertise but more profoundly is grounded in the controversial nature of medical causality. It is not enough to say that medical causality is multi-factorial in essence: it has to be figured out how physicians dissect these multiple factors and plot them against the chronological axis which frames their interventions.

Corinna Delkeskamp-Hayes and Mary Ann Gardell Cutter (eds.), Science, Technology, and the Art of Medicine, 153–162.

I will try to comment on Fagot-Largeault's paper by showing how in the 1970s modern physicians, being aware of Bernard's scientific dilemma, have shifted enthusiastically to a new explanatory mode, the immunological mode [31]. This mode throughout its historical development exhibits both features singled out by Juengst in his comment on Fagot-Largeault's paper: it is causal *and* teleologic, mechanic *and* functional, anatomic *and* physiologic. I will argue here that the recent development of the immunological style provides physicians with both a technological strategy for studying disease and an etiological framework for explaining disease. (On the prehistory of the concept of style (*Denkstil*), see [10]. For further general development see [14].) First, immunology suggests a respectable program for molecular research: coincidently, physicians have adopted a new ideological construct of the immune system. This concept of the immune system opens a new vista on the organism and can be referred to as a formal or definitional cause in the Aristotelian sense.

I. THE IMMUNOLOGICAL MODE OF EXPLANATION

The immunological mode of explanation, which developed after 1970, offers a syncretic approach to causality in medicine. This approach can be briefly presented by quoting Bernard's statement that "illness is determined by the nature of the thing ill" ([8], p. 14). The "immune system" (the term emerged in the mid-1960s) carries with it the historical connotations of "predispositions, idiosyncrasy, constitution ..." ([8], p. 14), such as have been discussed since Hippocrates' time, but interprets them in terms of "immune response". The immune system is a set of cells characterized by "competence" or the capacity to mount the immune response, in other words, to recognize molecules (defined as antigens) and dispose of them through various "effector" mechanisms. Obviously, the concept of the immune system stems from the historical intuition of the defensive role of cells in binding up bacterial toxins, as was recognized in the late nineteenth century (antidiphtheric serotherapy) ([29], [34]). But considerations on the immune system include, more broadly, all the ways in which the organism differentiates itself from the environment and maintains its identity ([18], [38]). The immune system is both a repository for biological memories and a powerful instrument for survival in the midst of the ever-present and potentially harmful agents both inside and outside the organism.

While the immune system provides the biologist with efficient causes to investigate, it may be at the same time viewed as an etiological construct. It has its own reality and substrata that can be (and have been) dissected. Cells belonging to the immune system bear some markers easy to identify with appropriate reagents. Some of these cells are located in particular organs and their genealogy has been thoroughly investigated ([19], [12]): much attention has been paid to the organs where they originate. Although these originating organs may be considered as the seat of immunity, the localization of new types of diseases labelled congenital immune deficiencies has been documented — most immunocompetent cells are circulating cells. The analogy with the circulation of blood makes it obvious that these organs are but the parts of a general system assigned to a new physiological function, namely, immunity.

Keeping this brief description in mind, the immune system displays the four characteristics of the Aristotelian paradigm. It may equally be considered as a formal cause (the system), a final cause (the purpose of immunity, the "integrity of the body" [5], a material cause (cells and factors), and efficient causes (the mechanisms of cytotoxicity, antibody production, etc., through which it does its work). But the immune system may be considered as well in the light of Juengst's opposition between causal and teleologic explanation, mechanical and physiological explanations ([17], p. 128ff).

II. THE MEDICAL ADOPTION OF THE IMMUNE SYSTEM

Since the concept of the immune system yields such a flexible instrument to account for causes of health and disease, the historical question may be raised why the immunological style of reasoning did not prevail among physicians until as late as the 1970s [30]. In fact, by the turn of the century, many physicians had already pointed out that the level of specific antibodies (or any other parameter of immunity) was but a poor indicator both for diagnosis and prognosis ([36], p. 681; [32]). Immunologically-oriented treatment was found to be unreliable and nonspecific in a number of cases where the outcome was submitted to what the bacteriologist Wilson phrased as "the hazards of immunization" [37]. These considerations led in the 1930s to a relative disfavor for immunological criteria in medical thinking.[1] Nevertheless, immunological practices, like serology, vaccination, and specific desensitization in allergies, did not disappear, but went apart from the scientific mainstream.

In recent times only, the immune system has been reconceived as a self-regulating system, composed of cells and secondarily factors, easily manipulated, either to simulate or inhibit the self's response to various antigens. It has thereby made plausible a new version of the natural history of disease [7], in which it is possible to describe the whole range of pathological situations by referring to the response of the immune system [22]. A pattern of four possible outcomes emerges which is applicable to nearly all cases of illness:

- the successful one: the immune system deals correctly with the perturbation and the disease is clinically unapparent;
- the immune system is active, but with some negative consequences for the subject: fever is high, symptoms are blatant. This is the evolution of acute disease;
- the response is weak or inappropriate (possibly harmful): viruses are able to multiply inside cells, or the initial damage is the starting point for further deterioration, leading to a state of chronic illness;
- don't let us omit death, either as an immediate hazard or as the ultimate end of evolution in the two last cases.

The response of the immune system is the key for evaluating the present status of the patient and discussing both diagnosis and prognosis. More interestingly, it yields some clues to the management of the medical care (through immunostimulation or immunosuppression).

The concept of the immune system has obviously been adopted on account of its versatility and its suggestions for treatment of disease. The mathematical formalization of the immune system has first been hardly influential on this success, since attempts in modelization have considerably lagged behind the widespread adoption of immunological criteria in medical terminology (such as "immunosuppressed" to indicate the status of the enfeebled or compromised host). The models invented after 1975 by theoretical biologists ([2] [3]) have so far indicated an abyss between constructs elaborated from restricted experimental data and the complexity of the real immune system which has to deal simultaneously with all kinds of external and internal challenges. Nevertheless, some attempts have emerged to estimate quantitatively the "vector of immunity" as the sum of different factors [21].[2] Such techniques are not commonly used, however.

But the adoption of the concept of immune system has a more far-reaching consequence; it solves the issue that was hereto the main conundrum of scientifically-oriented medicine, the unavoidable, unpredictable deviation of individuals from the norm: "the relation between the individual and the type" ([8], p. 113). Fagot-Largeault had admitted that "determinism (which authorized scientific prediction) finally dissolves, however, into idiosyncrasies of individual dispositions" ([8], p. 113). The immune system contributes to featuring biological individuality in a more positive and consistent way.

III. THE RATIONALE OF BIOLOGICAL INDIVIDUALITY

With the introduction of the concept of the immune system, individuality is no longer viewed as the main obstacle to the establishment of causal laws of (patho)physiology: it has been extolled to the dignity of the object of biological knowledge. Consequently, it is no longer linked to the personal expertise of the practitioner and concerns the biologically trained physician as well. In a nutshell, the concept of the immune system responds to the dilemma that plagued Claude Bernard and his followers. Biologists have not "reduced" individuality: they have raised it to the level of a full-fledged cause, knowable *per se*.

How is this shift to be described? Intuitions of the uniqueness of the individual had always been grounded upon religious and philosophical convictions; Descartes' *Cogito* is exemplar of this tradition. In the pioneering work of Alexis Carrel on transplantation, the singular essence of the organism had been presented as a formidable obstacle, biological in nature but of unknown constitution, preventing easy transfer of organs between individuals [33]. Later, Mendelian genetics had provided convincing arguments for the original character of every zygote. Immunogenetics [16] first developed as a descriptive science, based on classification of biological markers. But the contribution of immunogenetics was more decisive when some clues for the physiological implementation of individuality emerged from work on components of the immune response in the life and evolution of the organisms ([18], [25]).

The immune response is dependent on the genetic built-in determinants of the system and the pattern of these determinants is unique for each individual ([19], [20]). Only identical twins have a similar pattern for the major histocompatibility complex genes which rule the immune response. An interesting chapter is the discovery that some people might be (like mice among whom this phenomenon was first described) "high" or "low" responders to one specific antigen or set of antigens if they have inherited a particular set of regulatory genes (see the work of Biozzi, Benaceraf, and McDevitt in [18]). Also interesting, although still very controversial, is the suggestion that homologies between sequences of histocompatibility molecules or cells (as determined by the relevant genes) and antigens might explain the variation of response for this category of antigens [13]. These genes have been mapped for a restricted set of antigens such as synthetic polypeptides. We do not have yet a complete model of either the entire immune system or the regulatory genes that direct it, but the premise of such a model is contained within the concept of the immune system; enough work has been done to suggest that the variability of the immune response is the key to the broad spectrum of individual behaviors in health and disease.

The immune system also features the individual in a second, somewhat redundant manner. Not only does each person's immune system embody unique individual determinants, but the singular history of life, the sequence of events that goes with every individual, strongly influences immunological reactivity, creating a secondary pattern of unique determinations. For example, since sequential frameworks in the establishment of the immune response are settled on first meeting with the antigen ("antigenic sin" [9]), such chance infections may result in "blind spots" [39], particular molecular configurations which the individual's immune system is no longer capable of recognizing.

These two levels of explanation (genetic encoding and environmental determination) may be superimposed and articulated together, thanks to their mutual "transparence", as Juengst nicely suggested ([17], p. 139), referring to Wimsatt. But Juengst should have developed why transparence of explanatory levels is, in his view, intellectually safer than "dialectics of nature", which is another methodological device for mediation.

As is expected from these conciliatory comments, the resulting therapeutics by no means embodies an attitude of *laissez-faire*, of confidence in *Natura medicatrix*. Therapeutics develop along the two axes of ontological or mechanical commitment: first, "causal" therapeutics like non-specific immunostimulation or suppression: second, "mechanical" treatment intervening at the local level (specific receptor-targeting drugs).

IV. THE SHORTCOMINGS OF IMMUNOLOGICAL CAUSALITY

Our development made it clear why physicians adopted the immune system. All factors reviewed by previous work like nutrition, stress, depression, infection, can be plotted, if not quantitatively, at least qualitatively, in terms of factors of the immune system. Accumulated knowledge can be reformulated and stylized.

This all-encompassing approach, however, has in turn its specific shortcomings. Multi-factorial causality of disease does not facilitate a linear arrangement of causes. In other words, physicians now face a "web" of causation ([7], n. 15, p. 191; [28], p. 677), with no method for asserting the direction of causality, along the temporal axis: it is the stress of his exam that made him so badly depressed that he was so sick with ... or is it his sickness of ... which made ....

The physician has more or less to sneak into this network and try to assess his own intervention. He may find it difficult to appreciate the outcome of his treatment ([6], [15]). The present emphasis on the interaction between the immune system, the nervous system, and the endocrine system [35] makes it even more tricky to measure the impact of any tentative cure. Finally, the physician would be nearly tempted to give up intervening and argue in favor of an expectant attitude.

V. INDIVIDUALITY, A BIOLOGICAL CONCEPT TAILOR-MADE FOR THE PHYSICIAN

To summarize, the idea of the immune system provides a nice historical example of the complementarity of two modes of explanation, mechanical and functional, as described by Juengst and of a successful medical strategy. This strategy combines the quest of the laws of biological individuality with an empirical research of adapted management and cure of patients.

Not surprisingly, the word 'system' spontaneously appears in Juengst ([17], p. 132), when the "formal cause" is pointed to as "the effects of the system's functional organization", phrased elsewhere as "built-in functional tendencies to manifest disease" ([17], p. 131) or "essential property of a tissue" ([17], p. 131). The immune system can either be described in reference to its parts (morphology) or in reference to its goal-oriented mechanisms (physiology). Of course, again to cite Boorse as quoted by Juengst ([17], p. 131), "there is no way of finding a unique goal in relation to which traits of organism have functions...." We have seen the increasing emphasis on the interrelations with some other "systems" such as the nervous and endocrine systems, and some immunologist's have predicted their fusion into a unique "sensory system" [4]. Transdisciplinarity may be another way of circumventing multifactorial causality in medicine by pointing to the convergence of systems rather than focusing on a single one.

Fagot-Largeault has clothed explanatory modes in medicine with clothes from antiquity; Juengst has updated these venerable categories with the newfangled garments of the Anglo-Saxon school of analytic philosophy. Both have been tempted to account for the oddities of medical causality by reference to the issue of individuality, which cannot be framed by the lawlike logic of scientific determinism.

Fagot-Largeault in an indirect manner, Juengst more straightforwardly, point to the necessary complementarity of two modes of explanation in medicine, called either ontological versus causal or mechanical versus operational. They pave the way to a medical strategy based on the acceptance of this complementarity. Complementarity is used as a metaphor borrowed from Niels Bohr's principle in quantum physics, which inspired so strongly early molecular biologists for investigating new laws and phenomena in the domain of living beings. This complementarity in practice leaves the way open for alternative modes of treatment (for example, specific immunostimulation versus non-specific) but does not make the physician's assessment of his own activity easier. In treating explanatory modes of disease, let us not forget that the physician is also an operating force in the context of nature and candidate for medical causality, when he aspires to interfere with the natural history of the disease. If the levels of interpretation are transparent to each other, his own mode of intervening is far from being transparent.

Fagot-Largeault and Juengst both single out the issue of individuality as the frame where multi-factorial causality is displayed. This causality is accessible to probabilist analyses; more often, physicians refer to it as a theoretical construct, where all components of the response are ideally plotted, including individual determinants, and select such elements as fit their temporary views of etiology and treatment. This altogether idealistic and empirical strategy is in agreement with the tradition of medical concern for individuals and the lure of a treatment fitted for every patient. The dialectics between pragmatism and rationalism is as much characteristic of the medical philosophy and approach of causality as complementarity of explanatory modes is characteristic of the medical epistemology. It remains to be seen whether the ultimate quest for transparence might explain, in the future, a radical shift to some new style of causality like this cautiously predicted by Fagot-Largeault.

Centre National de la Recherche Scientifique
Paris, France, and
The Johns Hopkins University,
Baltimore, Maryland, U.S.A.

ACKNOWLEDGEMENT

I am greatly indebted to Dr. Harry Marks, from the Institute of the History of Medicine, Johns Hopkins University, for philosophical and stylistic comments.

NOTES

[1] No immunologist is listed among supporters of a New Biology between the two world wars [1].
[2] The evaluation of this vector was suggested for the management of medical care in hepatitis.

BIBLIOGRAPHY

1. Abir-Am, P.: 1987, 'The Biotheoretical Gathering: Transdisciplinary Authority and the Incipient Legitimization of Molecular Biology in the 1930's, *History of Science* **25**, 1–70.
2. Bell, G.I.: 1970, 'Mathematical Model of Clonal Selection and Antibody Production', *Nature* **228**, 739–744.
3. Bell, G.I., Perelson, A.L., and Pimbley, G.M. (eds.): 1978, *Theoretical Immunology*, Marcel Dekker, New York.
4. Blalock, J.E.: 1984, 'The Immune System as a Sensory Organ', *Journal of Immunology* **132**, 1067–1070.
5. Burnet, F.M.: 1964, *The Integrity of the Body*, Harvard University Press, Cambridge.
6. Canadian Multicenter Transplant Study Group: 1983, 'A Randomized Clinical Trial of Cyclosporine in Cadaver Renal Transplantation', *New England Journal of Medicine* **309**, 809–815.
7. Evans, A.S.: 1976, 'Causation and Disease: the Henle-Koch Postulates Revisited', *Yale Journal of Biology and Medicine* **49**, 175–195.
8. Fagot-Largeault, A.: 1992, On Medicine's Scientificity, Did Medicine's Accession to Scientific "Positivity" in the Course of the Nineteenth Century Require Giving up Causal (Etiological) Explanation?', in this volume pp. 105–126.
9. Fazekas de St. Groth, S.: 1967, 'Disquisitions on Antigenic Sin', *Annual Review of Medicine* **8**, 329–345.
10. Fleck, L.: 1979, *On the Genesis and Development of a Scientific Fact*, F. Bradley and T.J. Trenn (trs.), University of Chicago Press, Chicago (first German edition 1935).
11. Good, R.A.: 1983, 'Historic Aspects of Cellular Immunology', in J.I. Gallin, and A.S. Fauci (eds.), *Advances in Host Defense Mechanisms*, Raven Press, New York, pp. 1–42.
12. Good, R.A.: 1976, 'Runestones in Immunology, Inscriptions to Journeys of Discovery and Analysis', *Journal of Immunology* **117**, 1413–1428.
13. Guillet, J.G., Gefter, M.I. *et al.*: 1986, 'Interaction of Peptide Antigens and Class II Major Histocompatibility Complex Antigens', *Nature* **324**, 329–345.
14. Hacking, I.: 1982, 'Language, Truth and Reason', in M. Lukes and S. Hillis, (eds.), *Rationality and Relativism*, Oxford University Press, Oxford, pp. 48–66.
15. Hart, N.: 1986, 'Inequalities in Health: The Individual versus the Environment', *Journal of the Royal Statistical Society* **149**, 228–246.
16. Irwin, M.R.: 1974, 'Comments on the Early History of Immunogenetics', *Animal Blood Group Biochemistry and Genetics* **5**, 65–84.
17. Juengst, E.T.: 'Causation and the Conceptual Scheme of Medical Knowledge', in this volume, pp. 127–152.
18. Klein, J.: 1982, *Immunology, the Science of Self-nonself Discrimination*, Wiley, New York.
19. Loeb, L.: 1945, *The Biological Basis of Individuality*, Thomas, Springfield.

20. Mackenzie, D.: 1981, *Statistics in Britain, 1865–1930*, Edinburgh University Press, Edinburgh.
21. Marchuk, G.I.: 1983, *Mathematical Models in Immunology*, Organization Software, New York (first Russian edition 1980).
22. Marchuk, G.I. and Belykh, L.M. (eds.): 1983, *Mathematical Modelling in Immunology and Medicine*, North Holland Publishers, Amsterdam (first Russian edition 1982).
23. Marks, H.M.: 1983, 'Ideas as Social Reforms: An Ambiguous Legacy of Randomized Clinical Trials', unpublished paper.
24. Marks, H.M.: 1987, 'Notes from the Underground, the Social Organization of Therapeutic Research', in H.M. Marks, D. Long, and R. Maulitz (eds.), University of Pennsylvania Press, Philadelphia.
25. McDewitt, H.O. and Chinitz, A.: 1969, 'Genetic Control of the Antibody Response, Relationship Between Immune Response and Histocompatibility Type', *Science* **163**, 1207–1208.
26. Medawar, P.B.: 1958, 'The Immunology of Transplantation', *Harvey Lectures*, **52**,pp. 144–176.
27. Medawar, P.B.: 1981, *The Uniqueness of the Individual*, Dover, New York (first edition 1957).
28. Mike, V., and Good, R.A.: 1977, 'Medical Statistics and Ethics', *Science*, **198**, p. 677.
29. Moulin, A.M.: 1983, 'De l'analyse au système: le développement de l'immunologie', *Revue d'Histoire des Sciences* **36**, 49–67.
30. Moulin, A.M.: 1989, 'The Immune System: A Key Concept for the History of Immunology, *Journal of Philosophy and History of Life Sciences* **11**, 13–28.
31. Moulin, A.M.: 1991, *Le dernier langage de la médecine. De Pasteur au SIDA*, Presses Universitaires de France, Paris.
32. Moulin, A.M. and Lowy, I.: 1983, 'La double nature de l'immunologie: histoire de la transplantation rénale', *Fundamenta Scientiae* **3**, 201–218.
33. Nuttall, G.H.: 1904, *Blood Immunity and Blood Relationship*, Cambridge University Press, Cambridge.
34. Silverstein, A.M., and Bialasiewicz, A.: 1980, 'A History of Theories of Acquired Immunity', *Cellular Immunology* **51**, 151–167.
35. Steinberg, C.M., and Lefkovits, I. (eds.): 1981, *The Immune System, A Festschrift in Honor of N.K. Jerne*, Karger, Basel.
36. Widal, F.: 1986, 'Recherche de la réaction agglutinante dans le sang et le sérum desséché des typhiques', *Bulletins et Mémoires de la société des Médecins des Hôpitaux de Paris* **13**, 681.
37. Wilson, G.S.: 1967, *The Hazards of Immunization*, Oxford University Press, London.
38. Wilson, D.: 1972, *The Science of Self. A Report on the New Immunology*, Longman, London.
39. Zinkernagel, R.M.: 1982, 'How to Escape Immune Surveillance?', *Springer Seminars in Immunopathology* **5**, 107–112.

PART III

ART AND INTUITION IN MEDICAL DECISIONS: THE REGARD FOR KNOWLEDGE IN MEDICAL PRACTICE

WOLFGANG WIELAND

THE CONCEPT OF THE ART OF MEDICINE

The essays in this volume concern the nature and interplay among science, technology, and the art of medicine. All three of these human endeavors have to some extent been submitted to critical analysis. This essay will focus on the development of the concept of the art of medicine. Many have understood this concept as a designation for any irrational remnant in medicine that the onslaught of science has as yet failed to eliminate and which will someday disappear with the scientification of medicine. I will argue, however, that the proponents of that position have failed to appreciate the development of the concept of the 'art of medicine'. In doing so, I hope to shed light on the importance of this element of art in our understanding of medical practice.

If one speaks today about art, one usually refers to the realm of the aesthetic. In light of this usage, it makes sense to infer that accepting the designation of medicine as an art means viewing physicians alongside poets, musicians, and pictorial artists. Certainly, this is not bad company; and surely, the aesthetic perspective is of some importance in medical practice. Most will admit, however, that the physician's task cannot adequately be described from this point of view.

If one understands the concept of art solely in the aesthetic sense, one has fallen victim to a persistent and wide-ranging interpretation. One has overlooked the fact that the actual idea of art in the expression 'art of medicine' represents a relic of an older level of language ([27], pp. 83ff). The expression 'art' was originally not limited to the aesthetic sphere; rather, it represented in line with the Latin *ars* and the Greek *techne*, the human capacity for production by planned action. In this older sense, not only the statesman and the military person, but every competent craftsman practices an art. The breadth of this understanding of art is typified in the so-called liberal arts (*artes liberales*) of the Middle Ages that included disciplines such as arithmetic and geometry (which, according to contemporary usage, are not arts but sciences). The narrow understanding of art in the aesthetic sense is a relatively late product in this historical development. Until the last century when it was necessary to discuss the aesthetic realm, one spoke not of arts, but rather of the 'fine arts' [16].

Corinna Delkeskamp-Hayes and Mary Ann Gardell Cutter (eds.), Science, Technology, and the Art of Medicine, 165–181.

One must recognize, then, that the concept of the 'art of medicine' originally referred to everything that medicine represents in contrast to the natural sciences. The concept of the art of medicine reminds us that medicine is a practical discipline whose final goal lies not in understanding matters of facts and reasoning about them, but rather in acting reasonably or prudently.

Let us review in greater detail the history of the concept of the 'art of medicine'. At the beginning of this history we find the *Corpus Hippocraticum.* Here, the concept of 'art' (*techne*) is still a generic term covering medicine as a whole. It includes reference to the theoretical knowledge of the physician as well as to his judgmental capability and his practical skills. In the first aphorism: "Life is short, the art is long", the "art" of the doctor in no way contrasts with medical science. In this context, a study of the Hippocratic work, *On the Art*, is instructive. This work does not intend to determine the essence of the art of medicine; rather, through discussion of certain opposing positions, it serves to suggest that such an art does in fact exist. The author, who is probably not a physician, is concerned to prove that the doctor is not talking without meaning when he claims that his actions will produce very specific effects, namely, that they will free the sick from their sufffering. The opponents with whom the author argues had adduced cases where either the physician could be of no more help or where the sick person regains health all by himself and without being treated by a doctor. As the author tries to show, it does not follow at all from such cases that the art of medicine is ineffective and that every cure depends purely on chance. The fact that the art of medicine has its limits does not justify calling the existence of that art itself into question. The concept of art is thus not at all defined in contrast to the concept of science; rather, it stands in contrast to the concept of chance. In the *Corpus Hippocraticum*, it still includes the entire realm of medical knowledge, actions, and capabilities and skills.

The second current of the tradition that is important for understanding the concept of art proceeds from Aristotle. Aristotle undertakes to limit the concepts of art (*techne*) and science (*episteme*) more precisely as exclusive of one another (*Nichomachean Ethics*, Bk. VI, Ch. 3–8; *Metaphysics*, Bk. I, Ch. 1). The scientist is the theoretician; he strives for knowledge for its own sake. The object of this knowledge is not individual things or events, but universal essences and laws. This universality, which science wishes to discover, is always unaltered. It is not subject to mutation and change. In particular, it is not influenced by the scientist. Focused on this universality, science seeks to capture its discoveries in terms of statements that are capable of, and require, a foundation. In contrast to science, art is focused on activity. It wants to produce concrete results and to form and change its objects. In contrast to scientific understanding, art directs itself, as action does, toward the single and individual. This is the realm where art must prove itself, even while remaining oriented toward universally valid principles in utilizing the products of theoretical science. Thus, according to Aristotle, there exists a tension between the realms of science and art, because the realm of the individual and the changeable can never be completely apprehended by science. Nor can art (or practical reason in general) ever be reduced to theoretical science. In this sense, it is characteristic of the physician, Aristotle's favorite example of the practical man, that he deals not only with universally valid knowledge, but also most importantly with individual patients and specific situations. His art lies in his ability to do justice to that realm of the particular to which science does not extend.

Although the *Corpus Hippocraticum* proceeds from an all-inclusive conception of art, while Aristotle's concept proceeds from a strict distinction between art and science, no unbridgeable contrast exists between the histories of their influence. Medical tradition was able without difficulty to incorporate the Aristotelian distinction. Distinguishing between art and science in medicine was considered necessary only when reflections on methodology or theory of science would so require it. Thus, the Galenic tradition could distinguish between *scientia medica* and *ars medica* in a properly Aristotelian manner [22]. This distinction, however, did not yet possess any explosive polemical power. Except for contrasts motivated by methodological interest, the terms 'art' and 'science' were for a long time used in medicine without difference in meaning. Certainly the Aristotelian tradition continually emphasized the unknowability of the individual. The validity of the principle that no science arises from particulars (*de singularibus non est scientia*) was seldom attacked during the Middle Ages or during the greater part of modern times. This, however, did not impede talk about art and science from becoming commonplace. Users of these expressions no longer needed to think in terms of the Aristotelian categorical distinction. Thus one finds medical authors up to the beginning of the nineteenth century using the two concepts interchangeably when speaking of medicine ([4], [5], [21], [8], [15], [11]).

The absence of a polemic relationship between medical art and medical science during the eighteenth century is easily understood. Pre-eighteenth century medicine was a closed, dogmatic discipline. It was taught at universities, largely in the form of a commentary on the classical teachers Hippocrates, Galen, and Avicenna. Within the framework of this tradition, not yet haunted by any crisis regarding its foundations, a closed and dogmatic medical science and the art of (the practice of) medicine could be neatly joined to one another. The decisive turn occurred as research into the fundamentals of medicine developed. At least, since Harvey's discovery of the circulation of blood in 1616, one finds an open, inquiring, progressing science that must be distinguished from the practice of medicine. Other than within a closed, dogmatic science, the question of how to mediate theory and practice will now remain pressing. The concept of 'application' was designed to express the manner of this mediation. Rational practice results from the application of theoretical knowledge to the individual cases of practice.

The balance between the theoretical and the practical, which had characterized the Aristotelian tradition, is thereby disturbed. In general, the assumption of the practical disciplines' independence is gradually given up. This can be seen in the example of ancient political science [13]. The ancient practical disciplines, like medicine or politics, were oriented around the model of a closed, dogmatic science. After this model was replaced by the model of an open, progressing science, it was necessary for the old practical disciplines to redefine their position. The turn of the nineteenth century introduces an early peak of methodological reflection in the history of medicine.

In the course of further developments in medical thought, science and art increasingly contrast with each other. The question as to whether medicine should be seen as a science or as an art becomes pressing. J.W.H. Conradi responds to this question in 1828:

> Medicine is to be considered as a science (a science of healing, a science of drugs) insofar as it presents a mass of knowledge, traces this knowledge back to basic principles and derives it from them, insofar as it orders this knowledge and presents it in a systematic fashion. It is an art (an art of healing, an art of prescribing drugs), however, insofar as it consists in the capability of acting according to particular rules ([7], p. 8).

This distinction between knowledge and action is easily compatible with the Aristotelian distinction between universal and particular. Indeed, Selle, certainly the most important theoretician of medicine in the eighteenth century, complained that beginners are seldom given a proper concept of the difference between theoretical and practical medicine: "Understanding the particular is the essential object of practice . . . ; science always just deals with more or less universalized concepts. Understanding the particular remains the proper area of art and of immediate practice" ([24], p. 189). This is not attained by science. "The artistic insight for the particular can, after all, easily escape the learned and quick-witted doctor" ([24], p. 2; [25], p. 260).

In the context of modern science, this division of tasks between art and science in medicine leads to yet another difficulty. This arises from an imbalance in the division that reveals itself when one compares the certainty attained, or at least pursued by science, to the much lower degree of certainty that the physician can claim when practicing his art on individual patients. As Zimmermann writes in his "Von der Erfahrung in der Arzneykunst":

> An art rests very often on mere probability when it does not have irrevocable rules for all cases, when it is impossible to follow a certain prescription in all cases, when one's mind, while not having sufficient instruction, must proceed as though it had, when one can make judgments only on the basis of very changeable conditions and merely approach the truth rather than reach it. Statesmanship, the art of war, and the art of medicine are of this nature ([28], p. 282).

The question as to the degree of certainty possible in medicine thus becomes a central theme of methodological discussion. A major participant in these discussions was Cabanis ([6], whose thought influenced Ayrer [2]). People had become aware of the fact that the application of knowledge in practice follows less strict rules than does the acquisition of the knowledge. But already, Selle held that this is only a matter of passing deficiency. He hoped that some day this deficiency would be eliminated and claimed that we "... in time and with future experience must expect the perfection and exultation of the art of medicine into a science" ([25], p. 240). For many eighteenth century philosophers of medicine, the ideal state of affairs would be one in which the art of medicine will have been reduced to a medical science. The naive optimism of believing that this ideal might one day be realized was supported again and again. Nevertheless, developments in the nineteenth century are characterized by a rapidly increasing disproportionality between, on the one hand, the progress of research into the foundations of medicine, and on the other hand, the practice of medicine. The rapidly growing theoretical knowledge found at first only partial application in medical treatment. So-called "therapeutic nihilism" characterized a situation in which medical science and the art of medicine came to occupy an ever more antithetic and polemic relationship.

In 1879, Billroth makes a timid attempt to rehabilitate the art of medicine *vis-à-vis* medical science. Interestingly enough, this happens in a book bearing the title, *On Teaching and Learning Medical Science* [3]. The concept of the 'art of medicine' designates for Billroth that part of medicine which does not include abstract knowledge or a knowledge transmitted in writing, but rather a skill that is always bound up with the person of the individual physician. In addition, it is conveyed only through direct communication between teacher and student, between master and apprentice. Billroth recognizes that the prevailing opinion of his time considers this element of artfulness to be a flaw in medicine that has yet to be eliminated. "To render medical ability independent from personal tradition, to establish the art of medicine for all time so firmly in writing that it will be independent from the talent of individuals, and to transform it wholly into a science is the ideal goal of our current efforts" ([3], p. 4). Nevertheless, Billroth remains skeptical of such hopes. He believes that the concept of the 'art of medicine' designates an element of medicine that is fundamentally irreducible to the 'science of medicine'. As he puts it: "I doubt that this goal will ever be reached: it will at least not be reached by the art of medicine any sooner than the art of poetry dissolves into metrics, painting into color theory, or music into theory of harmony" ([3], p. 4). It is revealing that at this time the idea of art as limited to the realm of the aesthetic had become so current that the aesthetic arts provide the natural point of comparison for Billroth when he explains the concept of the 'art of medicine'. One must note, however, that the concept of the 'art of medicine' indicates nothing irrational for Billroth. On the contrary, Billroth holds that the art of medicine is concerned with a teachable and learnable discipline.

To be sure, the scientific nature of medicine was at this time never seriously doubted. The dispute solely revolved around the question of whether in the long run medicine would be able to reserve a certain area into which science could not intrude. The following often falsely quoted and usually misunderstood statement goes back to the clinician B. Naunyn: "Medicine will become a science or it will not be" ([19], p. 1348; [20], p. 3). Naunyn is usually cited as crowning evidence for an unbound scientification of medicine. One easily overlooks, however, the polemical point of his statement. It is directed toward permitting medical therapy to participate in scientific progress. Thus, the statement aims at overcoming 'therapeutic nihilism'. Naunyn challenges the view that therapy constitutes a realm of the art of medicine that cannot profit from scientific progress. He sees clearly that such an idea about therapy and the art of medicine represents *an asylum ignorantiae*. On the other hand, the scientification of medicine has a limit for him as well. "It will scarcely ever become a natural science, because every science places its boundaries around its capabilities: Medicine cannot accept such self-restriction, since it is too deeply involved with humanity" ([19], p. 1384).

Toward the end of the nineteenth century and with the sharpening of the polemical relationship between art and science, M. Mendelsohn published his *Ärztliche Kunst und medizinische Wissenschaft* [18]. To this little book we owe the sharpening of the contrast between the physician and the medical professional, as it is still prevalent today.[1] For Mendelsohn, the situation of the medicine of his time is characterized by tension between the art of medicine and medical science.

> ...That medicine, practical medicine, the practice of the art of medicine finds itself now in a period of decline, of inner decline, we as doctors must confess... [m]edicine is a science — this statement transmits itself like an eternal disease... And yet, this statement is absolutely false, at least in its generalization ([18], p. 15).

After all, medical practice has science only as its basis. In itself, it is, however, not a science, but an art.

> For our profession the one constant factor in the flight of phenomena is the art of medicine, which, in independence from all variable outlooks and theories of science, has one goal and one task: to lead each patient back to health by all possible means, but not only by those of so-called 'exact science' which represents only a small fraction of the many available means ([18], p. 15).

In his writing, Mendelsohn's goal was primarily to challenge critically the medical education system of his time. In his view, this system of education focused too extensively on scientific theory; it produced no practical physicians but only medical scientists. Thus, he was concerned to ensure that the practical, individual, and patient-oriented side of medicine be given appropriate attention in the education of the physician. "In practicing his profession, the doctor must reckon with imponderables of which the world of exact science does not even dream" ([18], p. 16).

It is only a small step from this view to medical irrationalism. Here the art of medicine is first established in an opposing position to science that is perceived as inhumane. It is here not the art of medicine, but medical science that must retreat in the presence of is adversary. The orientation of the art of medicine towards aesthetic art (for the interpretation of which the concept of genius becomes important) leads to the ideal picture of the physician-artist [23]. Even though the physician-artist will apply the results of science whenever this cannot be avoided, he mainly orients his actions around points of view that reside outside of the sphere of scientific rationality. The true physician no longer excels on the basis of his possessing scientific knowledge, nor even so much of his controlling teachable and learnable techniques. Rather, he is marked by the grace of a talent for which those who have not been similarly blessed by fate will only strive in vain.

Erwin Liek represents medical irrationalism in its most stark form [17]. For him the scientification of medicine is synonymous with the destruction of the soul of medicine. Thus, quite apart from teachable knowledge and skill, the personality of the physician is assigned the primary place in medicine. He even goes so far as to propose to the physician the role of a priest. Only the elect are called to this office: "One is either born a doctor or one is never a doctor. Benevolent gods laid gifts in his cradle, which can only be given, never sought after" ([17], p. 197). A more considerate representative of medical irrationalism is Diepgen, who restricts the role of the irrational. This realm coincides with that of the art of medicine and it should be secured and defended against every attack by science. "It concerns rightfully what is called intuition; that is, the insight which is gained through spiritual vision, an inner suggestion of the moment" ([9], p. 18). The trust that the patient places in the physician is founded on that irrational part of healing that belongs to the art.

The art of medicine as medical irrationalism is understood in such a way that it stands in contrast not only to all the sciences, including medical science, but also in contrast to everything teachable, learnable, and capable of being substantiated. In addition, the whole realm of proper medical routine and technique is excluded from the art of medicine. With irrationalism, the concept of the art of medicine appears disfigured almost to the point of caricature.

Having reviewed the history of these difficulties, let us now ask whether there are conditions under which one may still meaningfully employ the concept of the art of medicine today. The problems involved in this concept are more relevant today than ever because they concern questions as to the function and significance of science within the realm of medicine. We still remain interested in whether there are areas of medicine, particularly involving the actions of physicians, which cannot be sufficiently accounted for by science and research. Certainly, the products of scientific research are applied in clinical medicine, yet no program of basic research tells the doctor how he should deal with these data. The question concerning the art of medicine involves, after all, the question of how the hiatus between science and patient can be bridged by the physician's actions. In this context, the term "application" is generally favored. Yet, what can be sensibly intended by this term is a question that is more difficult to answer than at first appears.[2] Frequently, this word signifies only an *asylum ignorantiae*. But there is yet another problem: the physician is always destined or committed to action. He must also act even when the knowledge that he would like to apply and that could motivate and legitimize his action is not, or is not yet, available to him. And it is just such predisposing action not arising from medical science that is the subject of my essay about the art of medicine. Thus, questions concerning the art of medicine involve considerations about the practical character of medicine. One is led to suspect that a theory of medicine is not reducible to a theory of science and the analysis of the theoretical disciplines underlying its knowledge ([27], pp. 5ff).

Let us examine this supposition with respect to three areas of medical practice in which the concept of "art" is currently used. A clearly legitimate use of the concept of the art of medicine is to be found where one speaks of the manual dexterity and ability of physicians. Such skills differ substantially from the knowledge and understanding of medical researchers. Of course, such skills can themselves be made the subject of scientific research aiming at generally acceptable principles. But even the precise knowledge of this kind of research relieves no physician of the trouble of learning and training the required skills. It is a truism that proficiency in the performance of manual tasks can never be replaced by any number of sound principles. Rather, one says of a physician that he possesses the art of medicine when he can carry out these tasks skillfully. As long as medicine exists, there will always be the point at which particular manual tasks must be skillfully undertaken. It is improbable that scientific research will ever be able to free the physician from this necessity. The respective state of scientific research has, however, a powerful influence on the way, the degree, and the choice of the tasks that physicians undertake. Thus, even though the necessity of manual tasks will not disappear from medicine, one can clearly recognize progress in medical research gradually rendering the problems that need to be solved by manual skills ever more trivial.

A second area where the art of medicine is opposed to the science of medicine concerns clinical judgment. To be sure, an adequate medical judgment about an individual patient is impossible without training in the medical sciences. The art of medicine is required in order to bridge the gap between the concrete condition of the individual patient and the universally valid laws and rules of medical science that as such do not reach the individual case. The question arises here as to whether medicine necessarily requires a specific medical art or whether we can conceive of a stage in the progress of the basic sciences that is sufficient for covering the individual as well so that a special art would be rendered superfluous.

In view of this latter question, let us examine the structure of the laws and rules of medical science employed by physicians. Reduced to their simplest form, they have the shape of ordinary universal statements: $(x)\ (Kx \rightarrow Sx)$. This means: for any individual x, if x has the disease K, then x displays the symptom or symptom-complex S. This is the basic form of the laws underlying the diagnostic process. The basic form of the laws underlying therapeutic actions looks very similar: (x) (Kx $\rightarrow$ Tx). This means: for any x, if x has the disease K, then the therapeutic measure T is indicated for x. It must not be overlooked that here we are dealing with only the most simple form of this kind of rule. In this formulation we have neglected the probabilistic character of most valid laws of biology as well as all the possibilities for variation arising from the individual constitution and situation of each patient.

The diagnostic endeavor proceeds in such a way that for the particular objective or subjective symptoms and signs in a given case, a suitable disease K is sought which is characterized by a symptom-complex S containing the symptoms and signs that one already knows. In extreme cases, one must work through all of the known disease entities that fulfill the required conditions. Then it requires a so-called differential diagnosis to discover further symptoms so that finally only one disease entity remains that satisfies the conditions. In practice, however, this goal may not be reached in this fashion. On the other hand, the therapeutic process may appear simpler. If the physician knows that a particular disease is present, this knowledge leads to a particular course of action that must at most be modified by his knowledge of the patient's constitutional an individual constants. The difference in logical procedures provides an epistemological explanation for the fact that the diagnostic task is usually more troublesome than the choice of therapeutic measures. Many of the difficulties facing the physician as well as the medical scientist are connected with the fact that in our system of medical concepts, rules, and insights, there is not always a clearly defined chain of implications that leads directly from symptoms to therapy.

In each of the diagnostic or therapeutic rules made available by medical science, at least two concepts are joined with each other. The individual patient does not occur, however, in such rules. They contain no individual names, but rather only bound individual variables. But the doctor always deals with individual patients whom he must diagnose and treat. Thus, he must employ statements containing individual names when he wishes to describe and justify his actions. To be sure, one can individualize the general rules. That is, one must go from statements containing bound individual variables to the statements with individual names which they imply. But even then, one is not dealing with simple statements about individuals, but only with conditionals. The physician who wishes to describe and justify his actions requires, however, only simple statements. Even when advancing and documenting a finding, he makes use of this kind of simple statement.

Medical science can facilitate this task considerably for the doctor. It gives the physician orientations for action and helps in decision-making: it can help him substantiate each of his judgments. However, the knowledge of ever so many general rules and laws can never relieve the physician of the task of diagnostically evaluating each individual patient from — as it were — top to bottom.

To work with the diagnostic and therapeutic rules of medical science, the physician needs as a starting point simple statements about individuals that can themselves not be further derived. Henceforth, I will call these statements, which can be based either on subjective symptoms or on objective signs, *basic statements*. Their reference to the respective individual patient is evident from their very foundation. A further analysis of their meaning reveals that such statements refer to individuality in yet another sense: they reflect the views of an individual physician that are in addition related to a particular point in time. The production of such statements in the process of elaborating findings and taking a history appears to be a remnant of the art of medicine that can never be taken over by the universal rules of the sciences.

Here we are concerned with a problem that is traditionally known as the "subsumption problem of the power of judgment". This problem has to do with the circumstances under which we can subsume an individual thing or event under a concept. The request that we should let our judgment depend on the presence or absence of the characteristic corresponding to the concept hardly helps us further because whether a certain individual does in fact exhibit a particular characteristic is often disputed. One will always reach a point at which one can only appeal to the evidence of experience or of the power of judgment when rendering a particular diagnosis. For this reason, we require that the physician have practical training in addition to a theoretical scientific education. Even the well-prepared physician will encounter situations in which he requires not so much information about the state of medical science, but rather the judgment and advice of experienced colleagues. The art of medicine is in this sense a result of experience: experience underlies the very possibility of applying the universal rules of science to the care of the individual patient.

The concept of 'experience' is indeed ambiguous. Medical experience differs not only in content but also in form from experience obtained in the experimental sciences. The experimental sciences are concerned with supporting or rejecting general laws. The data of experience that contribute to this process are ideally simple, verifiable, and repeatable. In contrast, medical experience does not aim at general principles; it manifests itself in the physician's ability to evaluate correctly difficult and problematic individual facts and events. It is such experience that enables the physician to frame appropriate basic statements. It correlates with a disposition that is closely bound to the person of the physician. It cannot be separated from that person and transmitted directly. Thus we have arrived at quite another point of difference from experience gained within the experimental sciences. In the experimental sciences the result is abstract knowledge, not knowledge bound to a particular person. Experience in the experimental sciences can be recorded in writing and transmitted in that form. When acquiring this kind of experience, one may continue working from the point at which another leaves off. One is not forced to start over at the very beginning as one is when acquiring medical experience.

Nevertheless, medical experience is not inscrutably irrational. Its acquisition can be taught and learned through training. Such training takes time that cannot be substantially reduced. Still, we may observe that in the course of medicine's development, the area reserved for the art of medicine and understood on the basis of a specifically medical experience has been decreasing. This becomes clear if we subsume the various concepts found in the basic statements under various types of concepts, such as classificatory, comparative, and metrical. With classificatory concepts, the subsumption problem presents itself in its purest form. Medicine works with such concepts, for instance, when it deals with evaluating an exanthema, the results of auscultation, or a histological preparation, because these concepts are indispensable when one is concerned with morphological observations and methods. Here one is dealing with simple alternatives, such as yes/no decisions about whether an individual condition may be subsumed under a particular concept.

Leaving aside comparative concepts (as in evaluating the seriousness of coronary insufficiency), I will proceed to consider metrical concepts. Their use allows one to assign particular numerical values to objects under study. It must, of course, be established from the very start what kind of magnitude is to be measured in particular cases. Thus, the concern here is not whether certain objects are found to be present, as in the case of alternative classificatory decisions. Rather, it is the magnitude of a certain blood chemistry level (the existence of which is presupposed) that must be investigated in a certain patient.

Everyone knows that the development of medicine is characterized by metrical concepts taking precedence over classificatory ones. This has resulted in a replacement of morphological by functional analysis. Despite all declarations to the contrary, the modern physician has greater confidence in an exact laboratory value than in a classification gained through clinical bedside investigation. This shift of emphasis in the evaluation of clinical basic data is at least equally characteristic of medicine's scientification as is the progress of research in theoretical medicine.

For our present subject, it is important that this scientification of medicine occurs at the expense of the decreased significance of the 'medical art'. To be sure, the significance of individual basic statements that are themselves irreducible will not disappear as long as medicine continues to deal with and act on individual patients. However, one cannot determine *a priori* what kinds of statements these basic statements belong to. After all, the tendency in the development of medicine to replace classificatory concepts by metrical ones has rendered the problem of how to attain medical basic statements ever more trivial. The fixing of a numerical value on the basis of proper experimental and technical methods does not require the sort of experience that is indispensable for qualitative classification. In addition, the employment of metrical concepts brings with it more reliable results. Different diagnosticians can expect a higher likelihood of observer agreement. Thus it is no accident that basic statements employing metrical concepts are usually no longer acquired by the physician himself, but by his staff.

To be sure, this obvious trivialization of the "medical art" of evaluating individual cases is at present still limited. Nobody can predict with certainty whether in the future all basic statements will be gained in ways comparable to those of present-day clinical chemistry. For the moment, we are very far from this goal. Especially when taking a patient's history, classificatory concepts prevail, quite regardless of the introduction of questionnaires. The process of taking a history is thus still a refuge for the art of medicine. Nevertheless, we cannot be sure whether this state of affairs will not be altered one day by the introduction of ever more technical means of investigation.

On the other hand, one must not overlook the fact that a new kind of medical experience accompanies this development. I mean here an experience and an art of the sort needed when one wishes to apply prudently the new forms of technological and methodological assistance. It is the sort of experience that is required in deciding which measurements should be taken in any individual case and in evaluating the data obtained.

Thus a new field of action for the art of medicine has opened up. The issue is here not how to gain basic statements, but how adequately to interpret the information contained in them and how to utilize it in medical action. Neither the interpretation nor the utilization is sufficiently determined by the diagnostic or therapeutic rules medical science provides. After all, practicing physicians rarely, if ever, possess enough information about their patients' condition to render the application of such rules a simple matter. To be sure, progress in medical science has opened up the possibility of obtaining even greater quantities of verifiable data together with an increased understanding of their connections with one another. However, in comparison with these possibilities, the amount of data one actually obtains and uses becomes even smaller. The reasons are generally known: they range from the ethical to the economical. While medical progress has generally increased the amount of data actually used by the practicing physician as well as that of data potentially available, the quotient between both has beome even smaller. Up to the middle of the last century a physician could presume that all medically relevant data pertaining to each patient were really available to him. By contrast, today he is faced with the question of which information must, should, or may be sought in each particular case.

There is yet another aspect to the condition of insufficient knowledge: Even if a physician would want to, and was economically and ethically able to realize the potential of the theoretically available data, he would lack the time required for such a task. In this sense, not just the progress of medical science, but even more so this constraint is responsible for the irremediably insufficient information on the basis of which the physician must act. Thus, whenever one must act under conditions of insufficient information and scarcity of time, it, appears as though the art of medicine could claim a position of its own vis-à-vis medical science. Such conditions are more prevalent among general practitioners than among specialists.

In view of these conditions, an appeal is often made to the role played by intuition in medicine. Such appeals are favored wherever compensation must be made for a vexing lack of information. Yet some caution is in order. The concept of intuition is notoriously ambiguous and very likely covers up unclarities of thought. To be sure, no one denies that there are ways of gaining and utilizing information that remain below the threshold of consciousness. This is what is usually meant when one speaks of intuition in medicine. So long as one remains critical of the conclusions arrived at by the intuitive process, and as long as one resists the temptation of appealing to supposed undeniable evidence, nothing can be brought to bear against the claim that intuition plays an important role in medicine.

But even then the situation of having to act under conditions of insufficient information will not in the long run support the view that medicine remains in part an art. In the search for rational orientation when acting, given any amount of available information, medical science offers "normative decision theory" to replace the opaque notion of intuition.[3] Normative decision theory presents the physician and others with a formal instrument determining in individual cases, and under conditions of insufficient information, the probabilities of diagnostic alternatives as well as the values for risk assignment and risk assessment for particular treatment options. The usefulness of this instrument in medicine derives not only from its ability to make up for any actual lack of relevant information, but also from the fact that almost all laws discovered by medical science are statistical laws. This is reason enough for physicians' inability to guarantee the success of their measures. Thus in two ways physicians have to make their decisions under a risk in a sense relevant for decision theory. Normative decision theory presents a formal device for optimizing medical interventions in the face of such risks.

Does this mean, then, that this device can replace the art of medicine with sophisticated, accurate, and efficient calculations? Perhaps in the far distant future it may, but at present medical theoreticians and practitioners have much work to accomplish. To begin with, most areas of medicine have not been analyzed in terms of sufficiently reliable statistical laws. But even if medical science will have solved this problem one day, a difficulty of quite a different kind will remain. After all, each probability value must refer back to a scale that allows one to determine unambiguously the probable utility achievable through each possible action. Such kinds of utility scales have successfully been devised in the area in which normative decision theory was first used, namely, in games of chance and betting. Even in economics, this may work without problems. Devising utility scales in the diverse areas of medicine presupposes, however, a general consensus concerning how to solve the basic issues in medical ethics. The achievement of such a consensus presents a new set of difficulties. I cannot go into this matter here. But it becomes clear that decision theory has so far been utilized in medicine only for very small, circumscribed issues.

In conclusion, I have argued that the concept of art in contemporary medicine accounts for the fact that medicine cannot be reduced to the theoretical science of medicine. Art refers to the actions of the physician who is assigned the responsibility of performing manual tasks and of making decisions regarding patients in particular situations on the basis of limited information and within temporary contraints. This state of affairs uniquely structures the practical and personal relationship between physician and patient and is older than any application of science in medicine.

With the increasing scientification of medicine, this relationship has been considerably modified. It has to a large extent been pushed aside by anonymous, institutional ways of health administration — a tendency that will without doubt continue. Will patients in the long run accept their being deprived of a personal physician? Or will they force a change? If we consider the general tendency of modern society toward depersonalization, the former cannot be excluded. But this is a different problem.

What I hope to have made clear is that a medicine without art would be a medicine without a physician. This should be kept in mind by those who choose to claim that medicine is other than a science.

Universität Heidelberg
Heidelberg, Germany.

NOTES

[1] Mendelsohn's work is notable for the distinction it draws between doctor (*Arzt*) and medical man (*Mediziner*). The distinction is still made in the German language.
[2] See Feinstein for a critique of the concept 'application' ([10], pp. 27–28) and for a discussion of the physician's 'art' ([10], pp. 14ff, 37ff, 291ff).
[3] For a discussion of the foundation of normative decision theory, see Stegmüller [26]; for a consideration of its application in medicine, see Gross [12].

BIBLIOGRAPHY

1. Aristotle: 1941, 'Nicomachean Ethics', in R. McKeon (ed.), *The Basic Works of Aristotle*, Random House, New York, pp. 935–1112.
2. Ayrer, A.F.: 1799, 'Introduction', in P.J.G. Cabanis, *Über den möglichen Grad der Gewissheit in der Arzneywissenschaft*, Dieterich, Göttingen.
3. Billroth, Th.: 1876, *Über das Lehren und Lernen der medicinischen Wissenschaften an den Universitäten der deutschen Nation nebst allgemeinen Bemerkungen über Universitäten. Eine culturhistorische Studie*, Gerald, Wien.
4. Boerhaave, H.: 1708, *Institutiones medicae*, J. Van der Linden, Lugduni Bat.
5. Boerhaave, H.: 1744, *Methodus discendi artem medicam*, London.
6. Cabanis, P.J.G.: 1798, *Du Degré de la Certitude de la Médicine*, F. Didot, Paris.
7. Conradi, J.W.H.: 1828, *Einleitung in das Studium der Medicin*, 3rd ed., Krieger, Marburg.
8. Danz, F.G.: 1812, *Allgemeine medizinische Zeichenlehre. Neu bearbeitet von Johann Christian August Heinroth*, W. Vogel, Leipzig.
9. Diepgen, P.: 1938, *Die Heilkunde und der ärztliche Beruf. Eine Einführung*, Lehmann, München.
10. Feinstein, A.R.: 1967, *Clinical Judgment*, Krieger Press, Baltimore.
11. Frank, J.P.: 1817, *System einer vollständigen medicinischen Polizei*, Vol. 6, Ch. 1, Schaumburg, Wien.

12. Gross, R.: 1975, 'Ubersichten über diagnostische und therapeutische Entscheidungen', *Klinische Wochenschrift* **53**, 293–305.
13. Hennis, W.: 1963, *Politik und praktische Philosophie. Eine Studie zur Rekonstruktion der politischen Wissenschaft*, Luchterhand, Neuwied.
14. Hippocrates, Vol. II with an Emglish translation by W.H.S. Jones, London 1952.
15. Hufeland, C.W.: 1798, *Über die Kunst, das menschliche Leben zu verlängern*, 2nd ed., 2 Vols., Jena.
16. Kristeller, P.O.: 1951–1952, 'The Modern System of the Arts', *Journal of the History of Ideas* **XII**, 496–527; **XIII**, 17–46.
17. Liek, E.: 1936, *Der Arzt und seine Sendung*, 10th ed., Lehmann, München.
18. Mendelsohn, M.: 1894, *Ärztliche Kunst und medizinische Wissenschaft. Eine Untersuchung über die Ursachen der "ärztlichen Misere"*, 2nd rev. ed., Bergmann, Wiesbaden.
19. Naunyn, B.: 1905, 'Ärzte und Laien', in B. Naunyn, *Gesammelte Abhandlungen*, 2 Vols. 1862–1909, Vol. 2, Stürtz, Würzburg, 1909.
20. Naunyn, B.: 1900, *Die Entwickelung der inneren Medizin mit Hygiene und Bakteriologie im 19. Jahrhundert*, G. Fischer, Jena.
21. Ploucquet, W.G.: 1798, *Pathologie mit allgemeiner Heilkunde in Verbindung gesetzt*, Heerbrandt, Tübingen.
22. Rothschuh, K.E.: 1975, 'Konzepte der Allgemeinen Krankheitslehre. Ihre Bedeutung in Geschichte und Gegenwart', *Physis, Rivista Internazionale di Storia della Scienza* **18**, 219–238.
23. Schweninger, E.: 1926, *Der Arzt*, 2nd ed., Rohrmoser, Dresden.
24. Selle, C.G.: 1797, *Medicina clinica, oder Handbuch der medicinischen Praxix*, 7th ed., Hamburg, Berlin.
25. Selle, C.G.: 1787, *Studium physico medicum oder Einleitung in die Natur- und Arzneywissenschaft*, 2nd ed. Hamburg, Berlin.
26. Stegmüller, W.: 1973, *Probleme und Resultate der Wissenschaftstheorie und der analytischen Philosophie. Bd. 4: Personelle und statistische Wahrscheinlichkeit*, Springer, Berlin, Heidelberg.
27. Wieland, W.: 1975, *Diagnose: Überlegungen zur Medizin-Theorie*, de Gruyter, Berlin.
28. Zimmermann, J.G.: 1787, *Von der Erfahrung in der Arzneykunst*, new ed., Orell, Gessner and Füssli, Zürich.

RUDOLF GROSS

INTUITION AND TECHNOLOGY AS BASES OF MEDICAL DECISION-MAKING*

I. INTRODUCTORY OBSERVATIONS: ASPECTS OF MODERN MEDICINE

At least since the nineteenth century, medicine has held a precarious place between natural science and "helping and healing". Now, in addition to this widely discussed dualism, there are additional problems:

1. So far, there has been no uniform and generally accepted definition for either *health* or *disease*. In Germany this has been demonstrated in medicine especially by Rothschuh ([33], [34]), Schadewaldt [37], and Schaefer [39], among others, and by Gottschick [10], Gröning [11], and Lange [24] in jurisprudence. Indeed, it seems questionable even now whether uniform and generally valid terms for health and disease will ever be found. A similar difficulty exists with regard to many syndromes and symptoms, although this has not been discussed specifically in such cases.

Certainly there are methods by which to circumvent this problem:

(a) One might restrict oneself, following Sadegh-zadeh [36], to choosing arbitrary definitions for which merely semantic correctness is claimed. This method seems to be especially attractive since mathematicians have also given up axiomatics as an absolutely true system. They have replaced it by arbitrarily defined constructs which once established cannot be changed, but function as constants ([19], [20], [28]).

(b) In medical practice and in forensic medicine, we may suppose a tacit *consensus omnium* in a field with vague limits. For instance, for a long time legislation in the Federal Republic of Germany (and supposedly in other countries as well) made use of terms such as health, disease, and functional capacity in numerous laws and ordinances without ever providing a uniform and exact definition [10]. Civil law and insurance law assume a general concept of health and disease which they supplement in special ordinances. This is also true of medicine itself. As the general practitioner and philosopher Lukowsky says: "These are basic principles, basic categories of my science, which I know very well to be inaccessible to further resolution and declaration by means of my own discipline, medicine" ([26], p. 97ff).

Corinna Delkeskamp-Hayes and Mary Ann Gardell Cutter (eds.), Science, Technology, and the Art of Medicine, 183–197.

These ways of circumventing the conceptual problem are, however, not satisfactory. Decisions in medicine are often made not only on the basis of *insufficient information in an individual case*, but on the basis of *generally unclear notions*. Therefore, Gottschick's response to this may be appropriate: "If... experience and the power of the notion are not sufficient to define the terms health and disease according to strict requirements of formal logic, they can hardly serve for any concrete purposes within the range of experience" [10].

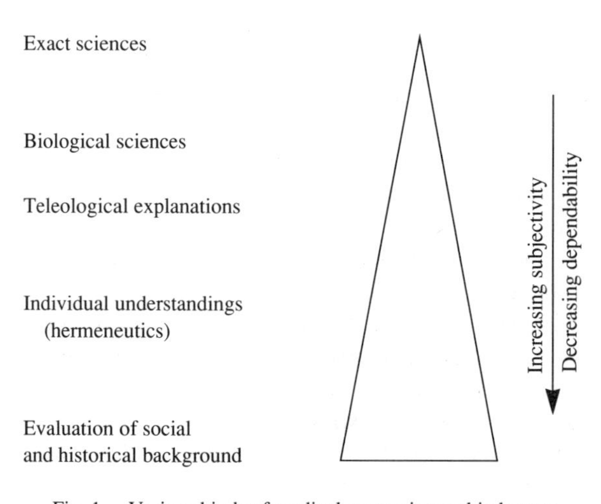

Fig. 1. Various kinds of medical perception and judgment.

2. Medical judgments carry varying degrees of scientific reliability. I have attempted to present this schematically (Figure 1) with reference to a monograph by Lukowsky [26]. Medical judgments can be based on the *methods of so-called exact natural sciences*, and thus reach the highest stage, i.e., they are not only exact (i.e., free of contradiction and reproducible) but also true (i.e., in accordance with reality) ([19], [20], [28]). For *biological results*, as well as most of those arrived at in medicine, variability is greater and truth is defined by appeal to some acceptable statistical level of coherence, relevance, completeness and unification. Biology prefers the weaker method of so-called *comparative induction*. We need only call to mind parameters that are important yet difficult to establish with the modern methods of technology, e.g., constitution, condition of the automatic nervous system, occurrence of death, and the like.

There are medical judgments which waive causal-analytical methodology and instead attempt to employ a *teleological contemplation* or an *understanding [verstehen] explanation.* Psychoanalysts prefer, for example, to work with methods of understanding which disregard the fact that Freud at the beginning thought that the human soul could be examined by the same methods as those which were applied to bodily organs (as is evident from Freud's consistent use of physical terms, such as mechanism, force, repression, etc., in his analyses).

We cannot deny that such methods are frequently necessary to enable us to make judgments concerning certain processes of life and disease. As the Basel pneumologist Herzog put it: "Relations with ... patients teach us at the same time how measurability can come to an end and how unmeasurable factors, at least intangible factors of the process, determine the process of a disease." According to Schaefer [39], there are no bridges from conscience to conscience. What happens in other minds must be found out hermeneutically. But we must always remain conscious of the relativity and susceptibility to error of such hermeneutical views. This relativity applies even more to *value judgments*, with which objective statements, subjective impressions of the examiner and, frequently, the projection of one's own standards and ideologies are mixed. This problem is often discussed in literature in that it plays an important role in medical decisions. Richard Koch [23] has already dealt thoroughly with these sources of error in medical decisions.

At least, the upper stages of medical judgement listed in Figure 2 should not be interpreted as representing an unavoidable sequence. One has the interesting cultural phenomenon that, with increasing formalization,

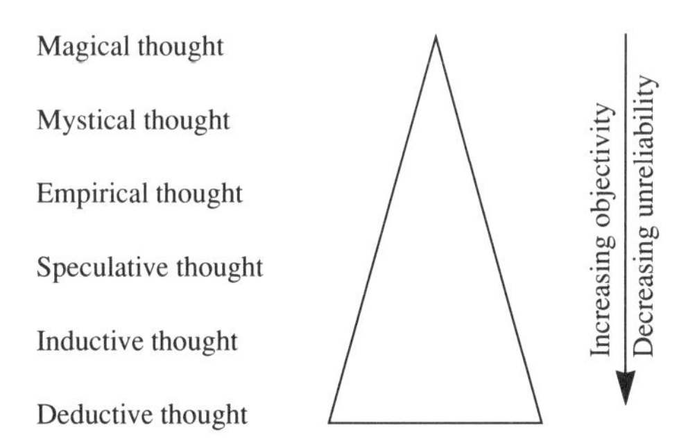

Fig. 2. Actual modes of thought about health and disease.

mathematization, and exact measurement of data, physicians are confronted with a flight into speculative, mystical, or magical imaginations. When we ask the disciples of such practices for their arguments in individual cases, they usually refer to their own *empiricism* or their *intuition*, instead of reproducible facts. In other words, empiricism and intuition for these people are indispensable as elements of medical decisions. However, the danger of error and abuse is great.

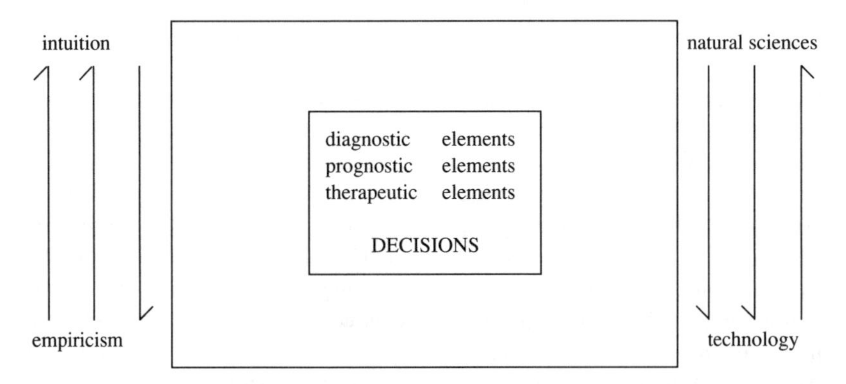

Fig. 3. Relations among the different elements of medical decisions.

3. This leads me to my last preliminary statement and, at the same time, back to the above-mentioned dualism. Figure 3 is an attempt to show the relations among the different elements of medical decisions: empiricism and intuition on the one hand, natural science and technology on the other hand.

Before I deal with the narrower themes of intuition and technology, I would like to hint at the interrelations of each side. On the left side, the relation between intuition and empiricism is represented. The emphasis is on *empiricism*. Leaving the divine illuminations of the scholastics and the flash of illumination of more recent philosophers (Figure 4) out of consideration, there is hardly any intuition without an empirical basis, whereas empiricism may exist without intuition. But let me make it clear, in no uncertain terms. This treatment of intuition does not concern Kant's account, according to which there is no real observation or formation of experience without categories independent of experience. There is neither a categorial (*a priori*) way of thinking in Kant's terms nor a purely *a posteriori* experience (in this con-

Enlightenment through divine revelation – e.g., Scholastics
Inner spiritual vision without reasoned insight – e.g., 16th to 19th century Philosophers
Unconscious transfer of cognitive processes – e.g., Bergson, C.G. Jung, *et al.*
Multi-dimensional conclusions based on premises which are unproven
Preferred organization of available information

Fig. 4. Development of the notion of "intuition".

nection, see M. Hartmann [18]). Medicine demonstrates this: I cannot think of any disease or syndrome I have never seen or experienced. I have, however, seen a group of symptoms, a disease, a dozen times already without having become aware of a new and special case of the interrelation of the symptoms. Yet, the Hamburg medical historian, Lichtenthaeler [25], has rightly pointed out the risk of our becoming pure empiricists given the abundance of disposable information, or what Hartmann calls "the avalanche of knowledge" [17]. Experience provides neither everything that is essential nor does it provide essential things only [10].

On the right side of Figure 3, things seem to be simpler: Technology is, even today, regarded as a product of the sciences. However, there are a number of scientific theorists who have pointed out an inversion of this mother-daughter relation (e.g., [19], [41]). According to them, technology is about to drive the sciences forward, not *vice versa*. Medicine gives striking examples of this: The problem-oriented specialist is, in part, replaced by method-oriented specialists. With a new methodology or a new device, very different groups of diseases are "exploited" more or less unsystematically. The harvest is gathered in the form of publications of "experience in case of..." and "experience with...". According to Schaefer, the sciences have become "hostile to intuition to the detriment of all" [38], and more and more withdrawn from the certain knowledge which, according to the judgment of even the most important mathematical logicians such as Russell [35], cannot be traced back exclusively to sense data. After all, logic guarantees only correct inferring, not true understanding.

II. MEDICAL INTUITION

The development of the concept of intuition is subject to the same rules as that of many other modern concepts. At first, only a small esoteric group is familiar with their use. Interaction with the general populace, however, introduces them into general use. As their use spreads, they lose their characteristic content, become flatter and more meaningless, generally losing their meaning altogether. With the increasing amount of contemporary literature on medical intuition (e.g., [4], [15], [14], [38]), we must ask, therefore, first of all what we wish to intend by intuition (Figure 5).

1. Usual associations of new and already-recognized sense impressions
2. Unusual and uncommon associations of impressions and thoughts ("lateral thought")
3. Unique associations of connections between new and earlier apperceptions

Fig. 5. Levels of intuition.

1. The combination of new impressions and experience with those already known from earlier observations or from the literature will probably play the major role in everyday medical decisions. This combination follows well-known laws of association (see de Bono [1], Eccles [7], and others). According to King [22], for example, every classification of an individual case into a diagnostic category or class, every selection from a number of possible treatments would be an act of intuition. This notion of intuition seems to me to go too far.
2. The unusual reclassification of information leads neuro-physiologically to so-called "lateral thinking" ([1], [2]), that is, to *a different linking of old and new thoughts* on the surface of memory.
3. When we disregard revolutionary new associations as rare, the central range (see the second item in Figure 5) covers what we would refer to as "occasional intuition".

This form of intuition is not or not so much an increment in information as a restructuring of the short- and long-term memory. The usual arrangement of intellectual information follows broken-in paths and patterns, it is always somewhat obsolete — at any rate, not optimal. In intuition, however, some stimulus causes a *modification of information arrangement*, in favor of the best possible or at least a comparatively better pattern. The stimulus must be a fortuitous one, since conscious effort would only cause an extension or intensification of the existing patterns. However, such ideas can also be pursued by certain thinking techniques ([1], [2], [12]). Logical thinking alone usually does not lead to that modification of sequence which leads to an intuitive rearrangement of information.

It need not be emphasized that this form of intuition plays a decisive role in the medical profession on the average level (item two of Figure 5), e.g., in making a diagnosis which does not offer itself implicitly, in selecting a therapy outside the usual schemes.

III. MEDICAL TECHNOLOGY

The change to medical technology, which is characterized by quite different possibilities and problems, requires nearly a conceptual leap. This can only be a brief summary of many thousands of quite different physical and chemical methods that have been introduced. The technical methods used today can be placed in different groups, usually coming from basic sciences and introduced to medical application (Figure 6).

Clinical-chemical, hematological, immunological, radio-immunological, micro-biological data
Mechanical measurement of pressures, functions
Determination of electrical readings (e.g., ECG, EEG, EMG)
X-ray and isotope data (organs, cavities, vessels)
Computer-tomography, sonography
Inspection and photography of body cavities
Cytological, histological, cytochemical, ultra-violet or fluorescent examination-of organ probes, secretions, cross-sections
Molecular Biology, Genetics
Controlled therapy with medication, radio-isotopes
Further tests

Fig. 6. Technical methods in clinical medicine.

Let me also illustrate by means of two examples taken from clinical chemistry what the advances of the last two decades have brought and where the problems are located. With regard to the technical *prerequisites* of modern clinical chemistry (Figure 7), I would like to restrict myself to a short graphic illustration. The most important *sources of error* (Figure 8) include the cost-increasing and, from my point of view, unjustified, indiscriminate ordering of batteries of tests, often on a daily basis. This not only increases costs but also reduces the reliability of the results. Most examiners still assume as a *standard limit* a purely arbitrary value of two standard deviations, although most values are distributed log-normally in biology or in a different asymmetry

Micro-methods
Instrumentation
Mechanization
Electronics
Self-recording measuring units
Plausibility control
Computer-printouts

Fig. 7. Prerequisites of modern clinical chemistry.

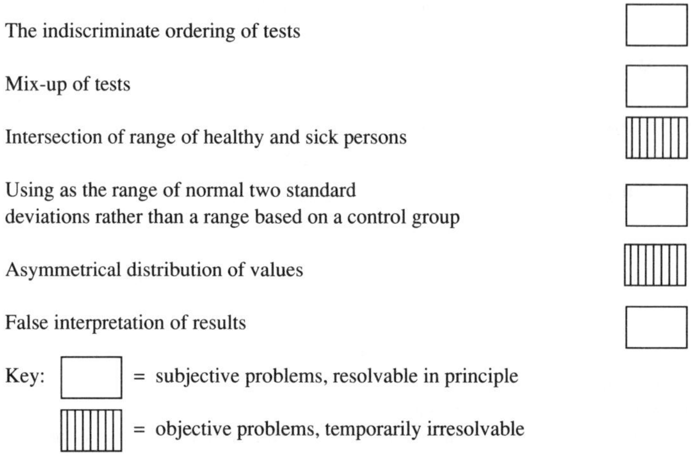

The indiscriminate ordering of tests

Mix-up of tests

Intersection of range of healthy and sick persons

Using as the range of normal two standard
deviations rather than a range based on a control group

Asymmetrical distribution of values

False interpretation of results

Key: = subjective problems, resolvable in principle

= objective problems, temporarily irresolvable

Fig. 8. Problems of clinical chemistry.

(e.g., being inclined to the normal values). For some determinations, there are examinations of sufficiently large *populations of healthy persons*, but not of *sick persons*, to establish a reference range of values. Also, ranges of values for healthy and sick persons overlap considerably, e.g., when determining uric acid or with regard to the number of leucocytes. A wide normal range increases the number of false positive results, a narrow range, the number of false negative results. When there is a considerable *overlapping* of really healthy and really sick individuals, three ranges become necessary, as I have already postulated [15]: certainly pathological, doubtfully pathological, and certainly normal.

These few brief comments are meant to show that even a highly developed and enormously expanding medical technology has its limits so that sources of error must be taken into consideration in the decision process.

IV. SYNTHETIC CONTEMPLATION: MEDICAL DECISION

Medical decisions can be either purely diagnostic (e.g., when a patient only wants to know his diagnosis), or purely therapeutic (e.g., in emergency situations when restoring homeostasis), but in most cases it is a combination of both. More and more it must be our aim to supplement decisions, which are often still purely empirical or intuitive in everyday practice, with mathematical, logical, and psychological elements.

	pk_1	$pk_2 \cdots$	$pk_i \cdots$	pk_m
a_1	u_{11}	$u_{12} \cdots$	$u_{1i} \cdots$	u_{1m}
			.	.
			.	.
			.	.
a_2	u_{21}	$u_{22} \cdots$	$u_{2i} \cdots$	u_{2m}
.	.	.	.	.
.	.	.	.	.
.	.	.	.	.
a_j	$u_{j1} \cdots$	$u_{j2} \cdots$	$u_{ji} \cdots$	u_{jm}
.	.	.	.	.
.	.	.	.	.
.	.	.	.	.
a_n	$u_{n1} \cdots$	$u_{n2} \cdots$	$u_{ni} \cdots$	u_{nm}

Fig. 9. Utility or decision matrix. Modified according to Schneeweiss [40]. See also [13], p. 41.

The *probability* of a particular disease, the *utility* of a therapy if used, and the *loss* caused by a failure to treat find their place in diagnostico-therapeutical decisions. I need not develop these points in greater detail here since I have treated them at length elsewhere [13]. Rather, I will restrict myself to mentioning two formalized procedures commonly recognized today: the so-called utility or decision matrix (Figure 9) ([13], [40]) and so-called decision trees (Figure 10) which are considerably simplified here ([31], [40]). In both cases, actions mix with probabilities, which are today regarded as 'subjective' but formalizable [13].

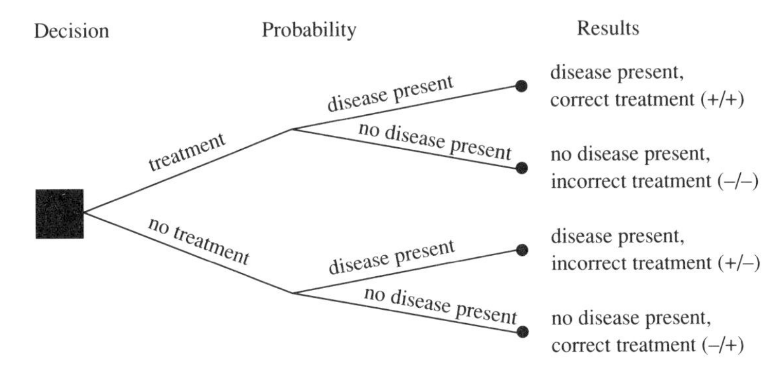

Fig. 10. Simplest model of a decision tree with four possible results.

I want to emphasize two *particularities of medicine* with regard to the mathematical criteria for decision-making:

1. Decisions in medicine are nearly always *decisions under risk* within the meaning of the theory of utility [5], [13], [40], in which different probabilities must be allocated to different events and different *formulae of utility* must be inserted (e.g., [13], [12]).

   Even then the decision-maker frequently does not foresee all possible future events, such as complications, individual reactions of the sick, etc. — sometimes not even all possible actions, i.e., diagnostic measures, treatments, etc. In medicine, this leads either to programs which are difficult to make and frequently changed or to an unjustified reduction to "simplistic schemes", as Feinstein appropriately notes [9].
2. In economics, for which the mathematical theory of utility was first developed and is far advanced, utility and loss are usually complementary. Not so in medicine: *riskier diagnostic actions* frequently promise a quicker and more reliable resolution. The more efficient and higher-dose *therapy* is usually at the same time the one richer in prospects and the more dangerous. The important indices of utility and loss as far, as I can see, have not been worked out systematically. This leaves a wide and fertile field for the theory of clinical medicine in particular, and the philosophy of medicine in general.

We want *hic et nunc* to restrict ourselves to the question of what *intuition* and *technology* contribute to clinical decisions. The cases in which a technical examination is the only basis for a diagnosis or makes at least the decisive contribution should today range from 10 to 80%. The enormous difference between these two limiting values is true of more or less all methods indicated in Figure 7. The large distribution results from whether one includes in the population any of the following:

1. patients who have no clinical complaints at all, i.e., so-called lanthanic diseases, following Feinstein's definition ([8], p. 145).
2. patients who see a doctor with complaints which are not typically a part of the final diagnosis; and
3. patients who come with complaints typical of the final diagnosis.

In each case, the technical methods fail in patients suffering from purely psychogenic or psychosomatic disorders. In practice, the exclusive function of technical methods is frequently a necessary but not sufficient part of diagnosing a psychosomatic disorder. The number of patients with psychogenic disorders varies enormously from one specialty to another, from one clientele to another. In internal medicine, however, above all it varies between in-

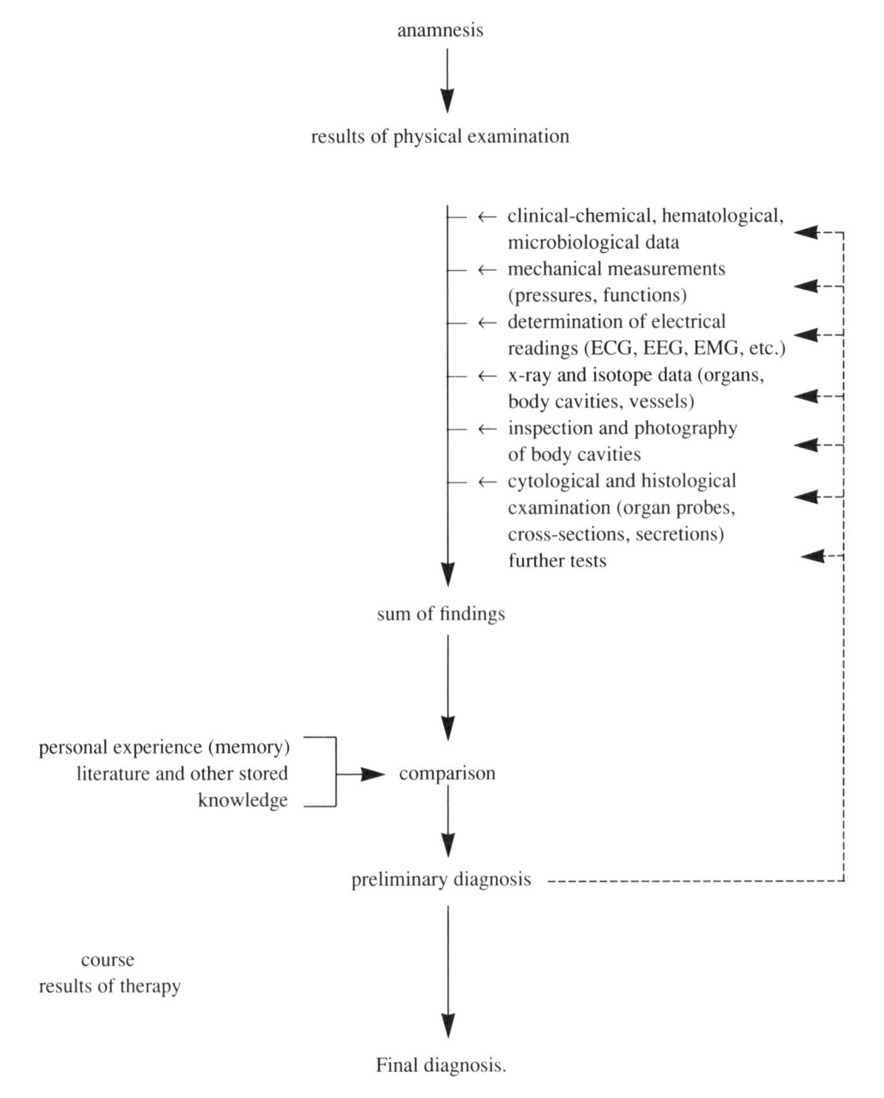

Fig. 11. Hypothetico – deductive diagnosis.

patients and out-patients. In the wards of the Medical University Clinic of Cologne, (entirely or prevailingly), psychogenic disorders constitute less than 10% of all admissions. In my own out-patient department, about 20% of the patients came with prevailingly psychogenic disorders, another about 40% had non-serious organic disorders but with a distinctly psychogenic overtone.

According to the experience of well-regarded American as well as German clinicians [15], 60–80% of all diagnoses can be made on the basis of a history and physical examination, the typical instruments of medical experience and intuition. But these figures too must be relativized: Such diagnoses are generally tentative *diagnoses*, which must be established or ruled out by scientific methods. At least in the clinics, they can today no longer be the only basis of far-reaching diagnostic, prognostic, or therapeutic decisions.

Let me close by addressing two interesting questions, which arise in reflection on the word 'and' in the phrase "intuition and technology".

1. Is a decision against the results of scientific examinations also made intuitively? I would answer: Hardly ever and only after careful consideration, provided that the laboratory results are reliable and stand up under review. If, however, the probabilities of certain diseases and the calculation of the utility of diagnostic or therapeutic actions within the framework of Figures 10 and 11 lead to rather similar results (indeed, this occurs quite often in the practice of medicine), the range of subjective or (if we wish) intuitive decisions is widened. In the geometric illustrations by Pauker and Kassirer [29], such decisions are near, or at, their threshold value limits.
2. Must technology be regarded as restricting freedom in medical decision-making? I would like to respond negatively to this question. Technology should be given its due place as an auxiliary, non-medical discipline. These points were also made by the judges of the Cartel Division of the Court of Appeal of Berlin — i.e., by a non-medical, non-philosophical group of German observers of these issues [21].

V. CONCLUSIONS

Instead of a summary, I would like to review a scheme I have developed [12], [16], and called, following a term of Francis Bacon, *hypothetico-deductive diagnosis* (Figure 11). The history (anamnesis) and physical examination lead to a preliminary diagnosis by analogy or, in more difficult cases, through so-called pattern recognition [13]. This process can be called more or less intuitive or on the basis of experience. The correctness of the conclusion as in every deduction, must be corroborated by new experiences (in this case

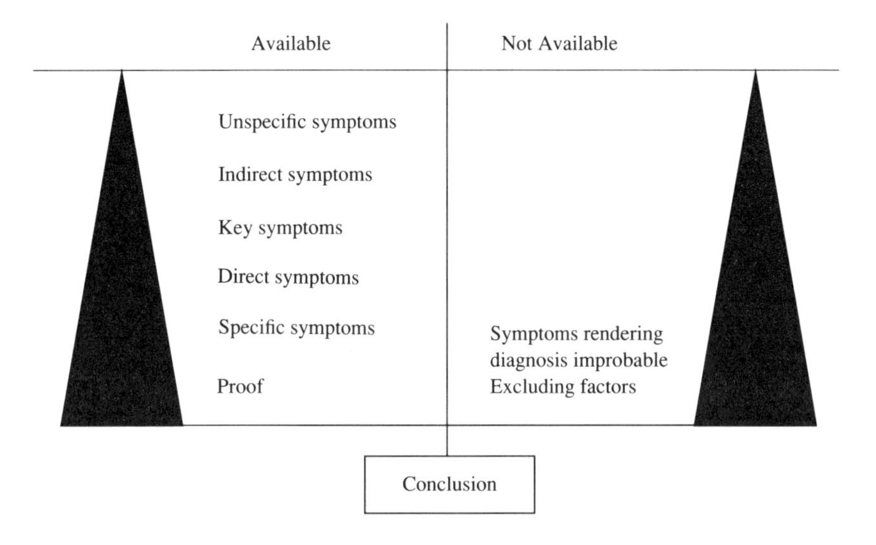

Fig. 12. Weighting symptoms.

technical examinations) or be falsified to use Popper's terminology [30]. By themselves alone the technical data can be pathognomic, highly reliable, relatively specific or entirely unspecific (Figure 12). In other cases, they have the weight of barring symptoms, symptoms which rule out the supposed diagnosis or make it improbable. Then, the process starts again in a sort of *recirculation*, with the diagnoses under consideration being restricted with every run, so that a sort of *diagnostic spiral* is formed. However, this does not preclude the physical examination or technical data from raising new questions for decision-making regarding the differential diagnosis or the differential therapy. In this context, I have not dealt specifically with *therapeutic decisions*. These can be simulated in terms of their desired and undesired effects before they are applied to human beings.

University of Cologne
Köln, Germany

NOTE

* This essay was developed from a lecture presented at the *Sixth Trans-Disciplinary Symposium on Philosophy and Medicine* that convened at Universitäts-Klinik, Hamburg, Germany, in 1977; the author wishes to express his gratitude to the Symposium's organizers.

BIBLIOGRAPHY

1. de Bono, E.: 1975, *Der Denkprozess*, Rowohlt, Hamburg.
2. de Bono, E.: 1975, *Laterales Denken*, Rowohlt, Hamburg.
3. Borch, K., and Mossin, I. (eds.): 1968, *Risk and Uncertainty*, Macmillan, London.
4. Catel, W.: 1976, 'Intellekt und Intuition', *Der Deutsche Apotheker* **28**, 458–464.
5. Chernoff, H. and Moses, L.: 1959, *Elementary Decision Theory*, John Wiley, New York.
6. Cramon, D. von and Backmund, H.: 1977, *Überlegungen zum diagnostischen Entscheidungsprozess am Beispiel der Neurologie*.
7. Eccles, I.C.: 1970, *Facing Reality: Philosophical Adventures by a Brain Scientist*, Springer-Verlag, New York; German trans.: 1975, *Wahrheit und Wirklichkeit*, Springer-Verlag, Heidelberg.
8. Feinstein, A.R.: 1973, 'An Analysis of Diagnostic Reasoning', *Yale Journal of Biology and Medicine* **46**, 212–264.
9. Feinstein, A.R.: 1967, *Clinical Judgement*, Krieger Press, Baltimore.
10. Gottschick, I.: 1963, 'Der medizinische und der juristische Gesundheits- und Krankheitsbegriff', *Deutsches Ärzteblatt* **11**, 475–480, 557–560.
11. Gröning, U.: 1974, 'Der Wandel des Krankheitsprozesses in der Rechtsprechung', *Deutsches Ärzteblatt* **11**, 475–480, 557–560.
12. Gross, R.: 1975, 'Zur allgemeinen Theorie der medizinischen Diagnostik und Therapie', *Zcitschrift für Geburtshilfe und Frauenheilkunde* **35**, 537–582.
13. Gross, R.: 1975, 'Über diagnostische und therapeutische Entscheidungen', *Klinische Wochenschrift* **53**, 193–305.
14. Gross, R.: 1975, 'Die Intuition in der ärztlichen Praxis und Forschung', *Deutsches Ärzteblatt* **72**, 3500–3502.
15. Gross, R.: 1969, *Medizinische Diagnostik-Grundlagen und Praxis*, Springer-Verlag, Heidelberg.
16. Gross, R.: 1973, 'Der Prozess der Diagnose', *Deutsche Medizinische Wochenschrift* **98**, 783–787.
17. Gross, R. and Fritz, R.: 1974, 'Die Wissenslawine und ihre Bewaltigung durch den Arzt', *Deutsches Ärzteblatt* **71**, 871–878, 950–958.
18. Hartmann, M.: 1956, *Die geistig philosophischen Grundlagen der Naturwissenschaften*, in M. Hartmann, *Gesammelte Vorträge*, G. Fischer, Stuttgart, p. 238.
19. Heitler, W.: 1972, 'Wahrheit und Richtigkeit in den exakten Wissenschaften', *Akademie der Wissenschaften und Literatur* **3**, 45–64.
20. Hilbert, D. and Ackermann, W.: 1972, *Grundzüge der theoretischen Logik*, Springer-Verlag, Heidelberg.

21. Kartellsenat d. Kammergerichtes Berlin. Urteil Kart. 32/74 vom 2.2.
22. King, L.S.: 1967, 'What Is a Diagnosis?', *Journal of the American Medical Association* **202**, 154–157.
23. Koch, R.: 1923, 'Irrtümer in der allgem. Diagnostik', in Z. Schwalbe (ed.) *Diagnostik*, Thieme, Leipzig.
24. Lange, R.: 1963, 'Der juristische Krankheitsbegriff', *Beiträge zur Sexualforschung* **28**, 1–20.
25. Lichtenthaeler, C.: 1965, *Unzeitgemässe Betrachtungen zur zeitgenossischen Medizin*, Lehmann, München.
26. Lukowsky, A.: 1966, *Philosophie des Arzttums. Ein Versuch*, Deutscher Ärzteverlag, Köln.
27. Lukowsky, A.: 1962, 'Das Unbestimmte im Denken des Logisters, des Juristen, des Arztes', *Verwaltungsarchiv* **53**, 25.
28. Meschkowski, H.: 1976, *Richtigkeit und Wahrheit in der Mathematik Mannheim*, Wissenschaftsverlag, Zürich.
29. Pauker, S.G. and Kassirer, J.P.: 1972 'Therapeutic Decision Making: A Cost/Benefit Analysis', *New England Journal of Medicine* **293**, 229–234.
30. Popper, K.P.: 1959, *Logic of Scientific Discovery*, Harper-Row, New York; German orig.: 1935, *Logik der Forschung*, Mohr, Tübingen.
31. Raiffa, H.: (1970^2), *Decision Analysis. Introductory Lectures on Choices under Uncertainty*, Reading, Ma, Addison-Wesley Comp.
32. Richter, H.: 1966, *Wahrscheinlichkeitstheorie*, Springer-Verlag, Heidelberg.
33. Rothschuh, K.E.: 1972, 'Der Krankheitsbegriff ', *Hippokrates* **43**, 3–17.
34. Rothschuh, K.E. (ed.): 1975, *Was ist eine Krankheit?*, Wissenschaftliche Buchgesellschaft, Darmstadt.
35. Russell, B.: 1976, Zit. n. Weischedel W.: *Die philosophische Hintertreppe*, Deutscher Taschenbuchverlag, München.
36. Sadegh-zadeh, K.: 1977, 'Krankheitsbegriffe und nosologische Systeme', *Metamedicine* **1**, 4–41.
37. Schadewaldt, H.: 1977, 'Grenzen von Gesundheit und Krankheit, historisch gesehen', *Medizinische Welt* **28**, 613–619.
38. Schaefer, H.: 1976, 'Intuition und Wissenschaft in Medizin', *Der Deutsche Apotheker* **28**, 1–5.
39. Schaefer, II.: 1976, 'Dcr Krankheitsbegriff', in M. Blohmke *et al.* (eds.), *Handbuch der Socialmedizin*, Vol. 3, Enke, Stuttgart, 15–30.
40. Schneeweiss, H.: 1967, *Entscheidungskriterien bei Risiko*, Springer-Verlag, Heidelberg.
41. Stent, G.S.: 1975, 'Limits to the Scientific Understanding of Man', *Science* **187**, 1052–1057.

STUART F. SPICKER

INTUITION AND THE PROCESS OF MEDICAL DIAGNOSIS: THE QUEST FOR EXPLICIT KNOWLEDGE IN THE TECHNOLOGICAL ERA

I. INTRODUCTION

Bertrand Russell once remarked that "Philosophy, like all other studies, aims primarily at knowledge. The knowledge it aims at is the kind of knowledge which gives unity and system to the body of the sciences, and the kind which results from a critical examination of the grounds of our convictions, prejudices and beliefs" ([10], p. 154). It is unlikely that Russell had medicine in mind when he referred to "all other studies", for it has always been a commonplace that attending physicians aim primarily at the restoration and preservation of health rather than at knowledge. As it is an indispensable part of this principal aim, however, physicians — since the days of the ancient school of Hippocrates of Cos — have been concerned with diagnosis, and in this sense at least *gnosis* still preoccupies members of the medical profession. For the practicing physician, the time between a patient's complaint and the initiation of a therapeutic regimen is occupied by a search for knowledge, though at this point it is a more utilitarian, pragmatic, or practical knowledge than Russell had in mind.

Professor Dr. Rudolf Gross, it is clear, shares with his fellow clinicians— attending physicians as well as specialists and clinical consultants — the concern for determining the accurate diagnosis or diagnoses pertinent to a particular patient's problems. For when Dr. Gross employs the terms 'clinical decisions' and 'medical decisions', he finds that, initially at least, he must focus on the problem the attending physician must face with each patient — the method of achieving an accurate and precise diagnosis, which will, in time, serve to determine the regimen and therapy for and management of the patient.

It is clear, too, that Dr. Gross is quite interested in the role *intuition* plays in this method of determining accurate diagnoses, especially in our medico-technological era. It should be noted here that the term 'diagnosis' is often used equivocally. It is a term with a variety of meanings, two of which are crucial to further discussion and so must not be confounded. The term 'diag-

Corinna Delkeskamp-Hayes and Mary Ann Gardell Cutter (eds.), Science, Technology, and the Art of Medicine, 199–210.

nosis' is commonly employed with reference to the *result* of a process of knowing — appearing as a category in the language of medicine, the disease nosology: "Mrs. Schmidt has aplastic anemia." But *diagnosis* frequently has as referent the *process* by which a physician arrives at a classification within the prevailing nosology: "Dr. Gross is a specialist in (is highly skilled in the process of) hematological diagnosis."[1] To avoid confusion, the term 'diagnosis' should be scrutinized each time it is used, so that these two principal senses be distinguished.

We call the *means* by which a diagnosis is established the *diagnostic process*. I take this to be the sense of *diagnosis* with which Dr. Gross is most concerned. He is concerned not so much with the results of medical decisions — that is, with the *classification* of diseases — as with understanding the intricate cognitive and conative *processes* that experienced physicans use when working toward establishing the appropriate disease category.

It follows from this interest that Dr. Gross is concerned also with those *errors* that surface all too often when something goes awry in the act of determining the correct disease category. I glean from his remarks that, generally speaking, Dr. Gross attributes error in clinical judgment to the confusion produced when clinicians carelessly relate (1) the objective findings revealed by rigorous scientific and technological tests with (2) the clinician's subjective impressions (the result of empirical and intuitive processes), and again with (3) the personal value system of the physician. Dr. Gross suggests the further problem that modern technological devices, constructed to provide objective findings, frequently do not serve to eliminate error in clinical judgment. Furthermore, he claims — and I have no good reason to doubt it — that technical instruments, tests, and methods frequently fail to help the physician determine the diagnostic category of patients who suffer from so-called 'purely psychosomatic' or 'psychogenic' conditions. In the process of determining the particular disease category of these patients — and there are many — technological advances in the medical armamentarium seem to give way to anamnesis, i.e., case histories and direct physical and psychological examination. Dr. Gross appears to conclude that anamnesis and such examinations do not require criteria and are not scientific in the pure sense, and he finds them subject to high observer variability. That is, the identification of a whole spectrum of clinical entities depends on the physician's empirical, experiential, or intuitive interpretation of audible, visible, and palpable clinical material.

Given this view, it is not inconsistent for Dr. Gross to conclude that psychosomatic diagnoses and classifications are largely "diagnoses of conjecture." Yet he goes on to say that they too can be "verified or refuted with scientific methods." Fundamentally, however, he believes that diagnostic classifications made primarily on the basis of case histories and direct examinations are chiefly the result of essentially *intuitive* processes, and that these processes are not as yet adequately understood by physicians or philosophers, in spite of the fact that the latter have proffered theories of intuitive knowledge since the Hellenic period, as indicated, for example, in Aristotle's *Analytica Posteriora*. The closing passage of this work reveals a surprising view of intuitive knowledge, one somewhat at odds with and in opposition to Dr. Gross's understanding.

> Now of the thinking states by which we grasp truth, some are unfailingly true, others admit of error — opinion, for instance, and calculation, whereas scientific knowledge and intuition are always true: further, no other kind of thought except intuition is more accurate than scientific knowledge... and since except intuition nothing can be truer than scientific knowledge, it will be intuition that apprehends the primary premises.... If, therefore, it is the only other kind of true thinking except scientific knowing, intuition will be the originative source of scientific knowledge. And the originative source of science grasps the original basic premise, while science as a whole is similarly related as originative source to the whole body of fact ([1], Bk. II, Ch. 19, 100b 5 –17).

It is clear from this passage that by *intuition* Aristotle is referring to an *immediate* mode of apprehension, not merely to a vague hunch, but to a cognitive process characterized by the absence of logical inference, as well as the absence of deductive justification. Moreover, for Aristotle, *intuition* does not refer to any mystical process, and he would not have us retreat into mystical or magical notions when using the term.

Although the philosophical tradition reveals a number of senses of *intuition*, only two need be distinguished here. First, *intuition* has been used to refer to a hunch or guess — that is, a belief not preceded by any inferential process. To be sure, we all recognize the existence of hunches, whether in ourselves or in physicians' assertions. This phenomenon is of particular philosophical interest and is thus in need of greater attention (to which I shall return shortly). The second sense, of course, is semise Aristotle's notion of intuition as *immediate knowledge of the truth of a premise, though devoid of any inferential process*. This view of intuition has also been problematic for philosophers simply because the paradigm of knowledge since Aristotle has been *inferential*. With inferential knowledge as the paradigm, it has been a puzzle to philosophers how knowledge can be obtained without having made oneself aware, through some process of inference, of a justification for a particular belief. After all, knowledge claims assume that particular beliefs held by certain persons — for example, physicians — are justified. Aristotle's view of intuitive knowledge is not, in this context, very comforting. For if we follow him here, we are compelled to accept the claim that the presence in the mind of the original starting points of knowledge is in principle inexplicable and must therefore be simply accepted without warrant. I do not know whether Aristotle was content with his account, but I fail to see why we should be. After all, in this view we would be committed to acknowledge a form of knowledge not preceded by inference, incapable of explanation, treated as if uncaused, and mysterious in a less than fruitful sense.

II. INTUITION AND THE ART OF MEDICINE

It is of some use, perhaps, to turn to a passage from Alvan Feinstein's *Clinical Judgment*. In his "Prologue", Dr. Feinstein reflects on his own medical education and remarks that

> a clinician performs an experiment every time he treats a patient. The experiment has purposes different from those of laboratory work, but the sequence and intellectual construction are the same: a plan, an execution, and an appraisal. Yet we had never been taught to give our ordinary clinical treatment the scientific 'respect' accorded to a laboratory experiment. Treatment was supposed to be an 'art', a humanistic application of established modes of therapy ([3], p. 14).

Feinstein goes on to discuss these clinical experiments and laments the fact that, some years before, he too had been taught to call such clinical experiments "art". In short, he was expected to "consign its intellectual aspects to some mystic realm of intuition that was 'unworthy' of scientific attention because it was used for the practical everyday work of clinical care" ([3], p. 14). At one point he notes that decisions regarding specific therapeutic maneuvers that proved beneficial to his group of patients "were based not on the intuitions of 'clinical experience', but on the analysis of quantified evidence" ([3], p. 6).

From these remarks, with which I hope Dr. Gross is in agreement, it is clear that intuitions in the sense of hunches prove of little value in medicine. Hence the difficulty I have with Dr. Gross's point of view is that he appears, in the end, to appeal to the traditional medical sciences — especially biology and pathology — in searching for the crucial data on the basis of which to proffer precise and accurate diagnoses. This is particularly true when he speaks of intuition as the opposite of precision and accuracy. Furthermore, in the end he seems to appeal to the same cognitive process of inductive reasoning that produces scientific or inferential truth for determining a patient's diagnosis. That is, he seems to agree with Feinstein that it is unfortunate that physicians are taught the application of experimental science in the laboratory, but not taught that these same methods are often quite "satisfactory for studying the live human beings who were the 'material' of our work at the bedside" ([3], p. 14). In *Clinical Judgment* and elsewhere [2], Feinstein has shown that *the clinician need not reconcile or integrate intuition and science* for he can always rest assured that there will continue to be improved scientific methods for observing and analyzing the intact human body.

Moreover, it is clear from his discussion that Dr. Gross is concerned with the following question: Can the thought (cognitive) processes underlying medical judgment, especially diagnostic judgment, be understood? It is also clear that he believes that the expert clinician's diagnostic judgment can be distinguished from the diagnostic judgment of less experienced clinicians on the basis of clinical intuition.

Because the precise meaning of the notion of 'clinical intuition' is itself unclear, one is not always able to provide a satisfactory answer to the question noticed above. For some, 'clinical intuition', even if a synonym for 'the art of medicine', does not refer to the opposite of scientific medicine, but is taken as identical to the 'thoughtful application' of medicine, and hence does not entail some mystical cognitive process forever unknowable. But for others, 'clinical intuition' is simply redundant, as when a critic claims that his doctor seems to arrive at diagnoses clinically, i.e., intuitively. The scientific or statistical process that leads to clinical diagnoses, therefore, is preferred to any intuitive process. Hence the formulation of intuitive judgments is generally rejected by those who advocate a statistical approach to clinical diagnoses, and who frame their analyses in terms of the concepts and functions of probability theory and the principle of uncertainty in virtually all matters clinical. This latter view is reflected in *Medical Thinking: The Psychology of Medical Judgment and Decision Making* by Steven Schwartz and Timothy Griffin [12].

They remark: "Those in the clinical camp believe that clinical intuition applied individually on a case-by-case basis will lead to the most sensitive decision making. Those favoring the statistical approach argue that case-by-case decision making is too easily biased." They prefer to set up *a priori* 'decision rules' that can then be applied in all similar cases" ([12], p. 8). An example of such a decision rule is: "Always order a throat culture when a patient presents with sore throat and swollen glands" ([12], p. 8). Such a decision rule is clearly a heuristic device, that is, a guide to inference, such as the classical maxim, *primum non nocere*.

Notwithstanding the extensive progress made in the accuracy of clinical diagnoses in the last few decades, in fairness to Dr. Gross one should point out that in spite of the presence of various cognitive biases among clinicians, "they have little effect on the *outcome* of actual clinical decisions" ([12], p. 81). Indeed, Schwartz and Griffin — citing the results of other cognitive researchers — conclude that even on the basis of cognitive intuition "... the outcomes of most of these doctors' diagnostic decisions were close to optimum. Moreover, since even misdiagnosed patients usually receive the same antibiotic treatment, for example, as those who are correctly diagnosed, judgment biases had little practical effect" ([12], p. 81). In short, intuitive and heuristic judgments in diagnosis more often than not do not lead to significant errors.

Hence even Schwartz and Griffin support, in part, Gross's thesis: "While these findings are encouraging because they suggest that clinical ['intuitive'] judgment is trustworthy, it should be noted that researchers typically observe only the outcome of clinical decisions, not the process by which they were reached." However, these authors would be critical of Gross's view since they also maintain that, "It is possible that doctors were making the 'right' decisions but for the 'wrong' reasons", and this is clearly unacceptable if the *process* and not merely the *naming* of diseases (also known as diagnosis) is to be founded on deductive and inductive reasoning.

III. INTUITION OR TACIT KNOWING

Dr. Gross's appeal to the usefulness of intuition in medical decision-making, diagnosis, and treatment has been anticipated by the extensive analyses of this mode of knowledge under the rubrics "subception", as striking confirmation of "tacit knowing" ([8], p. 143; [7], p. 7), "indwelling" ([7], p. 17), "sudden illumination" ([9], p. 96) or "hidden knowing" ([7], p. 21), terms employed by Michael Polanyi to signal the fact that "*we can know more than we can tell*" ([7], pp. 4, 23), but "by steps we cannot specify" ([8], p. 143). It is important immediately to add that by 'intuition' neither Gross not Polanyi is referring to the so-called supreme or immediate knowledge called 'intuition' by Leibniz, Spinoza, and Husserl. Rather, it is, in Polanyi's terms, "a work-a-day skill for scientific guessing with a chance of guessing right" ([8], p. 144).

Polanyi appreciated the fact that many view "the art of the expert diagnostician ... as a somewhat impoverished form of discovery," which, he adds, "we may put in the same class [as the] performance of skills, whether artistic, athletic, or technical" ([7], p. 6).

In distinguishing intellectual from practical knowing, wherein for Polanyi neither is ever present without the other, the clearest place to observe this is "in the art of diagnosing, which intimately combines skillful testing with expert observation" ([7], p. 7). It is important to note, however, that Polanyi does not suggest that we rely too heavily on intuitive or tacit knowledge, for it is limited by the fact that we may get to know a practical operation, but can not tell how it works ([7], p. 8). The enhancement of scientific medicine comes about when we come fully to understand the biochemical, genetic, and pathological mechanisms of disease, for example, and to more precisely understand "how" they work.

In short, tacit or intuitive knowing can only account for (1) the valid knowledge of a problem; (2) the physician's capacity to pursue it, guided by a sense of approaching its solution; and (3) a valid anticipation of the yet *indeterminate implications of the discovery or the more explicit knowledge* to be arrived at in the end, knowing, of course, that "a wholly explicit knowledge is unthinkable" ([8], p. 144). The discovery of tacit or intuitive knowing casts a pall, therefore, over the hope of discovering apodictically objective knowledge in any domain, including medicine, but this may prove a useful heuristic principle that leads to more modest epistemological claims by researchers and physicians alike.

The ideal of achieving a greater degree of *explicit* knowledge is, however, an ideal to be cherished and pursued by physicians. This is especially important because, for some, "diagnosis is almost a matter of opinion" ([14], p. 103). Since the "aetiology and pathology of most diseases is poorly understood," Dr. Henrik Wulff remarks, "the truth of the diagnosis is sometimes almost a matter of opinion" ([14], pp. 110–111). By relying on intuition in clinical diagnosis, Wulff warns, "We introduce more and more expensive diagnostic methods, but usually we have not proved their ultimate benefit to our patients" ([14], p. 115). Even a reliance on sophisticated probabilistic methods leading to the most probable diagnosis is not as weak as intuitive knowledge of the diagnosis, although the classical clinical maxim still stands: "When in doubt make a diagnosis" ([14], p. 118).

The force of this critique of the limits of clinical intuition, whether in the context of determining an accurate diagnosis or the most efficacious treatment option, is expressed by *the transition from empirical or experiential to rational diagnosis and treatment*. Whatever methods lead to greater stringency, such as various statistical methods, should be vigorously pursued. In short, as Wulff observes with respect to medicine in the mid-nineteenth century, "Clinical experience was unreliable, current therapy was useless, and consequently doctors retired from the bedside to the laboratory in order to study the causes of disease. This development was stimulated by the first results in experimental physiology" ([14], pp. 132–133). Interestingly, Wulff concludes that "It is important ... to realize that treatment can never be based exclusively on a knowledge of the causes and mechanisms of disease. Clinical experience still plays an important part" ([14], p. 133). The trouble is that all too often physicians are misled and deceived by uncontrolled experience. This concern has stimulated efforts to establish and execute carefully designed therapeutic trials to determine the most efficacious medical treatments for various diseases. In the end, such "controlled experience" often serves usefully to "*suggest* new forms of rational treatment" ([14], p. 134).

Indeed, Wolfgang Wieland, in "The Concept of the Art of Medicine" [13] cites the cogent view of the eighteenth-century theoretician, G. Selle, whose words are worth repeating: "Understanding the particular is the essential object of practice ... ; science always just deals with more or less universalized concepts. Understanding the particular remains the proper area of art and of immediate practice." This is not attained by science, Wieland adds. Furthermore, Selle continues, "The artistic insight for the particular can, after all, easily escape the learned and quick-witted doctor" ([13], p. 168).

Note that Wieland cites P. Diepgen, who maintains that physicians should defend the art of medicine against every attack by science. Thus, for Diepgen and Wieland the art of medicine "concerns rightfully what is called intuition, that is, the insight which is gained through spiritual vision, an inner suggestion of the moment." In short, *intuition* is identified with "spiritual vision" and may even refer to the "irrational part of the art of healing" ([13], p. 172). [It is interesting to note that Dr. Edmond A. Murphy, in his influential *The Logic of Medicine*, only once refers to 'intuition', and he employs it as synonymous with 'revelation' in his chapter on "Superstition" ([6], p. 162).]

Wieland continues, "To be sure, an adequate medical judgment about an individual patient is impossible without training in the medical sciences. The art of medicine is required in order to bridge the gap between the concrete condition of the individual patient and the universally valid laws and rules of medical science that as such do not reach the individual case" ([13], p. 173). "The art of medicine is in this sense a result of experience: experience underlies the very possibility of applying the universal rules of science to the care of the individual patient" ([13], p. 175).

But the point I wish to stress — having given what I believe is the proper, limited, and yet important place of the role of cognitive intuition in medicine — is perhaps best expressed by underscoring *four* considerations: (1) that medicine is necessarily an enterprise carried out under conditions of insufficient knowledge, and is thus always tentative; (2) that patients frequently present under duress when the benefits to be derived from extensive and sound biomedical and clinical research have not been determined, often because the research was not undertaken; (3) that intuitive or tacit knowledge is virtually immediate and insightful in its mode of expression, i.e., its speed far exceeds the processes of inductive and deductive reasoning; and (4) that information acquired by extensive clinical experience with particular patients tends to remain (as Polanyi has argued) tacit, hidden, or (as Wieland would have it), "below the threshold of consciousness" ([13] p. 107).

Thus we can at most safely conclude that the role of intuition in medical diagnosis, even in our era of sophisticated medical technology, is nothing less than an admission of our insufficient and tentative knowledge, but where, nevertheless, physicians are still compelled to act, and where what is understood as *intuition* is necessarily opaque. This in itself signals that we are perhaps obliged to move beyond tacit knowledge and intuition and compelled to initiate the painstaking, critical, rational, and controlled investigation of what is presently unclear or unknown in medicine.

IV. CONCLUSION

In my view, our technological era poses no special threat to medicine so long as it does not tend to disengage the clinician from his patient. Although technology can be abused, the real threat, as Feinstein puts it, is posed by those who give the label "science" to dissection of a patient's protoplasm but not to the careful description of his pain; by those who define a "disease" as the derangement of a molecule or the dysfunction of a cell but not the illness of a person; by those who perceive the patient as a reduced fragment and not as an intact whole; and by those who, maintaining false scientific dogmas and remaining intolerant of the dignity of the past, distract clinicians from attention to their ancient domain: the care of the sick ([3], p. 380). In short, clinicians of the future may come to appreciate that the process of clinical reasoning or clinical inference can be made a deliberate and articulate exercise and need not be construed as a nebulous process of intuition, taken in either of the senses noticed above.

Reliance on the basic sciences and the inferential knowledge gleaned from the comprehensive microscopic study of molecules and cells — their chemistry, physiology, and so forth — has resulted in the clinicians' belittling the scientific approach to symptoms, the relevance of epidemiologic factors, and even anamnesis. Feinstein and others worry, surely with reasonable justification, that schools of medicine in our time will be "converted from scholarly shrines for sick people to cenobitic citadels of cellular biology" ([3], p. 377). The balanced response is, of course, that there are quite warranted pedagogical problems that should rightfully worry us in this era of technological and scientific medicine. Equally, in seeking to understand the process of diagnosis, whatever its present enigmas, does not justify excessive pessimism even given the prevalence of physician variability in diagnosis and the limitations of empirical, "uncontrolled" experience.

The potential for inferential reasoning to which our particular species has become reasonably well adapted has already achieved a modest record of realization: We have learned how to incorporate intuition — so-called intuitive, tacit, or hidden knowledge — into the mainstream of inferential or explicit knowledge, dissolving its "hunch-like" and opaque characteristics in the process. In short, although I agree with Dr. Gross's general thesis that "empiricism and intuition are indispensable as well as equally valuable components of medical decisions" — that is, diagnostic decisions, I disagree with him that intuitive knowledge is "indispensable" in the sense that we shall never actually free ourselves from the naggings of beliefs that seem to resist all analytical justification. That is, though intuitive beliefs are "valuable components" of clinical diagnosis, as Dr. Gross suggests, I fail to see how intuitive beliefs can ever be relied upon to achieve the ends of accurate clinical diagnoses and the rational selection of treatments given the multiplicity of treatment options available to the clinician.

It is not so much, then, a matter of *connecting* or *integrating* intuition with logical reasoning or mathematical and scientific cognition, as it is a matter of *transforming* intuition into a systematic framework in which such "knowledge" is capable of being employed inferentially and explicitly. I see nothing in particular to be gained by appeals to associative psychology to unravel the structure of intuitive processes, or even to neurophysiology for final clues as to the structure of inferential and non-inferential reasoning ([4], pp. 131–132). On the other hand, I am not pessimistic about the recent appeal of information theory and the application of the computer in medicine, and I am not at all adverse to the idea that computer-aided diagnosis may be of great assistance in clinical decision-making, though as Wulff has indicated, "it is not possible to predict the future use of computer-aided diagnosis in clinical medicine" ([14], p. 119).

In conclusion, I remain skeptical that intuitive processes can be relied upon to provide new groupings of old information. There are better ways to explain, for example, the infrequent, rare, unusual, and yet correct diagnosis of a patient's complex set of signs, symptoms, and test results. For one thing, the clinician in such instances may have a true belief and yet not be said to have explicit knowledge. For although every case of knowledge is a case of true belief, the contrary proposition is, of course, false. An illustration provided by Bertrand Russell is apropos here: Consider the man "who looks at a clock which is not going, though he thinks it is, and who happens to look at it at the moment when it is right; this man acquires a true belief as to the time of day, but cannot be said to have knowledge" ([11], p. 154). Here there is a lesson for clinicians as well: One cannot always claim to have known because one turned out to be right. In such cases we should not be seduced into assuming that the clinician is simply an intuitive genius; we now simply know enough about the universe to chalk such "diagnoses" up to chance.[2]

Baylor College of Medicine
Houston, Texas, U.S.A.

NOTES

[1] The German language, less ambiguous than the English here, distinguishes between *Diagnostik* (the process of diagnosis) and *Diagnose* (the naming or classifying of a disease).

[2] I am very grateful to James Rundle, Ph.D., Emeritus Professor of English, Middlesex Community College, Connecticut, for his extremely helpful editorial suggestions on an early draft of this essay.

BIBLIOGRAPHY

1. Aristotle: 1963, *Analytica Posteriora*, G.R.G. Mure (tr), in W.D. Ross (ed.), *The Works of Aristotle*, Oxford University Press, Oxford.
2. Feinstein, A.R.: 1964, 'Scientific Methodology in Clinical Medicine', *Annals of Internal Medicine* **61** (1–4), 564–579, 757–781, 944–965, 1162–1193.
3. Feinstein, A.R.: 1967, *Clinical Judgment*, Robert E. Kreiger Publishing Co., Huntington, New York.
4. Gross, R.: 1976, *Zur klinischen Dimension der Medizin: Beiträge zu einigen Grundlagen und Grundfragen*, Hippokrates-Verlag, Stuttgart.
5. Gross, R.: 1992, 'Intuition and Technology as the Bases of Medical Decision-Making', in this volume, pp. 183–197.
6. Murphy, E.A.: 1976, *The Logic of Medicine*, Johns Hopkins University Press, Baltimore, Maryland.

7. Polanyi, M.: 1966, *The Tacit Dimension*, Routledge & Kegan Paul Ltd., London.
8. Polanyi, M.: 1969, *Knowing and Being: Essays by Michael Polanyi*, ed. by M. Grene, Routledge & Kegan Paul, London (*viz.* Chapter 11, "Tacit Knowing").
9. Polanyi, M. and Prosch, H.: 1975, *Meaning*, University of Chicago Press, Chicago, Illinois (*viz.* Chapter 2, "Personal Knowledge").
10. Russell, B.: 1959, *The Problems of Philosophy*, Oxford University Press, New York.
11. Russell, B.: 1962, *Human Knowledge: Its Scope and Limits*, Simon and Schuster, New York.
12. Schwartz, S. and Griffin, T.: 1986, *Medical Thinking: The Psychology of Medical Judgment and Decision Making*, Springer-Verlag, New York.
13. Wieland, W.: 1992, 'The Concept of the Art of Medicine', in this volume, pp. 165–181.
14. Wulff, H.R.: 1981, *Rational Diagnosis and Treatment: An Introduction to Clinical Decision-Making*, 2nd ed., Blackwell Scientific Publications, Oxford.

RAY MOSELEY

INTUITION IN THE ART AND SCIENCE OF MEDICINE

Persons practiced in the medical disciplines often simply see what is wrong with a patient and what ought to be done. They have the right intuition. There is a problem in assessing the significance of such claims. Do such persons simply have a non-inferential knowledge of what is at stake? Or are they so practiced that they, in an unconscious fashion, produce a proper inferential conclusion? Are such intuitions, in fact, simply unconscious inferences open to reconstruction and critical evaluation, so that in the end one would be able through a science of medical decision-making to improve upon their intuitions? What do intuitions have to do with physicians who say they are practiced in the art of medicine? Does being a good artist, rather than just a good scientist, require possessing intuitions? One's view of the significance of the intuitions of clinicians will have major implications for how one regards the science of medical decision-making as well as the contrast often drawn between the science and the art of medicine.

We have seen in the two preceding contributions by Drs. Gross and Spicker that physicians view 'intuition' as central to many medical diagnoses. Dr. Gross points out that as many as 50–80% of all medical diagnoses may involve an intuition by the attending physician. Unfortunately, it is not always clear what is meant by 'intuition' in the context of medical diagnoses.

In trying to unravel the meaning of 'intuition' in the context of medical diagnoses, it appears that at least three senses are present. I argue that in the end, however, there are only two proper senses and one mistaken sense of 'intuition' that may be found in these discussions of medical diagnoses. The mistake lies in falsely identifying some cases of inferential reasoning as cases of intuition. In the cases of the two proper senses of 'intuition', they generally do not, and more importantly should not, have a legitimate role in medical diagnosis.

The first and somewhat pedestrian sense of 'intuition' that may be found in discussions of medical diagnoses is, as Dr. Spicker points out, commonly used to refer to mere "hunches" or "guesses".[1] Intuition, as exemplified by mere guessing, is not a reliable or accurate method with which to make good medical diagnoses. The chance of picking the correct diagnosis from among the myriad disease possibilities is obviously quite slim.[2] I suspect, if this

Corinna Delkeskamp-Hayes and Mary Ann Gardell Cutter (eds.), Science, Technology, and the Art of Medicine, 211–218.

sense of 'intuition' is clearly distinguished from other possible meanings, then most physicians would reject, and wish to avoid, this characterization of the method whereby they make diagnoses. Similarly, it seems unlikely that patients would find such intuitive diagnoses very reassuring. This characterization of intuition clearly fails to capture the sense of intuition which Dr. Gross sees as intimately involved in many medical diagnoses. It fails to account for the fact that physicians appear to feel more certain about their diagnoses than a "guess" implies. Furthermore, they do a better job of diagnosis than mere "guesses" could possibly account for.

A second sense of intuition has been widely recognized and discussed by philosophers since the time of Aristotle. 'Intuition', in this sense, is the *certain*, and *immediate* apprehension of true facts about the moral or material world. Many philosophers have thought that intuition was a prominent method by which one acquired knowledge. For example, Duns Scotus held that men's souls were not restricted to merely observable material knowledge but thought that they were also capable of intuitive knowledge. In the nineteenth century, Wilhelm Dilthey postulated that *Verstehen*, i.e., intuitive understanding, was crucial for determining the true nature of reality [3]. 'Intuition' has also held an historically important position in moral philosophy. Ethicists such as Henry Sidgwick, H.A. Prichard, and W.D. Ross have all seen intuition as essential to understanding moral truths.[4]

Intuition, in this sense, does not depend on inductive or deductive reasoning processes for arriving at the correct conclusion.[5] Thus, one may claim that one intuitively *knows* that abortion is wrong, without conclusive supporting arguments, or that a patient has had a myocardial infarction, without knowing enough of the relevant presenting symptoms to make a probable diagnosis.[6] In this sense, an intuitive medical diagnosis is a certain and non-inferential apprehension of the true nature of a patient's problem. This may be what is meant by physicians when they claim that they have an "immediate grasp" of the correct diagnosis.

It has been suggested that an account of 'intuition' in this second sense when viewed in the medical context may be equivalent to that aspect or character of medicine which is referred to as "art".[7] This interpretation is based on the fact that the "art" of medicine is generally seen as being disconnected from any *quantifiable* aspect of medical science. However, this view overlooks the fact that the art of medicine is something that is learned from experience, namely, that of watching other experienced physicians diagnose and interact with their patients. Dr. Edmund D. Pellegrino has strongly argued that medicine as an "art" should not be understood in any such formal intuitive sense. This is because "medicine as art" involves inferential reasoning, "the knowing of what to do, how to do it, and why one does it" ([8], p. 190), just as "medicine as science" does. As Dr. Pellegrino points out, the "art of medicine" is essentially and inferentially based on experience, thus, it does not seem to be intuitive in the second sense.

It should be noted that 'intuition' in this second sense has not been considered the "paradigm of knowledge" acquisition. That honor, as Dr. Spicker points out, belongs to 'inference' as a mode of acquiring knowledge. The lesser status of 'intuition' is not merely due to a historical accident, as Dr. Spicker seems to suggest. Rather, it is because intuition in this traditional philosophical sense suffers from several metaethical and metaphysical flaws.[8] Two of these problems, in particular, are near fatal to intuitive medical diagnoses.

The first is that human beings simply do not seem to possess or "have" an intuition; its source cannot unequivocally be identified as "my intuitive faculty". As William Frankena has pointed out,

> If we had a distinct faculty which preceives what is right or wrong, and speaks with a clear voice, matters might still be tolerable. But anthropological and psychological evidence seems to be against the existence of such a faculty, as does the everyday experience of disagreement about what is right in particular situations ([5], p. 23).

In other words, it is simply not clear that we can identify when or if our judgments are the results of some sort of intuitive understanding.

Second, there seems to be no clear way in which satisfactorily to resolve conflicts among differing 'intuitions'. Philosophers have attempted to resolve this problem in the moral realm on the more general level of conflicting moral principles (where we might suspect that agreement would be most easily achieved) either by proposing an intuitively supported priority ranking of moral principles or by arguing that intuitively verified moral principles never in fact conflict with one another. These attempts at a resolution of the problem of apparent conflicts among intuitions rest on a further assumption, namely, that there will be agreement among reflective persons over the priority ranking or the lack of conflict between basic principles.[9] Such agreement simply has not been forthcoming.

This lack of agreement is an even more obvious problem when it comes to intuitive judgments about particulars. If Dr. Jones intuits the patient's problem as X and Dr. Smith intuits the patient's problem as Y, they are simply at an irresolvable impasse, because there is no way to adjudicate between the competing intuitions. The impasse may, of course, be eventually resolved when more symptoms manifest themselves or when the report of the pathologist arrives (although at that point it may be too late to benefit the patient).

The claim that physicians often rely on intuition in their medical diagnoses appears doubtful if we understand intuition as consisting either of a mere "guess" or "hunch", or as a certain and immediate apprehension of the truth. It seems that neither of these standard senses of intuition is the sense that is employed by Dr. Gross, or that is employed in the vast majority of medical diagnoses.

A third way in which intuitions seem often to be employed in the process of medical diagnoses springs from the incredible complexity of our inductive inferential processes and our often naive view of this process. Simply put, an inductive inferential process is one in which we begin with a particular set of evidence or clues and through a series of (inferential) steps based on knowledge concerning the probable causes and effects of these clues, arrive at a probable conclusion. For example, if a patient presents with severe substernal pain, which the patient describes as a crushing sensation, and complains of dyspnea and weakness, the physician may conclude that the patient has had a myocardial infarction.

Now it appears obvious that this example of an inference could not easily be mistaken for intuitive understanding in the previously discussed senses. This is because this inference is connected to empirical symptoms and signs, and on knowledge about the relationships between symptoms and their causes. An intuitive medical diagnosis in the first or second sense, on the other hand, does not rely on empirical data in the same way.

Physicians clearly depend heavily on this type of simple inductive inference for many of their medical diagnoses, and it seems that in these cases there is little problem in distinguishing inferences from intuitions. On the other hand, there are many diagnoses that do not seem to fit this simple model. The physician often cannot readily point to symptoms or tests on which he or she based a diagnosis. In such cases, intuitive understanding appears to be an appealing explanation of the process whereby the physician arrived at the diagnosis.

I believe we can account for often not being readily able to identify the data or the reasoning process used in arriving at many of our conclusions, and hence the intuitive explanation for that conclusion, with a fuller appreciation and understanding of the often complex process of inference.[10] To move toward a fuller account of the process of inference, it is useful to look at a distinction made by Michael Polanyi between two types of knowing.

> There are vast domains of knowledge — that exemplify in various ways that we are in general unable to tell what particulars we are aware of when attending to a whole, that is, to a coherent entity which they constitute. Thus we discover that there are two kinds of knowing which invariably enter jointly into any act of knowing a comprehensive entity. There is (1) the knowing of a thing by attending to it, in the way we attend to an entity as a whole, and (2) the knowing of a thing by relying on our awareness of it, in the way we rely on our awareness of the particulars forming the entity for attending to it as a whole ([10], p. 119).

These two ways of knowing are clearly distinct and seem to be quite often exclusive of one another. Thus there is the tendency for us to focus either on the entity itself or on the particulars that comprise the entity, and it is difficult to do both simultaneously. Polanyi's observation has been supported by Gestalt psychology. It has shown, especially in the realm of visual data, that when we focus our attention on particular empirical data we lose awareness of the whole, and conversely when we focus on the whole we lose awareness of particulars ([10], p. 119).

These tendencies, it would appear, are also at work in the process of medical diagnoses. The physician, when making a diagnosis, may tend to focus on the well-being of the patient as a whole. This focus on the patient's problem as a whole would, in many cases, if we may legitimately extrapolate from the Gestaltian observation, obscure one's conscious recognition of each particular bit of distinct empirical evidence on which the diagnosis was made. As Polanyi points out, "to diagnose a disease is to grasp the joint meaning of its symptoms, many of which we could not specify; so we know these particulars only by relying on them as clues" ([10], p. 123). This tendency does not mean, however, that in the physician's understanding of the patient's problem he or she has failed to base the diagnosis in a reliable way on distinct pieces of symptomatic data. The physician's inferences may very well be legitimate and yet the physician may be unable to easily identify precisely the inferential steps he or she made in reaching the particular diagnostic conclusion. Without the intermediate steps of an inference being readily apparent, the "intuitive" or "art" explanation for one's conclusion becomes quite attractive.

That which Dr. Gross calls 'intuition' should probably be understood to be cases of inferential reasoning in which the physician is not consciously or explicitly aware of the particular flow of his or her reasoning. If so, they are not truly intuitions, (if there are such things as intuitions) but only mistakenly labeled as 'intuitions'. In much the same way, it appears that medical diagnosis is often labeled "art" when, in fact, it is essentially an inferential, data-based science. Only with these understandings can we explain the ability of physicians to teach medical diagnosis and the recent successes of computer diagnostic programs such as Caduceus I and Caduceus II.[11]

A closer examination of Dr. Gross's analysis of medical diagnoses reveals that his description of medical intuition could quite reasonably be understood in terms of unconscious inferences. According to Dr. Gross, 'intuitions' are (1) "combinations of new impressions and experiences with those already known"; (2) "different linking of old and new thoughts"; and (3) "modification of information arrangement". Consider whether 'intuition' is an accurate description of these three processes that may be involved in a diagnosis. It seems that it is much more likely that the physician recognizes, although not necessarily consciously, various empirical clues which lead him or her through a series of inferential steps to a conclusion.

This interpretation is supported by three observations. First, it is clear that clinical experience is necessary and extremely valuable in the education of physicians. Practices in medical education strongly suggest that it is through clinical experience and not through certain and immediate apprehension that the physician learns to recognize important empirical clues and the empirical link between a particular clue and the patient's particular problem. Second, if one takes the time to make a conscious and reflective effort to examine closely an "intuitive diagnosis" or judgment, it seems that one can almost always trace the inferential steps that went into that diagnosis or judgment. Third, if no trace of inferential steps is apparent upon a formal analysis, such as with a Bayesian analysis, or if some new evidence counter to the intuitive conclusion is uncovered with a formal analysis, then at the very least the certainty enticingly offered by "intuitive diagnoses" will vanish.

Confidence in the ability of physicians to make accurate diagnoses is necessary. An 'intuitive' grasp (in the second sense) of the patient's problem suggests, at first glance, an attractive and reassuring certainty of diagnosis for both the physician and the patient. Medicine simply cannot afford this delusion.

Physicians' general efficacy in the practice of medical diagnoses will be enhanced and physicians will certainly be on better philosophical footing if they avoid the use of intuitions in the sense of a "mystical" certain and immediate apprehension of the truth. Describing medical diagnoses as 'intuitive' carries conceptual problems that will continue to haunt medicine. I agree with Dr. Spicker, that when there is legitimate doubt about a medical diagnosis, time is better spent by the physician in attempting to unravel the complex inferential process involved, rather than continuing to try to resolve conflicts at the level of impenetrable intuitions. Only in this way is one likely to learn how to avoid mistaken conclusions and to make better medical diagnoses.

University of Florida College of Medicine
Gainesville, Florida, U.S.A.

NOTES

[1] Dr. Spicker interprets Dr. Gross as understanding "intuition" in the context of medical diagnosis in this way. I argue in this paper that this interpretation may be somewhat harsh, since Dr. Gross may be interpreted as referring to unconscious inferences and not intuitions.

[2] This is not intended to imply that a medical diagnosis to be acceptable must be infallible. Medical diagnoses may necessarily involve some degree of fallibility. See Gorovitz, and MacIntyre [6].

[3] C.S. Peirce also made what might be identified as "intuition" as an important part of his theory. Peirce's conception of intuition, which he labels "abduction" is "weaker", than this second traditional sense of intuition. Although abduction is some sort of "immediate insight", Peirce does not also incorporate "certainty" into its meaning, but only what is "suggestive of a possibility" [9]. The notion of "certainty" that attaches to intuition, in this second sense, is a very important aspect of intuition's attractiveness since it is this "certainty" that sets it apart from the "merely" probable knowledge that inference offers.

[4] Although all of these philosophers held intuition to be very important, they did not all understand intuition as applicable to the same levels of moral inquiry. Sidgwick saw intuition as validating only the very basic moral principles, whereas Ross understood intuition as able to validate particular moral judgments ([11], pp. 20–21).

[5] This particular philosophical sense of "intuition" should be clearly distinguished from the specific technical sense of "intuition" that Immanuel Kant, for example, uses when referring to our apprehension of the categories of space and time.

[6] Intuition does, of course, rely on some empirical data in the sense that empirical data sets the basic parameters of the intuitive judgment, e.g., one could not intuitively know a person had cancer if he or she had never heard of the disease, cancer. Intuitions, in the second sense, as Dr. Gross points out, may also follow a series of inferential steps or be imbedded between a break in inferential steps.

[7] See, for example, Clouser [1], and Clouser and Zucker [2].

[8] See for a xbrief discussion of these problems Frankena [5].

[9] For example, one attempt was proposed by Sir David Ross. See Ross [11].

[10] Although this simple inference model belies the fact that most inferences are of varying accuracy and complexity, this does not mean that it is impossible to unravel these processes. Variations of the inference process have been well classified and analyzed. See, for example, Murphy [7]. Furthermore, very complex diagnostic inferences have been helpfully explained by the probabilistic or Bayesian models of medical diagnosis. See Wulff [12].

[11] The art of medicine seems to be centered in the doctor-patient relationship where the physician calls on his or her talents to communicate effectively, sympathize and understand the patient's needs. Computer diagnostic programs cannot and do not attempt to duplicate this art. As Alvin Feinstein says, "The computer cannot participate in the true art of medicine. It has none of the subtle sensory perception, intellectual imagination, and emotional sensitivity necessary for communicating with people and giving clinical care" ([4], p. 369).

BIBLIOGRAPHY

1. Clouser, K.D.: 1977 'Clinical Medicine as Science: Editorial', *The Journal of Medicine and Philosophy* **2**, 5–6.
2. Clouser, K.D., and Zucker, A.: 1974, 'Medicine as Art', *Texas Reports on Medicine and Biology* **32**, 267–274.
3. Dilthey, W.: 1883, *Einleitung in die Geisteswissenschaften*, Teubner, Leipzig.
4. Feinstein, A.R.: 1967, *Clinical Judgment*, Robert E. Krieger Publishing Company, Huntington, New York.
5. Frankena, W.: 1973, *Ethics*, Prentice-Hall, Inc., New Jersey.
6. Gorowitz, S. and MacIntyre, A.: 1976, 'Toward a Theory of Medical Fallibility', *The Journal of Medicine and Philosophy* **1**, 51–71.
7. Murphy, E.: 1976, *The Logic of Medicine*, Johns Hopkins Press, Baltimore.
8. Pellegrino, E.D.: 1979, 'The Anatomy of Clinical Judgments', in H.T. Engelhardt, Jr., S.F. Spicker, and B. Towers (eds.), *Clinical Judgment: A Critical Appraisal,* D. Reidel Publishing Company, Dordrecht, pp. 169–194.
9. Peirce, C.S.: 1940, *The Philosophy of Peirce*: *Selected Writings*, Harcourt, Brace and Company, New York.
10. Polanyi, M.: 1974, 'Faith and Reason', in F. Schwartz (ed.), *Scientific Thought and Social Reality: Essay by Michael Polanyi*, International Universities Press, New York, pp. 116–130.
11. Ross, W.D.: 1973, *The Right and the Good*, Clarendon Press, Oxford.
12. Wulff, E.: 1976, *The Logic of Medicine*, Johns Hopkins Press, Baltimore.

RAPHAEL SASSOWER

TECHNOSCIENCE AND MEDICINE

I

In an age that reconstructs itself historically and that attempts to deal with a postmodern condition in Lyotard's sense [8], one would expect a recurrent interest in the scientific credibility of all areas of research and practice. If science is to mean anything special at all, one could argue, then it must distinguish itself from all other discourses and deliver highly valued "goods" in an unqualified manner. In short, the reliance that can be expected of science and its attendant technologies surpasses in principle and in practice all other forms of expression generated culturally, from the film industry to religions. The privilege granted science and technology is a modern phenomenon, one that can be traced to the vision of the Enlightenment, where reason and empirical experimentation were joined to apprehend diverse data and cull specific results and predictions to protect and improve the human condition. The scientific revolutions that began in the seventeenth century transformed not only the intellectual arena, but also the broad cultural expectations of the eighteenth to the twentieth centuries. To a certain extent, and without disregard for numerous challenges — Romantic and Marxist — to the validity of science and its methods of inquiry, contemporary society remains enchanted and captivated by the great promises of the scientific era (especially since many of these great promises have been fulfilled, such as longer life expectancy, efficiency in food production, improved transportation, and communication techniques).

The love affair modern culture has had with science is neither uniform nor without its problems. To begin with, there is no "science" in a generic sense, but instead a variety of disciplines whose claims for scientific status have been examined and charted over the years. Following August Comte [1], there is even a codified hierarchy of the sciences, so that mathematics is all the way at the pinnacle of scientific certainty and credibility, followed by physics and then chemistry, all the way down to less prestigious disciplines such as sociology and political economy. Moreover, even when one focuses on particular disciplines, it is unclear whether or not they fulfill the criteria of

Corinna Delkeskamp-Hayes and Mary Ann Gardell Cutter (eds.), Science, Technology, and the Art of Medicine, 219–228.

demarcation between science and pseudo-science (e.g., in Popper's terms [13]): do they use induction or deduction?, can they quantify the entire spectrum of their date?, can they model their results and formulate precise enough predictions?, are their predictions testable, can such predictions be confirmed or falsified? In addition to methodological questions, those that can be classified as theoretical in nature, there are numerous questions whose answers rely on technological innovations, some of which, incidentally, impinge not only on practical matters but also on the very methodological principles that direct and test the research of the sciences as a whole. Herein lies my agreement with Peset [12]. These technical issues, as I have argued elsewhere [11], diffuse the ability to hold onto the binaries of theory and practice and of science and technology. For example, the development of the theory of probability has not only enhanced our understanding of gambling and the financial protection from overseas shipping trade [19], but altered the naive conception of induction attributed to Francis Bacon into a sophisticated probabilistic-induction associated with Carl Hempel ([6]: also, see Fagot-Largeault [2]). That is, the tools and instruments used by science, may they be Galileo's telescope or Hempel's probability, are not mere derivatives of scientific contemplations, but in fact bring about and influence scientific reconsideration and the continued revision of scientific theories and models so that, as Toulmin convincingly argues, the "modern", spectator vision of science is replaced with a "post-modern," participatory vision [18].

Perhaps the brief backdrop I painted above is superfluous for those familiar with the history and philosophy of science and technology; yet I think it is a useful heuristic to answer my first question: why focus on medicine when one asks questions about science and technology? Medicine provides a fascinating case study for the deconstruction and reconstruction of the discourses of science and technology because of the way in which the development of medicine problematizes the categories of science and technology. First, there is a history (even a genealogy) of debates concerning the classification of medicine as either science or art. Galen's treatise of the second century surveys many of the issues that haunts contemporary commentators to the present in relation to the scientific basis of medical practice: empiricism vs. rationalism, experience vs. speculation, diagnosis vs. prognosis, individual reports vs. generalizations, principles of operation vs. laws of practice [3]. Second, medicine has always benefited from technological advances so that the entire practice of medicine has been transformed from, for example, surgical instruments to the ability to transplant organs. In this respect, theory (science) and practice (technology) are intimately intertwined; separating the two is a violation of the historical record and a misappropriation of discursive binaries from elsewhere. Third, since medicine at once relies heavily on scientific methods and data and depends on intuition (i.e., personal and speculative insights), it cannot simply situate itself in a comfortable position along Comte's hierarchy of sciences. I use the term 'intuition' to distinguish between following the rules of scientific methods from a personal insight, instead of Gross's sense of intuition as a conceptual leap of faith [5]. In this respect, medicine draws from the studies of biology and chemistry, psychology and sociology, yet deals with social ills (culturally defined and expressed) that manifest themselves in individual patients. The issue here is not (as in my first comment) whether medicine is science or not, but how scientific it is in fact (assuming that the balance between or mixture of science and its cultural context is shifted towards science along Comte's conception of the hierarchy of science).

The reasons given above for focusing on medicine would make it seem that, if anything, one should avoid discussing the scientific and technological aspects of medicine and limit one's examination to particular instances in which medicine has been successful in eradicating a disease or treating an illness. My own view is that especially because of its perennial ambiguities, medicine is a prime candidate for the examination of broad philosophical and practical issues endemic to the discourse of technoscience (the diffusion of the binary of science and technology), as Latour calls it ([8], pp. 174–175). Instead of putting medicine under the microscope, why not look from medicine's perspective through a prism that would turn, so to speak, the microscope into a telescope with which to view the technoscientific horizon? It is in this sense that I agree with Tsouyopoulos [20] that medicine engages in an active pursuit of goals.

II

Assuming that medicine provides an interesting focus for the examination of traditional views concerning science and the roles it plays in our culture, one may ask more specifically: what difference has technology made both in the practice of medicine and in its conceptual models? That technological innovations have made a difference in the practice of medicine is clear from any of the instruments that appear in a doctor's office (e.g., ophthalmoscope), a hospital (e.g., X-rays machines), or a specialized clinic (e.g., MRIs). What sometimes escapes public notice is the manner by which such technical innovations have disrupted and transformed conceptual models. Here Wieland [22] targets an important issue.

Historical examples abound, from surgical techniques that changed the model of the circulation of blood to laboratory capabilities that first established a model concerning the immune system of the body. These historical examples portray outdated and superseded models as if they were mere superstitions and speculations concerning, for example, the efficacy of bloodletting as a technique to drain poisons from the body. Medical practices, based on technological innovations, took hold so permanently in these respective areas of research that attendant conceptual transformations have been accepted without much ado, as if these transformations were "natural" developments, proceeding from one model to another.

With contemporary examples, the narrative is not so easily reconstructed or brought to a closure (at least in the sense of a consensus concerning the circulation of blood or that the body has an immune system, however defined). Perhaps a reconstruction of contemporary examples in medicine is more difficult because of the realization that the reconstruction may continue indefinitely. I wish to mention briefly two contemporary examples, one more complex than the other, that will illustrate how technology changes conceptual models in medicine. The first example concerns the definition of death. According to U.S. law, the only person who can pronounce a human being dead is a physician. Moreover, when the courts are involved in cases that require a determination of time of death, for example, they refer back to the medical profession and not to the police, requesting an autopsy and a medical confirmation of time and cause of death. Now if science deals with empirical certainties, and if medicine is scientific, and if human death is an empirical fact, then medicine should be able to be certain about one's death. How certain can medicine be about one's death?

Different methods of examination have been used historically, from mirrors to salts, from smelling a body's decomposition to the use of a stethoscope. But what about certainty? To be sure about death, any physician would agree even today, a body should be left alone long enough to observe its decomposition. This level of certainty has a price tag: allocation of space, terrible odors, and a lengthy observation process. So, some shortcuts have been proposed. If blood is the lifeline of one's body, providing oxygen to every part of the body, and since oxygen is necessary to sustain the body, then when blood stops flowing the body dies. All one needs to examine, then, is the flow of blood, either by pressing on veins or by listening to the heart pump blood to the rest of the body. A similar model of empirical testing can be constructed in terms of lungs (and oxygen) and their essential function in the maintenance of the body.

Medical models of the role of blood and lungs capture a certain view of the human body and a certain view of human life. They frame not only methods of treatment but also the very foundation of the cycle of life, the way a religious doctrine or a scientific theory would express it. These models in effect determine not only the parameters of epistemological concerns, but also the epistemic or ontological and metaphysical assumptions with which the medical community works. Put differently, and with Schäfer [17], epistemic assumptions are influenced by non-epistemic ones (e.g., treatment techniques, cultural contexts, values, and political goals).

What happens to a person who is hit by a car and is rushed to a hospital in an ambulance, having been connected on route to various machines that pump oxygen into the lungs and ensure the circulation of blood through a transfusion of liquids? Upon arrival at the hospital, the patient's lungs "breathe" and the blood "flows". Is that person dead or alive? The intervention of machinery is no longer incidental or ancillary but primary: human frailty is enhanced by equipment designed specifically to overcome bodily weaknesses, temporarily or permanently. Given the models of the role of blood and lungs, the person in question is "alive," plain and simple. But what about the intrusion of equipment into the life cycle of the human body? Is the intrusion natural or artificial? If artificial, does it not put into question whether or not the patient is in fact alive? Just because the equipment is labeled artificial does not disqualify the judgment that a patient is alive, in a similar fashion that wearing spectacles does not disqualify the observation of a near-sighted person. So the debate cannot rest on the question of the character of the equipment, namely, that it is "artificial." What other tests would help decide the seemingly simple question of death?

I would make the claim that technological innovations have rendered the blood and lungs models useless and have therefore required (and not merely enhanced, as I have argued elsewhere [16]) a reconception of the criteria by which medicine judges a body to be dead or alive. The criteria have shifted to the testing of the brain and its functions. Death, as is currently defined in accordance with several sets of criteria (e.g., Harvard or those defined nationally [14]) is defined in terms of the brain. A detailed account of the historical, clinical, epistemological, and ethical concerns that surround the definition of death can be found in [16]. No matter what physical condition one's lungs and heart are in, if there is a certain amount of brain damage (paralysis or dysfunction), then a patient can be pronounced dead. Back to our accident patient.

The patient is hooked to various machines, the blood circulates, the lungs move and breathe, but the brain is considered "dead" because it failed certain tests (undertaken by different machines, e.g., an EEG). At that point the physician is in a position to pronounce the patient dead and remove all equipment from the room. The removal of the equipment becomes unproblematic only if one realizes that these machines were in fact *organ*-sustaining and not, as so commonly perceived, *life*-sustaining. For all we know, the patient died at the scene of the accident, and it was only the demands for "certainty" in the pronouncement of death, involving extra brain tests (for insurance purposes or for the possibility of harvesting some organs [4]), that verified what was true all along: the person is dead! If science raised the specter of certainty, and if the sacredness of life (given adherence to sacred texts) demands that its cessation would be pronounced with certainty, it is technology that enables medicine to fulfill its cultural obligations, whether these obligations are understood legally or ethically (also see Wieland [22]).

The example of the determination of death in light of technological developments concerning brain functions is simple by comparison with the configuration of DNA structures or the genetic coding of humans. I will only mention in this context that the "human genome project" [21] will not merely aid in treating specific diseases, but will require a rethinking about evolution models of the species and more narrowly of genetic hereditary (also see [7]). Given the extent of human genetic variation and the large number of people who will be found to have disease-related genes, how can we understand the concept of genetic disease, genetic disorder, and abnormality? The age-old debate of nature versus nurture may be laid to rest once a full genetic "map" is drawn and tested and human interference in "natural phenomena" for purposes of averting (medical and personal) problems would become not only a concern of speculative philosophers but of the public at large.

Genetic engineering, as it is being called, puts into question the very foundation of the medical profession: is medicine supposed to cure and treat patients' diseases and illnesses or is it supposed to prevent these diseases and illnesses from ever appearing? Is medicine a caretaker or a controller of the body? What authority should medicine command? Whose political agenda should medicine follow? Some of these questions were brought up during the Nuremberg Trials after World War II when the world encountered the horrifying experiences of Nazi experiments in the name of (scientific) medicine [9]. These questions were thought to have been sufficiently answered so that the abuses of medicine would never be repeated. However, the genetic revolution has reintroduced these questions in different guises but with the same urgency. The promises of science (and therefore of medicine) overshadowed for the longest time possible fears and anxieties. But these anxieties are never fully forgotten, and their reappearance cannot be ignored.

III

The third question I wish to raise concerns the cultural attitude that should be embraced in the face of contemporary medical developments: are there reasons for optimism in light of the progress made in medicine? While technological developments seem to increase our options and open opportunities to rethink medical practices, both in research and the clinic, it seems that the sophistication and complexity of medical technologies have also had the opposite effect by forcing medical training to become overly specialized and narrow-minded with its focus on the latest scientific research. Not that one would want physicians to be poorly trained or not keep up with the latest developments in their fields, but at what costs [15]?

If medical education is a bellwether of attitudes pervasive in the profession, then there is a great deal to worry about. Though there is some broadening of the curriculum and the addition of social and ethical issues into the first two years of medical school, there is still a prejudice concerning the importance of mastering materials that may be never referred to in one's clinical settings. In short, though the apprenticeship method is valued and of great importance in the education of physicians, there is no clear indication of the priorities of training, so that a great deal of time and energy is still spent on the "scientific" basis of medicine and very little on its *techne*.

The promises of science and medicine are tempered, in my view, by the possible loss of broad perspective that comes with the emphasis on the technological gadgets that continually appear on the hospital's doorsteps. Not that these gadgets cannot save lives; they often do. But the use of these gadgets does not guarantee what possible conceptions and models can be conjured; for this to happen, a broad education, including sociology and history, psychology and philosophy, must be pursued alongside one's learning about science. My concern here is with some sort of technophilia that overlooks the potential and the hazards associated with the technological apparatus of medicine. It is a concern not limited to medical pedagogy, but one that extends to the role of physicians as experts and public servants at the same time.

Because, as I said earlier, medicine provides an interesting case study of a discipline with blurred boundaries with regard to the conceptions of science and technology within the natural and social sciences and the humanities, it should — as a discipline — stop playing "catch-up" with the scientific establishment (that includes, for instance, astrophysics and biochemistry). Instead of craving for scientific recognition and credibility, it is in a unique position to challenge the very criteria by which science is evaluated. That is, it could shift the seat of authority from the scientific hierarchy as established by Comte, and introduce alternative parameters according to which to evaluate a discipline or an area of research. Consider, the medical establishment could argue that successful treatments of diseases, for example, depend on a successful implementation of new technologies or old technologies in new ways. This dependence may be unique to medicine or may be shared by other disciplines, just as the apprenticeship method of training may be uniquely fit for medicine but not as much for other areas of research and practice.

New technologies bring with them the possibility of undermining traditions and hierarchies and questioning positions of power so readily granted. In judging medical practices differently from other forms of scientific activity (from experimentation in quantum mechanics to radon testing and launching space shuttles), there is an implicit recognition that not all scientific endeavors follow the same rules of conduct nor that they all fulfill the same criteria of (scientific) demarcation. The contextualization of judgments and references to specific sets of rules and criteria would diffuse the power relations and structure that seem to pervade the scientific community in general and the medical community in particular, a community still held spellbound by scientific rhetoric. In following this advice, the pressure to emulate methods of inquiry that may be inappropriate for medical research and clinical practice would be eliminated.

The optimistic note with which I would like to end relies on a transformation of orientation, a change of heart, so to speak. This is quite a bit to ask of those heavily invested in positions of authority that appeal to preconceptions about the validity and power of science. Yet, the possibility of a change of orientation can be realized through the training of physicians. Instead of providing students with ready-made answers to medical questions, additional effort should be made to suggest novel and unproven answers, and to encourage the continued questioning not only of specific clinical results but of the entire medical edifice. Technoscience, as it conditions and is conditioned by medicine, reaches the main themes that characterize the postmodern condition, such as a relentless skeptical and critical engagement with any claim to a permanent foundation and an appeal to authority. Medicine can lead the way in construing different models of interpretation and practice instead of following the scientific pack.

University of Colorado
Colorado Springs, Colorado, U.S.A.

BIBLIOGRAPHY

1. Comte, A.: 1975, *Auguste Comte and Positivism*, ed. by Lenzer, G., Harper & Row, New York.
2. Fagot–Largeault A.: 1992, 'On Medicine's Scientificity – Did Medicine's Accession to Scientific "Positivity" in the Course of the Nineteenth Century Require Giving Up Causal (Etiological) Explanation', in this volume, pp. 105–126.
3. Galen: 1985, *Three Treatises on the Nature of Science*, M. Walzer and M. Frede trans., Hackett Publishing, Indianapolis.
4. Gaylin, W.: 1974, 'Harvesting the Dead', *Harpers* **249**, 23–30.
5. Gross, R.: 1992, 'Intuition and Technology as the Bases of Medical Decision-Making', in this volume, pp. 183–197.
6. Hempel, C. G.: 1952, 'Fundamentals of Concept Formation in Empirical Science', *International Encyclopedia of Unified Science*, Vol. II (7), 1–93.
7. Holtzman, N. A.: 1989, *Proceed With Caution*, Johns Hopkins University Press, Baltimore.
8. Latour, B.: 1987, *Science in Action*, Harvard University Press, Cambridge.
9. Lifton, R. J.: 1986, *The Nazi Doctors: Medical Killing and the Psychology of Genocide*, Basic Books, New York.
10. Lyotard, J.-F.: 1984, *The Postmodern Condition: A Report on Knowledge*, G. Bennington and B. Massumi, University of Minnesota Press, Minneapolis.
11. Ormiston, G. and Sassower, R.: 1989, *Narrative Experiments: The Discursive Authority of Science and Technology*, University of Minnesota Press, Minneapolis.
12. Peset, J. L.: 1992, 'On the History of Medical Causality', in this volume, pp. 57–74.
13. Popper, K. R.: 1935/1959, *The Logic of Scientific Discovery*, Harper & Row, New York.
14. President's Commission for Ethical Problems in Medicine: 1981, *Defining Death*, Government Printing Office, Washington, D.C.
15. Sassower, R.: 1990, 'Medical Education: The Training of Ethical Physicians', *Studies in Philosophy and Education* **10**, 251–261.
16. Sassower, R. and Grodin, M. A.: 1986, 'Epistemological Questions Concerning Death', *Death Studies* **10**, 341–353.
17. Schäfer, L.: 1992, 'On the Scientific Status of Medical Research: Case Study and Interpretation According to Ludwik Fleck', in this volume, pp. 23–38.
18. Toulmin, S.: 1981, 'The Emergence of Post-Modern Science', *The Great Ideas Today*, Encyclopedia Britannica, Chicago, pp. 68–114.
19. Todhunter, I.: 1865, *A History of the Mathematical Theory of Probability, From the Time of Pascal to that of Laplace*, Macmillan and Co., Cambridge.

20. Tsouyopoulos, N.: 1992, 'The Scientific Status of Medical Research: A Reply to Schäfer', in this volume, pp. 39–46.
21. U.S. Department of Health and Human Services, National Institute of Health, National Center for Human Genome Research, and the U.S. Department of Energy, Office of Health & Environmental Research, Human Genome Program: 1990, *Understanding Our Genetic Inheritance: The U.S. Human Genome Project, The First Five Years, FY 1991–1995*, National Technical Information Service, Springfield, VA.
22. Wieland, W.: 1992, 'The Concept of the Art of Medicine', in this volume, pp. 165–181.

PART IV

OBLIGATIONS TO PATIENTS: THE PURPOSE OF MEDICAL PRACTICE AND ITS CONSEQUENCES FOR KNOWLEDGE

STEPHEN TOULMIN

KNOWLEDGE AND ART IN THE PRACTICE OF MEDICINE: CLINICAL JUDGMENT AND HISTORICAL RECONSTRUCTION

Two symbolic figures have dominated the development of Western epistemology: that of the geometer, and that of the physician. From Aristotle the biologist right up to Karl Jaspers the psychiatrist, the experience of medicine has challenged the philosopher's analysis of the relations between *theoria* and *praxis*. From classical Greece on, indeed, medicine has presented philosophers with a peculiarly rich and close alliance of mind and hand, theory and practice, universal and existential. The art of medicine demonstrates that human reason is practical as well as theoretical, existential as well as universal; that is, reason is concerned not just with abstract, but also with *flesh and blood* issues. (In this context, the word "concrete" is inappropriate.) Accordingly, the central question about medical epistemology is: "How can medicine sustain this paradoxical combination of contrasted features in the first place?" What, then, is this *doctrina* or *mysterium* — this "medical knowledge" — which paradoxically combines within itself pairs of characteristics (practical and theoretical, universal and existential, and so on) that have to be so sharply distinguished and separated in other contexts?

This question is especially significant at the present time, because so many aspects of medicine and the medical profession are at present under critical review. Physicians in the United States, and to some extent in Europe also, have become the objects of disappointed hopes, even of suspicion and envy, and at times there is an outright *politicization* of medical issues. And this disillusion goes rather deep. This is not merely an extension to medicine of those general anti-authoritarian attitudes prevalent during the late 1960s. It is also a response to the move by many physicians to redefine medical knowledge and practice — to transform the traditional craft into a modern science, and to give the enterprise of medical care more and more the character of "biological research". As a result, contemporary uncertainties about the ethics of medical care, the social role of the medical profession, and the proper scope of medical knowledge, go hand in hand, and need to be dealt with all together. Let us, therefore, try here to make explicit certain assumptions and problems lying behind the current tensions between physicians and their clientele; and, in doing so, hazard a *definition* of medicine rich enough to

Corinna Delkeskamp-Hayes and Mary Ann Gardell Cutter (eds.), Science, Technology, and the Art of Medicine, 231–249.

remain valid today, despite the new incorporation of so much new scientific material into medical practice during our own generation.

The current debates about medicine and medical practice may seem to involve three distinct and separate controversies — ethical, socio-political, and philosophical. Yet, a single deeper epistemological issue underlies all three controversies; and the protagonists' positions over all three questions display similar differences of opinion about the epistemological character of medical knowledge itself. So we have three tasks to perform here. First, we must recall the terms in which the current related controversies are commonly discussed. Next, we shall see how attempts to handle any one of them in isolation from the others distort the issues in debate. And, finally, we shall recognize why all three disputes can be fully resolved, only if our philosophical account of medicine does justice to the complexity of the physician's vocation, and to the consequent richness and subtlety of medical knowledge.

II

Let us glance in turn at:

(A) the social and political critique of professionalism in medicine associated with (among others) the name of Ivan Illich [2];

(B) the intellectual claim that, in the course of the last hundred years, the practice of medicine itself has been transformed into "biomedical science";

(C) the current ethical challenge to medical practitioners, namely, that they are increasingly inclined to treat their clients as "cases" rather than "persons", objects of theoretical knowledge rather than subjects for clinical understanding.

(A) In the argument about professionalism, and the supposed need to "deprofessionalize" health services, Ivan Illich is only one social critic of contemporary medicine among many. His extreme claims have naturally attracted particular attention; but they are correspondingly easy to ridicule, and so make him an unsatisfactory subject for analysis. So let us here state the essentials of the dispute in general terms, aiming to do justice to the opposed positions without making either side look merely silly.

The second half of the twentieth century has seen two seemingly contrary organizational changes in the health care field. On the one hand, extraordinary developments in, for example, pharmacology and medical technology have made sophisticated medical practice much more technical, and have drastically reduced the need for clinical management of many childhood and adult illnesses. (Only those of us who grew up before the 1940s still recall the long weeks of rest and nursing associated with a simple streptococcal infection in the days before antibiotics.) Thus, since the 1940s, the focus of medical care has shifted from the bedside at home to the hospital clinic and ward; and meanwhile the traditional "general practitioner" has apparently lost importance as compared with the hospital physician and specialist.

This change has been perceived in two different ways. On the one hand, many physicians have been delighted by the change, and have welcomed the resulting intellectual and practical challenges. In their view, criticizing the growing technicization of medicine means ignoring the inevitable, and all those with the medical welfare of patients at heart should join them in welcoming it. On the other hand, many people — chiefly, but not entirely, lay people — have been saddened by the consequent separation between physicians and their patients. The old style general practitioner (they argue) at any rate took continuing responsibility for the personal care of patients: now, patients all too often do not know where to look for real "medical attention". Home visits from the "doctor" are no longer available; but, if they present themselves at a hospital, they quickly come to feel like packages, being shunted from department to department until they chance (if lucky) to end at one that can deal with their current problems. On this alternative view, the general public are obliged by the current technicization of medicine to take responsibility for their own health care back into their own hands; and they must organize it in ways that bypass the formal profession of medicine, e.g., in "self-help groups" and "community health centers". The physicians are (on this view) not just pricing themselves out of the market. Even more, they are failing to deliver what the public rightly wants, namely, medical attention directed at their individual needs.

Recall some practical developments of the latter 1970s: for instance, the cooperative gynecological and child care clinics set up as offshoots of the Women's Movement. Many of these have operated successfully with no more technical involvement than that of a part-time nurse practitioner, with at most a fully trained physician on call. And the advocates of extreme deprofessionalization argue, quite simply, that all routine medical attention could be channeled through such "self-help centers" — and would be the better for it, since the members of such cooperative groups can give each other a kind of personal care no longer forthcoming from the medical profession proper.

(B) Similarly, with the current debate about how far medicine is a science: the main focus of this debate again grows out of mid-twentieth-century developments. Until quite recently, the craft of medicine owed very little to scientific biology, and many practicing physicians shared the craftsman's traditional anti-intellectualism. Only since the 1860s have there been major changes in this respect, as the science of physiology began to make progress in the universities of Germany and the hospital laboratories of France. The large-scale transformation of medical practice, involving new modes of diagnosis and treatment with roots in science, has in fact been entirely a twentieth-century affair.

There are substantial differences of opinion today about the lengths to which this transformation should go. Many working physicians, particularly in medical schools, see this conversion of medicine into a branch of science as the source of all real good in the physician's armamentarium. On their view, the apprentice physician can best learn whatever the sciences of physiology, biochemistry, and pharmacology have made possible; so the most important result of medical education is a proper grasp of these new "scientific" techniques. By contrast, all the care and comfort provided by the traditional "doctor" was, at best, a collection of harmless spiritual placebos, at worst the source of false hopes. Rather than perpetuating the training of doctors as "secular priests", we should be turning out physicians who take pride in being applied physiologists and biochemists.

In the opposite camp are those, both inside and outside the profession, who deplore the human consequences of this transformation. The misplaced emulation by physicians of "science" is, in their eyes, the real reason why the general public is alienated from professional medicine. Inevitably and properly, the focus of the natural sciences is on the general rather than the particular, the universal rather than the existential: this is equally true of any "biomedical science". So, instead of seeing individual subjects as "patients" afflicted with various "ills" (i.e., *people* with individual *complaints*), biomedical scientists legitimately regard them as "cases" of general syndromes or conditions: as people, their subjects are therefore interesting only incidentally, to the extent that they exemplify some pathological entity that is interesting *in itself*. Rather (the argument concludes) the physician should model himself on, for example, the gardener, for whom the individual plant is not just an indifferent thing to be studied and if necessary dissected for botany's sake, but a beloved creature to be tended and cared for so that it grows and flowers to best advantage.

(Goethe[1], of course, levelled similar objections against the Newtonian science of his time. In studying the Nature of Life, Goethe argued, the physiologist paradoxically begins by anatomizing the living creature, so destroying the very Life whose Nature he is investigating; in studying the nature of white light, the physicist similarly starts by passing the light through a prism, so abolishing the very whiteness he is studying; and so on. Instead, true understanding would come only by adopting quite different methods: by giving one's objects of study the individual, detailed, and loving concern of a physician, a nurse, or an intimate friend. In this way, Goethe called for natural science to become an art on the model of medicine, rather than medicine seeking to model itself on the abstract, detached natural sciences.)

Finally, we should glance at (C) current controversies about the ethics of clinical medicine. Here, the most powerful criticism comes from those within the medical profession who are responsible for training apprentice physicians in the clinical arts. When faced with critical problems in, e.g., the management of terminally ill patients (they argue), these physicians behave in ways strikingly unlike those they would have to adopt if they fully accepted those dying patients as "persons"—strikingly unlike, say, their conduct toward their close friends or their own families. The formal training of physicians seemingly promotes the formation of individuals with "split" personalities, having specialized "part selves" for handling technical problems in medicine in isolation from normal personal relations. As a result, clinical practice is damaged both technically and ethically. Technically, physicians with such fragmented and "defended" personalities often overlook the slight and subtle indications to which a more psychologically "open" physician will be sensitive; while the failure to accept patients as full "persons" also offends ethically, directly against the basic Kantian obligation of respect for the dignity of one's fellow humans as persons.

In opposition, the question is raised whether too personal an involvement with individual patients does not unsettle a physician's judgment. Only by preserving his detachment from their anxieties, wishes, and feelings can he achieve a properly objective and "clinical" view. (So, the word "clinical" has by now come to mean, colloquially, something close to "pitiless".) The greatest enemy of objectivity and accuracy is sentimentality: far from demonstrating disrespect for their patients, physicians who deliberately preserve a psychological "distance" are doing them the best service they can. In short, technical, scientific physicians will be more efficacious physicians, just because their heads are clearer and cooler, and their judgments steadier.

To put this last controversy in a nutshell: how far do legitimate ethical demands require the physician to view every patient fully as a person, and how far is it acceptable to see him, rather, as a "case" of some pathological condition? Surely (we might respond) both attitudes have a place. Evidently, for the purposes of successful therapy, the patient's condition needs to be diagnosed as falling under the generalizations of medical science; yet the physician who is interested in a patient's condition *only* as exemplifying some pathology he already understands is neglecting important resources available to him in his personality and experience, and missing the chance to explore new challenges to his clinical judgment.

Let us notice two points about all these three controversies. First: all three issues focus on genuine problems of medical practice that oblige us to strike a delicate balance. In the social organization of medicine, professional autonomy and lay participation both have legitimate claims. A totally autonomous medical profession would almost certainly develop authoritarian attitudes toward the lay public; but a system of totally deprofessionalized self-help groups would, with similar certainty, become dangerously ineffective. Again, the practice of medicine must balance the proper demands of science and art. A medicine pursued only as "science" would have no reason to be interested in patients having routine or intractable conditions, without direct significance for "biomedical" research; but a medicine that cultivated care to the exclusion of science would scarcely prove useful to its patients. Finally, the clinical situation demands a corresponding balance from the physician between personal and technical attitudes. A physician who perceived his patients only as cases, never as persons, would be lacking in practical sensitivity; but one who perceives them only as persons, and never at all as cases, would be incompetent.

Secondly: to argue in terms of extremes, or polar opposites, will in all three cases generate ideologically-toned exaggerations. Attacking all professional organizations of medicine as jealously authoritarian is as exaggerated as condemning all self-help groups for being dangerously ineffective. Again, accusing all biomedical scientists for being personally indifferent to their clinical subjects is no less a caricature than depicting all oldstyle practitioners as bumbling purveyors of false comfort. And, finally, blaming individual physicians for adopting in actual practice the attitudes of "clinical" detachment they learned as apprentices is no less an exaggeration than portraying the physician's personal concern for his patients as inescapably sentimental.

III

At this point, we may shift our attention to a more obviously philosophical plane, and bring to the surface the epistemological issues underlying these practical controversies. First, we may focus on the relations between science and art (*episteme* and *techne*) as illustrated in the differences between Plato and Hippocrates; and, later, we may move on to discuss the significance for medicine of the idea of "history" as developed in Giambattista Vico's polemics against the Cartesian philosophy.

The twin models of knowledge (geometrical and medical) are already at work in Plato's cosmology. Plato's account of the Myth of Timaeus, for instance, offers two alternative, complementary accounts of the human frame and its bodily organs. On the one hand, these organs work as they do for geometrical reasons — e.g., the gullet can channel food to the esophagus just because of its conical shape. (It *acts like* a funnel, because geometrically it *is* a funnel.) On the other hand, a deeper understanding requires us to attend to the physiological *functions* of these structures — the gullet *has to* funnel food to the esophagus, so that the digestion may maintain the organism's life and activity. So, in the *Timaeus*, the continual interplay of physiological form and function is already evident, if only on the level of *theoria*. While Plato moves away from geometry in a functional direction, physiology remains of concern to him simply as one more generalizable *episteme*: not as *techne*, still less as individual *praxis*. In physiology as much as geometry, Plato perceives the universal, not the particular: the general functions of different bodily organs, not the particular disorders afflicting this or that individual patient. (Compare Plato's famous political analogy between the State and an Organism. The proper concern of political theory is, likewise, the general functions of social structures and relationships. Particular cities, nations or societies may go astray in their own idiosyncratic ways; but Plato's general blueprint for a well structured and properly functioning "republic" remains always the same.)

The classical account of medical *praxis* and its ethical obligations was of course given not by the Athenian philosophers but rather by the physicians of Cos. The name of Hippocrates is associated with the first fully organized school of medical practice, and physicians today still appeal to the Code of Hippocrates as expressing their moral loyalty to the traditional doctor-patient relationship. At the very least, then, Hippocrates and his colleagues deserve the credit for articulating one crucial feature of medical knowledge — namely, that the focus and object of medical practice and knowledge are alike to be found in the *individual patient*. In that respect, medical knowledge can make no pretense at being a general and universal Platonic *episteme*: rather, it is intrinsically a variety of particular, existential knowledge.

This is not to imply that Hippocratic medicine was merely *pastoral*: i.e., a "helping" or "caring" profession giving personal advice to particular individuals. On the contrary, some of the twentieth-century ethical tensions, between the physician's obligations to individual patients and to the collective professional art, are already recognizable in the Hippocratic Code. The physician is exhorted to acknowledge a duty to the art of medicine, as well as to his patients: he must act so as to promote the art, and take care to transmit it faithfully to his apprentices. So, Medicine had a claim on the classical Greek physician also as a variety of *generalized* knowledge; but still as *techne*, rather than as *episteme*, because it was a practical and not a theoretical kind of knowledge.

However, even the generalized principles of the medical art could be learned and exercised only as applied to, and embodied in, the condition of particular human beings. And indeed, today also, the best way of grasping and refining the general principles of medicine is to go on "rounds". For all their abstract generality, these principles are *abstracted from* the symptoms and syndromes presented by individuals, and a proper understanding of them can be achieved only by seeing how their relevance and applicability to those individuals depend on their specific backgrounds, lives, constitutions, and other idiosyncratic features — in a word, on their *histories*.

This characteristic duality — both general and particular, both universal and existential — carries over from medical knowledge to the physician's vocation, and so to the "virtues" proper to the physician's modes of life. Since this vocation is to put the general body of knowledge about health and disease to use, for the benefit of particular human beings, it defines also the *Lebensform*, or "way of being in the world", within which the professional physician must do his work.

In this respect, medicine is the paradigmatic case of an *applied* art or science, and can be cited to demonstrate to us one important point about all applied arts and sciences. However much the pool of general knowledge on which medical practice draws may be ethically *neutral*, all specific applications of this knowledge in medical practice necessarily involve estimates about the individual patient's "good". So — contrary to Heidegger's claims about technology — the modes of thought and action of the applied sciences are *not* purely instrumental, i.e., concerned merely with "means" and never with "ends". On the contrary, the proper application of general medical knowledge to individual human beings demands an accurate appreciation of their particular needs and conditions; so that the task of medicine — however "scientific" it may become — remains fully *ethical*.

IV

Is this analysis of medical knowledge still relevant to twentieth century medicine? Given all the new modalities and facilities for medical intervention, and all the new roles of biomedical technology in the contemporary physician's practice, may not the vocation of medicine — and with it the *Lebensform* and the virtues of medicine — have changed also?

We must not underestimate the force of this question. After two-and-a-half millennia during which the central focus of medicine remained its *pastoral* function, the last hundred years have seen several new foci of medical practice emerging. The development of bacteriology established the *preventive* function of medicine, initially associated with the sanitation movement and the inauguration of public health services during the late nineteenth century. The rise of scientific pharmacology has created a genuinely *curative* function for medicine, beginning with, for example, Ehrlich's work on salvarsan but reaching its spectacular peak only during the last thirty years. More recently, medicine has begun to help people, in addition, to enhance their natural appearances, faculties and abilities, by procedures ranging from cosmetic surgery to sex therapy — so establishing what may be called *augmentive* medicine. And finally, underlying all these new foci of medical practice, is the *research* function, which is continually extending and validating the armamentarium of medical and surgical procedures.

A parallel development of new medical facilities has led to the creation of new types of hospitals and laboratories, radiological and psychiatric clinics, public health operations, mental health centers, "encounter groups", and so on. Correspondingly, there is a great variety of novel social roles available in the health care system: for physiotherapists and laboratory pathologists, x-ray technicians and marriage counselors, sports physicians, and the rest. And one might well raise the question: "Surely, within this new, late twentieth-century situation, the traditional Hippocratic conception must be rethought and revised; and surely, also, this new situation calls for equally new analyses and definitions of medicine, and of the physician's role?"

Attractive though the appeal to modernity always is, we can give a conservative answer to this question. Some basic features of the human situation are slower and harder to change than we may suppose. For all their novelty and consequences, all these new medical functions and vocations still fall under the same general heading of "applied art and science". As such, they still share the same epistemological character — notably, the traditional duality of general and particular, universal and existential. (A public health official, for instance, may not treat individual patients; but he applies the general principles of bacteriology and epidemiology to the specific problems of his particular community using the same kind of particular judgment that a personal physician brings to the traditional single doctor/single patient situation.)

What does create some ambiguities, however, is the way in which the needs of any single patient are often split up and parceled out among different specialists. "In a world of internists and pathologists, radiologists and other specialists," we may ask, "Who then is, truly, *the physician*?" With this question in mind, we may now turn to a second general epistemological argument. For, if this paper is intended to underline and emphasize any one central point, it is the fundamental importance of *the idea of a history* for all medicine:

> However specialized and fragmented the technical practice of medicine may become, the true physician remains — as he has always been — the person who *takes the patient's history.*

Let us now set about explaining and justifying that central statement.

The application of general principles to particular individuals in medicine (we remarked) always requires us to understand what exactly it is about the present patient that is peculiar, particular, idiosyncratic — what makes this specific individual the "individual" he is. But it requires us, equally, to judge what makes this patient the individual he is *at this particular moment.* A patient is not just what he is, he is even more particularly what his past life has made of him, what he has made of himself, and to some extent also what he now perceives himself as being and becoming. (The English and German languages, in fact, lack a useful distinction present in some Romance languages — between, for example, the two Spanish verbs *to be*, *ser* and *estar*. It will be self-evident to a Spanish physician that he must understand not just what any patient *es*, but also what he *esta* — not just his constitution, but also his current condition.) The clinical appropriateness of any medical judgment will often depend in crucial ways on historical facts about the patient in question. To understand your patient is to understand his "history" — to understand who he is, and how he now perceives himself to be.

What is the *epistemological* significance of this fact? Let us again go back in Western thought, this time to the seventeenth and eighteenth centuries A.D. In his role as the philosophical prophet and spokesman for the modern scientific movement, René Descartes called for a science having the same general, abstract, mathematical basis that Plato had dreamed of two thousand years before. (In this respect, Newton was both the executor and the beneficiary of Descartes' methodology, as Euclid earlier had been of Plato's.) But Descartes' scheme of scientific priorities also had two less fortunate consequences. First, *medicine* was in his eyes as theoretical as any other science. The practice of medicine was of serious intellectual concern only as promoting the development of a scientific physiology. Descartes's sole concern with medicine, in short, was with the general mechanisms of respiration, blood circulation, and the rest: not at all with the art of ministering to the particular needs of individual men. Secondly, Descartes regarded *history* too as intrinsically uninteresting. Historical knowledge was a compendium of particular, contingent facts bound together by no general, intelligible principles: as such, it could not be taken seriously by the scientifically-minded philosopher.

This scornful attitude towards history and historical knowledge was challenged, in the early eighteenth century, by Giambattista Vico of Naples. Vico argued — as Dilthey, Croce, and Collingwood were to do much later — that historical knowledge is, in its own way, as serious and profound as knowledge of natural science. Moreover, (he argued) we find the course of human history directly intelligible *from within*, by a kind of historical empathy, since we too can recognize the human significance of the problems confronting our predecessors. To that extent, historical knowledge might even be *more* serious and *more* profound than, say, our understanding of the fall of heavy bodies, the progress of chemical reactions, or the flow of blood through the arteries — all of which are accessible to us only *from outside*, as external onlookers.

So, far from regarding historical understanding as epistemologically deficient, by comparison to scientific, we should see matters the other way about. Even the abstract principles of theoretical physics and physiology can be given a full and final epistemological basis, only by relating them to the particular empirical observations by which they were arrived at; and the reports of those observations are just as much "timebound chronicles" as any other historical reports. Thus, from Vico's point of view, history — with its timebound, narrative knowledge — truly deserves the epistemological primacy that Descartes had mistakenly claimed for the "clear and distinct general ideas" of the mathematical sciences.

For our own present purposes, there is no need to go all the way with Vico and his modern followers. Nor need we take sides, here, between Descartes and Vico at all. There is in fact something a little fruitless about insisting *either* that physics is "intellectually more serious" than history, and physiology "intellectually more profound" than clinical medicine, *or* the other way around. Both science and history have their proper places and purposes. Developing a general scientific understanding of natural processes by observational means, and interpreting particular historical episodes by the exercise of human sympathy: these are two complementary intellectual enterprises, each with its own problematic, methods, and criteria of success. To praise the one and decry the other is, at best, to declare one's own personal priorities: at worst, it is to put oneself in blinders. For our purposes, it is more helpful to accept both the general understanding provided by science (about natural processes occurring in any place and at any time) and the particular understanding provided by history (about specific human episodes occurring here-and-now, rather than there-and-then), as equally serious and valid intellectual goals when pursued in appropriate situations.

If we do take this even-handed position, we shall then be in a position to recognize the true source of the last epistemological feature — the "existential" character of medical knowledge — that we remarked on at the outset. This character arises directly from the involvement of medicine with history as well as science. Historical knowledge of human affairs is always knowledge of the here-and-now: knowledge of human beings as they are, as they have been shaped by their past, as they have made themselves, and as they perceive themselves to be. We might, indeed, say that all historical knowledge is, in a sense, *clinical* knowledge; or, at the very least, that all clinical knowledge is, essentially, *historical* knowledge. As applied to specific states of affairs, the physician's understanding refers always to the condition, of some one particular subject, in his, her, or its present condition, with all the consequent timebound, local, and particular features. So, the physician always brings the general principles of medicine to bear on the transient existential peculiarities of his subject, and the specific "medical situation" with which he has to deal thus constitutes one particular intersection of nature and history.

V

The reasons for advancing the central thesis of this paper, namely, that *the true physician is the person who takes the patient's history*, should now be clearer. Only the person who has taken this history, who has probed it sympathetically, and tested it in dialogue with the patient, can truly know "the medical situation", as here defined. Only he can judge with any confidence the medical significance of this-or-that feature or the likely consequences of this-or-that course of treatment; for only he can have any reliable sense of the strengths and weaknesses of the patient's medical record, *judged simply as an historical narrative*. (The weaknesses of this narrative are, of course, as important as its strengths. A woman gynecologist from Venezuela remarked to me once: "I have learned from experience that, when I take a patient's history, the things she does *not* tell me are very often quite as important as the things she does tell me.") And, since the physician's central vocation has always been to recognize what general principles are relevant and applicable to *this* individual, given his condition *here and now*, that judgment of relevance and applicability is essentially one for the person who best understands *the individual's history*.

Even in the most complex and differentiated hospital situation — with pathology labs, radiology clinics, psychiatric support services, and the rest — one specific physician is usually responsible for taking, mastering, understanding and refining the "history" of any individual patient. It is he who sends out specimens for tests, interprets the radiology reports, takes account of the psychiatric assessments, and so on; and it is he who decides, in each case, just what weight any specialist opinion deserves, when seen against the background of the patient's entire history. Or, at any rate, that is how it *should* be: for, once this specific locus of understanding and responsibility is lost, patients can indeed complain that they are failing to get the individual personal attention their conditions require.

Notice: as used in this context, there is nothing *sentimental* about the notion of individual, personal attention. The case for such individual attention is in no way concerned with *Schmaltz* or *Schwärmerei* — it is not a matter of advocating warm, human fellow feelings between patients and their attending physicians. The argument has simply to do with the conditions required to meet the central needs of medicine. All that is being claimed is that a full, accurate, and psychologically revealing "history" must be developed, if the physician is to judge with any real accuracy and objectivity just what truly is the patient's "condition", and what prospects can be reasonably foreseen from different courses of treatment. Specialist consultants of different kinds all have their value and function; but the traditional locus of medical knowledge, judgment and attention — today as much as ever — still lies in the attendant physician. He alone can put all the specialists' reports together, interpret their significance for this individual here-and-now, and test his judgment in the confidence of (and in dialogue with) the affected individual. If the *medical situation* is considered as "one particular intersection of nature and history", then the locus where this intersection is actually grasped and understood is, more specifically, the *mind of the attendant physician*.

To pull the threads of this discussion together: our two central epistemological distinctions, between Plato and Hippocrates (personifying theoretical and practical knowledge) and between Descartes and Vico (personifying scientific and historical knowledge), should make it clear just how far the fusion of medicine with biological science can afford to go, if it is not to destroy the essential character of medical practice and understanding. Though medicine may continue to draw on, and profit from, an ever larger pool of general scientific knowledge, the essential demands of the physician's vocation will still require that the particular, idiosyncratic, and existential features of every single "medical situation" (in a word, its *historical* features) should preserve a central place in the medical enterprise. To that extent, the special kind of narrative we call a *medical history* will retain its vital role in the organization of the physician's understanding.

With this point in mind, we can see how our three initial controversies are related; and how we can find our way between the polemical extremes that are currently presented to us. On the one hand, to make medicine over entirely into a natural science might indeed turn doctors into "authorities on biomedical science" — so giving them all the professional autonomy and moral privilege of other natural scientists, and seemingly entitling them to view their clients as experimental subjects rather than as personal recipients of responsible care. But, in that event, there would indeed be a strong case for taking the responsibility for our health care out of the hands of the medical profession, and looking for personal attention and the traditional physician's understanding elsewhere.

On the other hand, if we acknowledge and respect the essential differences between scientific and medical knowledge — notably, the physician's complex but indispensable fusion of the theoretical and the practical, the general and the particular, the universal and the existential — we can avoid being driven to these extremes. But we can do so only at a price. (1) In the first place, in order to escape the extremes of deprofessionalization, physicians must accept the need — both as individuals, and through their collective organizations — to take lay people more into their confidence, and to concede them the right to have a say in the urgent socio-political problems surrounding the provision of health care today. Though professionalism may be here to stay in medicine, the day of the "authoritarian personality" in medicine is past. (2) In the second place, the intellectual analysis of medical issues may continue to borrow more and more ideas from natural science — from biochemistry, from physiology, even from nuclear physics — but their relevance to medicine must continue to be judged, in the future as in the past, in terms of their clinical/historical application and human significance. (3) Finally, however much future medical training takes place within a professional context and focusses in on fruitful new scientific ideas, the teaching of clinical judgment will call more than ever for physicians to grasp the individual, personal, and essentially *ethical* character of clinical problems, and the dangers of becoming split-minded technicians without a proper feeling for matters of personal individuality.

The resolution of all three disputes will, likewise, best come through linking them together. The characteristic intellectual feature of medicine as a *techne* — its dependence on establishing an individual historical narrative — is bound up with the characteristic ethical feature of medicine as an art — its concern with persons regarded as "ends in themselves", rather than "as means only"; and these two features of present-day medicine alone justify the professional claims of physicians to autonomy, self-regulation, and public respect, and the modes of professional organization for medicine developed on the basis of these claims. It is also this same rich, complex, and existential character that makes medicine a *philosophically profound* field of study, and so calls for a fuller and more accurate epistemological analysis. So, it becomes clear why, from the outset, Hippocrates the Physician was entitled to a place in the Pantheon of Human Understanding — alongside Plato the Astronomer and Aristotle the Zoologist — among "the masters of them that know".

VI

Going beyond these conclusions, some fresh lines of thought are worth opening up. These have to do with:

(1) the light that the philosophy of medicine can throw on the philosophy of natural science;
(2) the significance (if any) of the current distinction between organic and psychosomatic diseases; and
(3) the consequent relations between somatic medicine and psychotherapy, especially psychoanalysis.

(1) Seeing how recently medicine has become a field of epistemological inquiry, writers on the philosophy of medicine have — naturally enough — turned to the philosophy of science for initial guidance and hypotheses: asking how far the general patterns of analysis already developed in the philosophy of science throw light on the particular character of medical knowledge. If our present analysis is correct, however, more significant suggestions may well be arrived at by moving in the opposite direction: by asking, rather, how far an understanding of the particular character of medicine throws light on the nature of scientific knowledge, more generally.

The reasons for believing this have to do with the special status of *physics*, which is the science most studied by philosophers. Most theoretical discussions in philosophy of science (or, to use the happier plural French phrase, *la philosophie des sciences*) have been dominated by examples from physics, especially from astronomy: other sciences have commonly been treated as philosophically serious only to the extent that they approximate to physics. Our present conclusions, by contrast, suggest that physics should be regarded not as the prototypical science, but as a special case. The physical scientist places himself in a peculiar position, quite foreign to all our experience with our fellow humans. His objects of knowledge are quite unresponsive to the fact of being studied, and his understanding of them is purely external. As a result, the theoretical relations in terms of which he discusses those objects are entirely products of his own construction. They are not given to him in the course of interactions with his objects of knowledge: rather, he brings them to those interactions from outside. Accordingly, the "objectivity" of the physical sciences is not just the objectivity of undistorted accuracy; it is, in addition, the objectivity of intellectual detachment.

In medical practice, on the other hand, the physician acquires real clinical knowledge, only to the extent that he engages in personal interactions with his patients. The interpretations he arrives at emerge out of this two-way relationship with the patient; and indeed, if the physician holds himself back from such two-way interactions — in the manner of a "medical astronomer" — he may well miss indispensable signs. (To return to my Venezuelan gynecologist friend: she reports the case of a patient who came in for a normal gynecological exam, but was not apparently reassured on being told that everything appeared entirely in order. Sensing this unease, my friend remarked that the patient had somehow gone through the entire examination, without exposing the upper half of her body. Only then did it come to light that the patient had undergone a radical mastectomy, and was afraid of possible metastases.)

It is true, of course, that this personal side of the medical enterprise is temporarily suspended for the technical purposes of the laboratory, in pathology tests, x-ray photographs, and the like; but that suspension is still only temporary. The actual relevance of such technical investigations to clinical understanding remains a matter for personal judgment, in which all the psychic resources of the physician's personality — all his capacity to interact and feel with the patient, on a basis of mutual understanding — may need to be brought into play. By comparison with a physician's knowledge of an individual patient, therefore, the physicist's knowledge of his own objects of study is, at best, partial and abstract. So, rather than attempting to explain the epistemological status of "medical knowledge" as a species of the broader genus, "scientific knowledge", we would do better to regard the physicist's theoretical knowledge as itself a special case — arrived at by placing special restrictions on the richer and more concrete (or "flesh-and-blood") understanding exemplified in, for example, the practice of medicine.

(2) With this first point in mind, we can go one step further. The recent moves toward making medicine an exclusively technical, or scientific field has reinforced the distinction between organic or somatic illness, on the one hand, and functional or psychosomatic illness, on the other. From that viewpoint, the physician's primary business is to deal with the pathological manifestations of organic disorders. The psychological concomitants of these disorders are not his direct concern; and the vicissitudes of psychopathology (e.g., functional disorders that actually mimic organic disease) merely place obstacles in the way of this primary medical task.

Our present analysis suggests an alternative view of the matter. From our viewpoint, the distinction between "purely somatic" and "psychosomatic" illness is misleading. Rather, all sickness has a psychic component, which plays a larger or smaller part in the physician's actual diagnosis, but is never wholly negligible or absent. Indeed, the typical initiation of a medical interaction is for the patient to come to the physician with a "complaint"; this *act* of complaining is an essential component in the clinical situation, whose significance the physician has to figure out. What is the nature and basis of this complaint, i.e., of this psychological act of "complaining"? Does it lie in some specifically organic disorder? Or does it spring, rather, from some other source, e.g., anxiety about the consequences of some earlier medical episode? (Recall the Venezuelan gynecologist's case, once again.) Thus, the psychological component in clinical medicine does not represent the external intrusion of issues irrelevant to the study of pathology: rather, the discovery of organic disorder puts a physiological slant on a clinical interaction that initially takes place on a psychological plane.

From this viewpoint, it is desirable to reconsider also the role of *intuition* in medical knowledge. In this context, the terms "intuition" and "intuitive" are seriously ambiguous. They have both an *intellectual* sense — so that we may speak of a scientist's "intuition" permitting him to leap, as by a shortcut, to some hypothesis which subsequently proves to be well-founded — and also a quite other *empathetic interpersonal* sense — so that we may speak of one person having an "intuitive" feeling for another person's state of mind. For those who see medicine as essentially scientific, the skilled diagnostician's "medical intuition" is naturally to be understood in the first sense: from an interpersonal viewpoint, however, the physician's "intuitive" capacity is, rather, one of insight or empathy (*Einfühlung* or *Mitgefühl*) for the inner character of the patient's complaint.

(3) We may end by pursuing this direction of thought one last step. In terms of its content (our present argument suggests), all sound clinical *knowledge* tends to the condition of history, rather than to the condition of natural science: the clinician's central responsibility is to arrive at a sound historical interpretation. In terms of its method, by contrast, all sound clinical *practice* tends to the condition of psychoanalysis. The physician's understanding of a patient's complaint requires him to "get inside" those features of the patient's history that have led to his act of "complaining".

These two theses are entirely compatible. The epistemological problems of psychoanalysis closely resemble the problems of historiography: in both cases, the crucial question is, how an historical reconstruction is to be justified — either, the reconstruction of some significant episode in the past history of a nation, or alternatively, the reconstruction of some significant episode in the early history of the analysand. The central task of psychoanalysis is, in fact, to enable the analysand — in empathetic collaboration with the analyst — to work his way back to an accurate historical reconstruction of his own lost past, and achieve the self-understanding necessary for him to accept, and live with, that past.

To conclude on a personal note: the most puzzling aspect of the epistemology of medicine is, to me, why so many physicians resist these present conclusions. Seemingly, these physicians *want* clinical medicine to become an impersonal technology. The importance we have allotted here to medical history and personal empathy is something they find unwelcome, and even reject. Why is this? It is as though, having become physicians, they then resisted the need to become conversant with their patient's histories and states of mind: even, in some cases, as though they had some deeper psychic interests, which were served by adopting the detached attitudes more appropriate to a natural scientist. Yet, if one does not wish to accept some real psychic involvement with sick people and is not really willing to involve one's whole personality in that interaction — and it is not just a case of the physician treating the patient as a "whole man", but rather one of the physician himself, as a "whole man," dealing with the patient as a "whole man" — then, I would ask, *why be a physician at all*?

University of Southern California
Los Angeles, California, U.S.A.

NOTE

[1] The key documents are in Goethe's collected works, notably in [1]: some 15 volumes have appeared to date. After 1945, there was a reappraisal of Goethe's scientific claims by a series of scholars and scientists, e.g., Agnes Arber, who treat him with more sympathy and perception than his nineteenth and early twentieth century critics: see, particularly, [3].

BIBLIOGRAPHY

1. Goethe, J. W. von: 1947–, *Die Schriften zur Naturwissenschaft*, ed. R. Matthaei, Böhleau, Weimar.
2. Illich, I.: 1976, *Medical Nemesis*, Pantheon Books, New York.
3. Sepper, D. L.: 1988, *Goethe Contra Newton*, Cambridge University Press, England.

MARY ANN GARDELL CUTTER

MEDICINE: EXPLANATION, MANIPULATION, AND CREATIVITY

I. INTRODUCTION

There is a common view among the authors in this volume that it is misleading to understand modern medicine simply as a science, as a technology, or as an art. Some ([28], [20], [9], [13], [11], [27], [29]) hold that although medicine embraces all three components, it favors one over the other. Others ([24], [1], [9], [13], [11], [27], [21]) hold that medicine is not reducible to these components and that it is "more than the sum of its parts". This essay shares this latter view and continues the dialogue regarding the complexity of medicine by reflecting on the dynamic interplay among science, technology, and art in modern medicine [3]. Moreover, this essay reflects on the grounds and the limits of the medical profession's legitimate authority and its obligation toward patients through an analysis of the role of values in medicine.

II. EXPLANATION (SCIENCE)

Clinical science employs explanation (L. *planare*, to make plain) to render the world intelligible. Intelligibility is tied to intersubjectivity and agreement among cognitive agents. Intersubjectivity grounds the possibility in medicine of usual and customary standards of care and formally articulated indications for treatment. Agreement makes possible action, which in medicine involves the caring and are curing of human disease and illness. In the history and development of medicine ([7], [5]), explanation has a two-tiered character, that of 1) the *clinical* understanding of disease as a syndrome, or constellation of signs and symptoms; and 2) the *laboratory* or *basic scientific* account, which interprets the clinical elements of a syndrome in terms of pathoanatomical and pathophysiological correlates with unacceptable pain, disability, dysfunction, or death. That is, with developments in seventeenth and eighteenth century medicine, accounts of the constellation of signs and symptoms provided the basis of clinicians' understanding of disease. This is seen, for example, in the clinical categories of Thomas Sydenham [26], François Boissier de Sauvages [23], and William Cullen [4]. Sauvages, for instance, organized the clinical world in terms of ten major classes: *vitia*

Corinna Delkeskamp-Hayes and Mary Ann Gardell Cutter (eds.), Science, Technology, and the Art of Medicine, 251–258.

(defects), *febres* (fevers), *phlegmasiae* (inflammations), *spasmi* (spasms), *anhelationes* (difficulties in breathing), *debilitates* (weaknesses), *dolores* (pains), *vesaniae* (insanities), *fluxus* (fluxes), and *cachexiae* (constitutional disorders). The *clinical* understanding of disease provides, in short, a phenomenological approach to organizing disease, resulting in descriptive accounts of disease in terms of patient signs and symptoms.

With the advent of nineteenth century pathological construals of human illness, clinical signs and symptoms were reinterpreted in terms of the basic laboratory sciences, which have in turn delivered new clinical descriptions, and which are themselves submitted to further revision and given further changes in basic scientific appreciations of disease. Fevers, pain, and fluxes were no longer considered diseases in their own right, but symptoms associated with underlying pathoanatomical, pathophysiological, or pathopsychological processes. Clinical complaints that had not been previously associated could now be brought together under one rubric. Phthisis, scrofula, consumption, and Pott's disease, for example, could be reorganized as manifestations of tuberculosis. Etiological accounts of the origin of the pathoanatomical and pathophysiological findings had their impact as well on distinguishing previously undiscriminated problems such as anemia, which could now be distinguished as general, chronic, and pernicious. The laboratory or basic scientific account of disease provides, in short, theoretical bases through which to organize and reorganize descriptive accounts of patient complaints in terms of underlying pathoanatomical, pathophysiological, and pathopsychological correlates.

The interaction between the two levels is dynamic. Clinical facts and scientific theories of disease interplay in such a way as to require multifactorial accounts of clinical findings (also see Fagot-Largeault [9], Juengst [13], and Moulin [18]). On the one hand, the basic scientific level tends to interpret the world of clinical complaints in terms of a number of pathoanatomical, pathophysiological, and pathopsychological correlates so that often no single factor stands out as both necessary and sufficient [25]. On the other hand, the clinical level tells the laboratory scientist what is pathological and thus brings direction to medical problems. This two-tiered, dynamic account of clinical problems provides, in short, ways in which to organize and predict clinical reality.

On this analysis, clinical problems are not simply discovered but are in part the product of the creative enterprise we call medicine. This appeal to an interplay between discovery and invention is indebted to Immanuel Kant (1724–1804), who, in the *Critique of Pure Reason*, grounds the character of the known in the characteristic of the knower:

> Now all experience does indeed contain, in addition to the intuition of the senses through which something is given, a *concept* of an object as being thereby given, that is to say, as appearing. Concepts of objects in general underlie all empirical knowledge as its *a priori* conditions. The objective validity of the categories as *a priori* concepts rests, therefore, on the fact that, *so far as the form of thought is concerned, through them alone does experience become possible*. They relate of necessity and *a priori* to objects of experience, for the reason that only by means of them can any object whatsoever of experience be thought ([14], B126) (my emphasis).

The world is not simply discovered, but rather created as well. In medicine, clinical reality is constituted for practitioners in the sense that the world is seen through prevailing theoretical frameworks ([10], [16], [24]). One can never know truly, undistorted by particular frameworks of interpretation and understanding (see also Fleck [10], Kuhn [5], and Schäfer [24]). As Kant reminds us, "What the things-in-themselves may be I do not know, nor do I need to know, since a thing can never come before me except in appearance" ([14], A277–B333). Reality is theory-laden and fashioned within a particular framework. In medicine, there is not a determinable number of clinical problems existing apart from clinical knowers. Clinicians' observations and goals guide the construction of disease nosologies and nosographies, which change with time. Clinical reality is created in the sense that it is seen and interpreted within the embrace of theoretical frameworks, or visions of reality.

In short, there are no unique lines in reality in which clinical definitions, nomenclature, or classifications correspond. Rather, explanations of clinical reality are created, not simply discovered, through endowing certain findings with significance, signs, or value.

III. MANIPULATION (TECHNOLOGY)

Insofar as medicine is tied to action for the sake of treating patient complaints and achieving goods, instrumental values direct clinical explanation. Herein lies a claim of significance, or value. Clinicians explain clinical problems in terms of certain goals, e.g., maximizing benefits and minimizing bur-dens (e.g., costs and harms) to particular patients. Such consequential or outcome language underlies the enterprise of manipulation or technology in medicine.

Technology in medicine is devoted to the achievement of goals through means that are developed for effectively and efficiently obtaining the goals [21]. Moreover, the technological dimension of clinical explanation is dynamic. This dimension interprets models and modes of examining medical phenomena in terms of therapeutic goals, such as providing more encompassing explanations of disease phenomena and in the end more useful forms of treatment. Seventeenth and eighteenth century medicine delivered, at best, symptomatic treatments (e.g., herbs, purges, cold baths) for a clinical problem. The clinical problem (e.g., pains) was that which was to be manipulated and was seen in the context of the patient's whole person.

With developments in nineteenth and twentieth century medicine, a critical dialectic was initiated between the earlier world of symptomatic treatment and the later world of etiologic intervention. Medicine no longer simply specified that which was to be manipulated, the basis for medical intervention. Rather, medicine offered various accounts of that which manipulates, treatments that were supported by theory, by etiologic interpretation. Such accounts are seen to vary and include pharmaceutical (e.g., penicillin), surgical (replacement valve surgery), and preventive (e.g., immunization) intervention. An interplay between clinical practitioners and biomedical scientists emerges [12] in which the basic sciences provide the bases for realizing certain ends (the treatment or prevention of patient complaints) by structuring the character of manipulations, of that which manipulates.

Since all clinicians and patients do not share the same views regarding which clinical goals are to be achieved and why, and since cooperation among physicians, patients, and third parties requires agreement concerning understandings about what are 'good' goals, ethical (Gr. *ethike* [custom] or moral (L. *mos, mores* [belonging to manners]) values (those concerned with assignments of praiseworthiness and blameworthiness) are tied to the formulation of clinical explanation and manipulation.

Major ethical values arising in the clinical setting include those tied to 1) autonomy and 2) the individual and/or societal welfare or interest [1]. Allowing an individual to choose to live his life as he chooses insofar as he does not disrespect another reflects a commitment to autonomy (i.e., self-determination or individual freedom). The procedural expression of this moral principle is found in the action of treating another with respect ([15], [7]). Respect for person provides the basis for free exchange among members of the moral community. In civil societies, one respects another by (not) doing unto another as he or she has (not) contracted to do. In medicine, respect for person takes on the procedural expression of (not) doing unto another that which one has (not) contracted to do. In the 'postmodern' tradition and given the unavailability of any single universal account of the good life, authority in the clinic and elsewhere at best derives from respect for person, or mutual consent. (Also see [6].)

Concerns regarding the welfare of another or others take on the procedural expression of doing good to another or acting in another's best interest (beneficence). In civil societies, acting in another's welfare raises numerous issues, such as a) *What* is an individual's welfare? b) *Who* decides what is a particular individual's welfare? c) *How* does one resolve conflicts regarding another's welfare? and d) *How* ought benefits and burdens be distributed in a good society so that welfare is promoted? In medicine, questions arise such as a) Are there general criteria for a patient's welfare? b) Who decides the welfare of a patient? c) How does one resolve conflicts between a patient's determination of welfare and alternative determinations of that welfare? and d) How ought benefits and burdens in medicine be allocated in a good society so that "patient" welfare is promoted? Herein lies not only deep concerns we may share about the welfare of our fellow beings in general, and our friends in particular, but also the practical and moral limits of the medical profession's obligation towards patients.

These and related questions and concerns are the focus of sustained discussion in the philosophy of medicine. They emphasize the central role manipulation or technology plays in our understanding of clinical problems. They emphasize the role values play in our understanding of clinical reality. They emphasize the practical and moral grounds and limits of the medical professions' obligation toward patients.

IV. CREATIVITY (THE ART OF CARING AND CURING)

In addressing the illness of men and women, we will need to look, then, not only at how scientific facts are appreciated simply by leading physicians and biomedical scientists but at the ways in which illnesses are viewed by individual practitioners, individual patients, and interested others. Fashioned by a given community's understanding of medical science and disease and illness, the facts of clinical problems may appear quite differently to patients than they appear to a well-trained physician. Facts, theories, and values are always interpreted within particular contexts. Expressions of particular understandings and manipulations of clinical problems constitute the art of medicine.

Art (L. *ars* [*artis*], root *ar*, to join) focusses our attention, expresses knowledge, and has a goal within particular contexts [3]. In this way, art is neither theory nor practice, but both, dynamically interacting. Art produces a change in the viewer's perspective (theory) as well as in our world (practice) without a concern about the distinction between the self and the world. For the artistic consciousness, the world is for the self simply what is here and now experienced. Within this interchange between the self and other, and in knowing the self through the world, one makes the self through the world, which is in turn made through the self. Yet, the world is not made 'out of nothing'. Similarly, the self is not made 'out of nothing'. The artist is not God, not a solipsist, but instead a finite mind. The artist makes the self and the world 'out of' what is presented to her, e.g., colors, sound, appearances, action. The expression of this level of consciousness, this interplay between and among the self and the world, is art [3], that which joins perspectives.

Medicine as art is theory and practice — interacting and in the making. Medicine produces a change in the understanding we have of ourselves (through, e.g., what it means to be healthy and diseased; normal and abnormal; functional and dysfunctional; ordered and disordered) as well as in the world of the manipulation of patient complaints (therapeutics) often without a concern about any sharp distinction between medical knowledge and medical therapeutics, diagnosis and therapeutics. (Simply going for a physical may be therapeutic for a patient. Then again the gap between diagnosis and therapeutics in contemporary genetics is evident.) Our understanding of ourselves as anatomical, physiological, and psychological beings influences and is influenced by the ways in which we manipulate the various aspects of our being — through, for example, surgery, pharmaceutical drugs, and or psychotherapy. Diagnosis and therapeutics — theory and practice, science and technology, explanation and manipulation — are inextricably interwoven within medical thought and action. In knowing ourselves, we make and manipulate ourselves. By making and manipulating ourselves, we come to know the nature and limits of the self. Yet, the world of medicine is not made 'out of nothing'. Members of the medical community (e.g., clinicians, concerned parties, patients) are not omnipotent beings, not solipsists, but rather finite minds. The world of medicine evolves out of interpretations of clinical reality or appearances that are experienced by various individuals. The expression of this dynamic evolution — in theory and in practice — constitutes the art of medicine.

It is against this backdrop that Bernadine Healy, Director of the National Institutes of Health, makes the following claim. In addressing the critical role creativity plays in medicine, she says:

> ... if medicine is to succeed, the Mozarts in biomedical research must be allowed to flourish. ... What we seek is a synthesis like that proposed by Alfred North Whitehead for dealing with the natural tendency for "those who are imaginative to have but slight experience and those who are experienced to have feeble imagination." Whitehead's solution was to wield together imagination and experience. ([12], p. 33).

This vision is not only one in which scientists accept the evolutionary character of their discipline, but also one in which clinicians and biomedical scientists, the clinic and the lab, work hand-in-hand, casting their creations. Such is artistic or creative expression. How unfortunate that this vision is seen by many in contemporary science and medicine to be inaccurate or threatening as opposed to heuristic. (Also see Toulmin [26].)

V. INTERPLAY: SCIENCE, TECHNOLOGY, AND ART

Medicine is an interplay among three axes of interpretations: science, technology, and art. Each axis of interpretation is necessary. Without science, there would be no episteme, no theory, no framework. Without technology, there would be no manipulation, no practice, no action in the world. Without art, there would be no individual expression and no cast and hand of medicine. The result is more than the sum of its parts. Within this new enterprise called medicine, there emerges a multiplicity of values or signs of significance, e.g., instrumental, aesthetic, and moral. Through such values, medicine comes to recognize its legitimate authority and its obligations to patients.

University of Colorado
Colorado Springs, Colorado, U.S.A.

REFERENCES

1. Beauchamp, T. L., and Childress, J. F.: 1991, *Principles of Biomedical Ethics*, 3rd ed., Oxford, New York.
2. Blois, M. S.: 1990, 'Medicine and the Nature of Vertical Reasoning', *New England Journal of Medicine* **318** (13), 847–851.
3. Collingwood, R.G.: 1978, *The Principles of Art*, Oxford University Press, London.
4. Cullen, W.: 1772, *Synopsis nosologiae methodicae*, Charles Elliott, Edinburgh.
5. Cutter, M. A. G.: 1988, *Explanation in Clinical Medicine: Analysis and Critique*, doctoral dissertation, Georgetown University, Washington, D.C.
6. Engelhardt, H.T., Jr.: 1991, *Bioethics and Secular Humanism*, Trinity Press International, Philadelphia, Pennsylvania.
7. Engelhardt, H. T., Jr.: 1986, *The Foundations of Bioethics*, Oxford University Press, New York, Ch. 5.
8. Engelhardt, D. von: 1992, 'Causality and Conditionality in Medicine Around 1900', in this volume, pp. 75–104.
9. Fagot-Largeault, A. M.: 1992, 'On Medicine's Scientificity: Did Medicine's Accession to Scientific "Positivity" in the Course of the Nineteenth Century

Require Giving Up Causal (Etiological) Explanation?', in this volume, pp. 105–126.

10. Fleck, L.: 1935, *Entstehung und Entwicklung einer wissenschaftlichen Tatsache: Einführung in die Lehre vom Denkstil und Denkkollektiv*, Benno Schwabe, Basel; English translation: 1981, *Genesis and Development of a Scientific Fact*, F. Bradley and T.J. Trenn (trs.), University of Chicago Press, Chicago.
11. Gross, R.: 1992, 'Intuition and Technology as the Bases of Medical Decision-Making', in this volume, pp. 183–197.
12. Healy, B.: 1991, 'The Nurture of Creativity', *Journal of NIH Research*, **3** (8), 33–34.
13. Juengst, E. T.: 1992, 'Causation and the Conceptual Scheme of Medical Knowledge', in this volume, pp. 127–152.
14. Kant, I.: 1929 (1781/1787), *Critique of Pure Reason*, St. Martin's, New York.
15. Kant, I.: 1990 (1785), *Foundations of the Metaphysics of Morals*, L.W. Beck (tr.), Macmillan, New York.
16. Kuhn, T. S.: 1970, *The Structure of Scientific Revolutions*, 2nd ed., University of Chicago Press, Chicago.
17. Lie, R. K.: 1992, 'Ludwik Fleck and the Philosophy of Medicine: A Commentary on Schäfer and Tsouyopoulos', in this volume, pp. 47–54.
18. Moulin, A. M.: 1992, 'The Dilemma of Medical Causality and the Issue of Biological Individuality', in this volume, pp. 153–162.
19. Nozick R.: 1974, *Anarchy, State, and Utopia*, Basic Books, New York.
20. Peset, J. L.: 1992, 'On the History of Medical Causality', in this volume, pp. 57–74
21. Sass, H.-M.: 1992, 'Medicine — Beyond the Boundaries of Sciences, Technologies, and Arts', in this volume, pp. 259–270.
22. Sassower, R.: 1992, 'Technoscience and Medicine', in this volume, pp. 219–228.
23. Sauvages de la Croix, F. B. de: 1763, *Nosologia methodica sistens morborum classes juxta Sydenhami mentem et botanicorum ordinem*, 5 vols., Fratrum de Tournes, Amsterdam.
24. Schäfer, L.: 1992, 'On the Scientific Status of Medical Research: Case Study and Interpretation According to Ludwik Fleck', in this volume, pp. 23–37.
25. Schaffner, K. F.: 1980, 'Theory Structure in the Biomedical Sciences', *Journal of Medicine and Philosophy* **5**, 55–97.
26. Sydenham, T.: 1848, *The Works of Thomas Sydenham*, R.G. Latham (tr.), Vol. I, The Sydenham Society, London.
27. Toulmin, S.: 1992, 'Knowledge and Art in the Practice of Medicine: Clinical Judgment and Historical Reconstruction', in this volume, pp. 231–248.
28. Tsouyopoulos, N.: 1992, 'The Scientific Status of Medical Research: A Reply to Schäfer', in this volume, pp. 39–46.
29. Wieland, W.: 1992, 'The Concept of the Art of Medicine', in this volume, pp. 165–181.

HANS-MARTIN SASS

MEDICINE—BEYOND THE BOUNDARIES OF SCIENCES, TECHNOLOGIES, AND ARTS

The media, the public, the politicians have increasingly become interested in medicine. This time, however, the interest is not so much related to traditional public interest in medical breakthroughs or progress in healing, as with such issues as the allocation of scarce resources, withholding of treatment which on medical or technical grounds could easily be provided, paternalism of the physician and consumerism of the patient, active or passive euthanasia, genetic diagnosis, *in vitro* fertilization. These issues are neither discussed as purely scientific or technical matters, they are discussed instead as value-related issues, as moral, cultural, political or religious challenges. Among the dozen odd questions of everyday medical practice discussed by v. Troschke ([32], p. 4) are the following: Shall terminally ill patients be fully informed about their diagnosis? Shall life be prolonged under any circumstances using all available technology? Shall the physician get involved in human experimentation in order to promote medical progress? Shall the physician do abortions on his or her patients? None of these questions is a scientific one. But they enjoy high public visibility and prominence and have to be answered either by the medical profession or the public, i.e., the regulators, the patients, insurers, self-help groups, etc. Definitely, medicine is not one of the hard sciences, when we look at these issues.

Hans Schaefer [26], in an article on the ostensible and underlying reasons for public criticism of medicine, identifies (1) the prevailing concept of medicine as a natural science; (2) the failure to address the causation of chronic diseases; (3) superficial contact with patients; and (4) an overestimation of the effectiveness of medicine as the four major reasons for public questioning of medicine. The issue addressed in this volume: Is medicine science, technology, or art ?, therefore is not just a theoretical question within the philosophical history of ideas and theory of science discussions; it is an eminently value-related question of highest practical and public consequences.

In this paper, I will briefly discuss three aspects of the question within the cultural dimensions of Western culture: (1) the historical framework of discussions in the theory of science; (2) contemporary challenges to the normative understanding of medicine and the medical profession; (3) a new form of triage in medicine: moral triage.

Corinna Delkeskamp-Hayes and Mary Ann Gardell Cutter (eds.), Science, Technology, and the Art of Medicine, 259–270.

I. HISTORICAL PERSPECTIVE

Most questions are posed not directly with regard to what the "reality per se" is, but with regard to what we, for personal, professional, historical or cultural reasons, believe or have been made to believe the "real issues" are. In everyday life as well as in the sciences and professional activities, questions and issues are predetermined by cultural or other value-related attitudes or customs. Within the sciences, we have additionally learned to differentiate between internal factors and external factors causing scientific progress or stagnation. The possibility of raising questions such as "Is Medicine an Art?" is rooted in the special conditions of the Western intellectual tradition [23].

Epistemological questions regarding medicine are predetermined by the way "science" "technology", and "art" have been defined since the Age of Reason. Descartes's limitation of the concept of truth to only and exclusively one form of truth, that which any or nearly any individual subject would be able clearly and distinctly to perceive — *illud omne esse verum quod valde clare et distincte percipio* (Descartes [5], Med. III) — excluded other forms of understanding and establishing "objective" knowledge and obligation concerning reality. The continuous process of sharpening the Cartesian concept of objectivity and truth and the increase of practical applications of the new methodology in understanding, changing, and influencing reality reinforced and strengthened each other immensely. At the same time, the traditional cognitive concept of the liberal arts as *cognitio sive representatio sensitiva* ([2], p. 17) pauperized itself into the stylish concept of *expressio*, of subjective or Romantic self-expression of the single creative and innovative individual, the genius. The sciences establish or verify the laws of nature, while the arts and humanities create or describe ideas; in the words of Windelband [37], the sciences are nomothetic, the humanities ideographic. And between those strictly separated kingdoms of property, there was nothing, not even a no-man's land. Vast properties of human concern and human activity fell into this black hole of no-existence or were held hostage in either one of the two campsites.

Alfred Ayer, e.g., having fixed himself into this Procrustean bed of modern Western definitions, consequently put human activities, such as moral argumentation, religious experience and aesthetic statements, into the one basket for "expressions of emotions," denying that there would be another basket for "judgements of value" (Ayer [1], p. 136). There have been more discriminating positions in the theory and history of science and arts (Baumgarten [2], Schiller [16], Otto [10], Rickert [14]), but the mainstream forcefully went from Bacon to Descartes to analytic and neopositivist positions like that of Ayer.

Notwithstanding the intellectual debates among philosophers and theoreticians of the sciences, medicine did quite well in considering itself to be on the side of the sciences. In fact, medicine, using natural scientific and experimental approaches, achieved all the major break-throughs which various other methods and concepts of medicine had never even dreamed of: surgery under anaesthesia, sulfinamids and antibiotics, chemotherapy, genetic screening and genetic engineering (Gerok [8], Wolff [38]). Virchow's thesis that "life is nothing more than a form of mechanisms, even though the most complicated one" [23] was more or less the commonly accepted ideational and conceptual platform for medical research and medical treatment in the nineteenth and twentieth centuries.

Consequently, the Brockhaus encyclopedia of 1971 defines medicine as "the science of the healthy and sick forms of life, the causes, symptoms and results of diseases, their diagnosis, therapy and prevention" ([13], p. 322) and presents the term "ars medicina" and "Heilkunde" (healing wisdom) as synonymous with the "science of the healthy and sick forms of life". This definition is not only incorrect and unproductive, it is not responsive to the difference between *ars* and *scientia*.

While the Brockhaus encyclopedia stresses the aspect of science in medicine, its American counterpart, the Academic American Encyclopedia [12], emphasizes the societal role of medicine as being "responsible for the maintenance of good health..., promoting hygiene, preventing and detecting disease, curing disorders, decreasing pain, improving public health, encouraging safe water, nutritious and non-toxic food and unpolluted air" and the fact "that society demands greater access to health and greater accountability from the healing professions."

Understanding medicine as either science or art or social contract is a tempting simplification which leads into deprofessionalizing of the essentials of the healing profession. Confusing medical activities with those involved in the development and application of the natural sciences can lead into a scientification which replaces persons with data and which loses any understanding of human dignity. Confusing medicine with arts might lead into impermissible professional paternalism. Confusing medicine with political or social contracting might lead into an unprofessional and incompetent, as well as unconstitutional, political or social paternalism (totalitarian) which recognizes neither the realities of medical technologies nor the rules of free societies, not the rights of people [24].

Toulmin [33] has underlined that the historical model of Vico provides a better orientational framework for analyzing and performing medical activities than does the Cartesian geometrical model. More precisely, it is the concept of diagnosis that does not fit into the geometrical classification scheme. The progress in medicine from Sydenham to the present can be described along the lines of the diminution, even disappearance, of the substantialized concept of disease (*species morbosa*) and the move towards flexible data collections and their flexible interpretation (Wieland [36], pp. 106, 119); and within the "history" of the person, the person's history of well-feeling (not well-being primarily) and history of data collected, which are then interpreted by comparing them to what, as a result of experience, generally is understood to be normal or regular rather than abnormal or irregular. Various patterns of data are linked together in the diagnostic process. This process represents itself as a history of using various tools for analysis and for linking these tools together: finding all about the history that led to the actual state of health (anamnesis), checking data (Gross [9]) obtained in different tests (blood tests, X-rays, urine tests, genetic screening, electrical readings, cytological and histological examinations, temperature and circulatory information, etc.) against average (i.e., non-extreme, or normal) findings, and then determining and applying tools in order to influence the future history of the healthy/unhealthy person in the desired direction (therapy).

For diagnosis as well as for therapy, there are no ontological classification schemes available for distinct and clear definitions of disease and health. On the other hand, the Cartesian requirement of clear and distinct classifiable knowledge is neither the only *conditio per quam* nor the only *conditio sine qua non* in the process of influencing future developments of life forms, i.e., medical treatment, which can be initiated or continued.

Gross rightly states that so far there has not been a uniform and generally valid definition for either health or disease ([9], p. 183). This is not only true with regard to the various different parameter schemes used in forms of medical reasoning and healing generally accepted in academic circles. It is also true with regard to those models of medical reasoning, explanation, and treatment generally not widely accepted, as represented by various schools of Western or Asian holistic medicine (Blohmke [4], Schwanitz [29], Polkert [18]), psycho-social (Siegrist [30]) or environmental reasoning (Perger [17]) with regard to subclinical intoxications or regulatory disorders. Diagnosis, according to Wieland, is a "temporary singular statement"—*zeitgebundene Singulaeraus-sage* ([36], p. 171)—integrated into a context of treatment.

Medicine's main goal is singular treatment of individual persons, not the establishment of general rules, nor the classification or integration of "disorders" or "irregularities" into parameters or classificatory schemes, even though various classificatory schemes play auxiliary roles in diagnosis and therapy.

Classifications of various forms are used not for finding the "truth" but for predetermining the problem and its final resolution. Various sciences, such as radiology, hematology, pharmacology, even sociology (Begemann [3]) and psychology, and various technologies and techniques, such as surgery, electrocardiography, anaesthesia or message, are applied in medical activity. However, the auxiliary use of applied sciences and applied technologies in medicine does not make medicine a science; neither does it make it a technology, nor an applied science or an applied technology. Modern efficacious medicine will not disrespect clinical experience and scientific data (Schuster [28], p. 446) nor will it disregard basic standards in technology and its applications like accuracy. But the proper application of the natural or social sciences and various technologies, their methods and results, does not by itself make good medicine. The *essentialia* lie beyond the boundaries of the applied sciences, technologies, and arts.

As the confusion of medicine with science or technology or art is not even half right but all wrong, the history of medicine suggests that we ought to look somewhere else when we want to identify the *differentia specifica of medicine*. We want to look into the traditional attitudes of healers towards the persons in need of help; and we want to look into the perceptions and expectations of the individual patient and of the general public with regard to what the responsibilities and the duties of the medical person and the medical profession are, or ought to be, i.e., into the professional ethos.

Traditionally, the members of the medical profession have not simply been identified by the level of their scientific knowledge or technical expertise, but rather by whether or not they had taken an oath to respect certain standards and values with regard to the understanding and implementation of their professional activity, to their peers and their patients, and with regard to a certain relatedness to the cosmic harmony of life. The distinctive value was professional responsibility, not truth, not efficiency. The ethos of a very special professional responsibility rules and governs in the Hippocratic Oath and all subsequent formulas that govern scientific insight, technological expertise and social and political power and focus them to one and only one goal: comforting, healing, and protecting the well-being of the patient, not scientific truth, technical efficiency, or political power (Wolff [38], p. 127f).

II. CONTEMPORARY CHALLENGES

Pellegrino and Thomasma identify four ingredients that are essential in each and any medical event: responsibility, trust, decision orientation, and etiology ([15], p. 69). No one of the ingredients alone would establish the medical event. All four of them have to come together in an integrative manner in a situation in which one person is seeking help and the medical person is providing that help. Pellegrino and Thomasma identify, without intending to present a watertight definition of medicine, various aspects of medical activity: a cognitive aspect appears in diagnosis; a predictive aspect in prognosis; and an operative aspect in therapy. But there also is an affective aspect in the relation between patient and medical person; an affection of trust, of friendship, of gratitude, of justice, of beneficence, of hope, of respect. Also, cultural values are involved: changing perceptions about the nature or benefit or harm of certain statuses of health; the goal, the limits and the means of health care and curing; perceptions about the social relevance of or the societal responsibility for health care ([15], p. 69).

Though the preliminary phenomenological descriptions of the medical event, as presented by Pellegrino and Thomasma, are far from a watertight definition of medicine, the question is whether working for such a clear-cut definition is adequate or beneficial or achievable at all. If medicine is something other than a science or a technology, even though it applies sciences and technologies, such a definition seems not to be a prime concern in the theory and practice of medicine. And the question of truth or precise classification seems not to be a prime concern in medicine either, nor does the efficiency in applying each and every available technical possibility to fight disease or termination of life or some of its functions in any case. Medicine, therefore, is a very important case in the theory of science, important because it shows that the philosophical debate concerning conceptualization in the theory of science and of technology has failed to recognize areas of activity in which the sciences and the arts and the technologies are not only applied but which are beyond the limits of each of them.

Facing the multitude of applications of technical and scientific expertise in medicine calls increasingly for not confusing medicine with science, technology and arts, but rather identifying it with its own ethos and its own tradition of true professionalism. For the time being we would hold that in medicine *responsibilitas governs possibilitas*, that applied sciences and applied technologies and applied arts play important auxiliary roles in medical activity, and that there is no urgent need for establishing a Procrustean bed of definition of medicine, but rather an urgent need to continue to increase the dialogue concerning the ethos of healing inside and outside the healing professions.

Such a dialogue would have to assess two sets of data, the scientific and technical data such as x-ray and blood tests, and the non-technical (cultural and moral data) such as risk acceptance, value priorities, goals of well-being and well-feeling. Accentuating both sets, rather than suppressing one or the other in diagnosis or therapy, would help to (a) leave behind the narrow-minded natural science concept of medicine; (b) decrease outdated professional medical paternalism; (c) prepare a healthy and communicative physician-patient relationship; and (d) lead to an educated and experienced, not just informed, health management [25]. Such an approach will finally restore and preserve the specific professional authority of the healing professions.

Among the contemporary challenges to such a transformation of healing authority into the modern world of technology and value plurality are: (1) the fashionable professional retreat into defensive ethics and defensive medicine (refusnikism); (2) the shift of emphasis from medicine towards economization, legalization, politization, and consumerization of genuine medical issues (confusionism); (3) the loss of moral and professional authority to the forces of value pluralism, consumerism, and the heteronomous regulation of health matters by public authorities (deprofessionalization).

1. The most influential position regarding the moral assessment of new technology in Germany is Jonas's concept of defensive ethics [11]. It is Jonas's thesis that technological developments have outgrown human capacity to handle the most recent developments and results of technological progress and that therefore the progress and the application of technology have to be curtailed politically and culturally. Morally refusing to recognize the multiple benefits modern technology could provide for healing and comforting as well as for other human needs or goals has a strong support in the contemporary culture of technophobia among intellectuals. If the intellectual trend of generally promoting defensive ethics spills over into the healing professions, we will face serious medical and moral problems, among them: non-application of available treatment, non-development of better knowledge, non-development of better treatment, non-disclosure and non-discussion of available options for improving well-being or well-feeling, and finally the loss of moral authority and moral responsibility in stewarding medical sciences and technologies for the sake of the patient and for the improvement of health and health management. Given the rise in technical breakthroughs in

medicine and the likelihood of further good progress, quite the opposite of defensive ethics would be required: an offensive and aggressive ethics that would fight harder than ever before to guide and to safeguard and steward the natural sciences and the various forms of technologies for the better cure of the patient and for better health management options. Offensive medical ethics will have to protect medicine and the medical professions from becoming a part of the intellectual technophobic mania, from violating the codes of good medical conduct that require the provision of the best available help to the sick and the suffering. Concentrating on aggressive and offensive medical ethics would include an increase in medical and public moral argumentation, more research and teaching in medical ethics, an increase in risk competence and risk responsibility, and an anticipatory debate regarding upcoming moral options opened up by morally appreciable progress in medical technology and medical sciences ([23], [24]).

2. A certain loss of moral authority by the medical professions has already caused or at least is accompanied by an oversupply of legal, economical, political frameworks dealing with health matters. Economic considerations have led to lists of diagnosis-related or age-related selections of patients and forms of treatment or non-treatment. In the future, the successes of health management in a nation might be more dependent on legislative or administrative moves and insurance policy than on good medicine or good health management. Debates on the structure and development of public health care systems might replace case-oriented sound physician-patient relationship or at least interfere severely with those relationships, as well as with good personal health information and health management. Consumerism on the part of patients and a self-understanding of physicians who see themselves as nothing else than scientists will develop new forms of consumer-provider relationships. The negative side of such a development involves the confusion between medical and economic issues and the confusion between the right to personal autonomy in an open society with the presumed right to place the unintended but predictable results of lifestyle choices on others, the taxpayers, the society. The best available strategy against such a value confusion is an increase in moral argumentation in public debates to which the healing professions have a great deal to contribute. Engelhardt has proposed understanding bioethics as the exercise grounds and tournament fields of educating, strengthening, and supporting the dialogue among educated citizens, educated and health risk competent patients, and risk competent health professionals [6]. From a different point of view, Fuchs has called for a careful study in translating traditional religious and moral values into new situations

created by the abundance of technology and plurality in values in the modern society [7], thus transforming good and proven values into new situations in post-Enlightenment times, characterized by personal responsibility emancipated from heteronomous forces and by technology not available to previous generations of physicians and patients.

3. Finally, the traditional concept of medicine is challenged by not balancing the principle of informed consent against any other values or principles. Surely, there is no place anymore for outdated forms of professional paternalism over uneducated and unemancipated people as in former times. Rather, professional authority has to be transformed into the new world of a pluralistic society based on free choice and technology. The prescription given for such a transition by Engelhardt — "full scale development of bioethics as the language of secular policy planning for the health care professions and the professional ethics of health care" ([16], p. 236) — and Fuchs — the reassessment of traditional values for "responsible manipulation of life and human life" ([17], p. 241) are indispensable tools for undertaking such an effort. But most importantly, this will need to lead to a new value-oriented understanding of health and disease. As Roessler puts it, "Health is not the absence of obstructions, but the capacity to deal with them" ([21], p. 63). If this is an acceptable definition, and if we follow Roessler in this, then health education and health risk education will have to play a major role in any society based on freely-contracting individuals and on free and responsible individual choices. It is in this context that the health professions will have to exercise their professional authority in educating the public and the individual patient as to the choices and the costs and benefits involved in making good or bad choices. Such an approach to health education and health management cannot be a narrow one that focuses only on medical intervention in crisis style management. It has to address the long-term implications of lifestyle choices, as well as the implications of new forms of genetic and other early forms of diagnosis.

Translating Rawl's understanding of justice into the medical setting, Rendtorff describes professional medical responsibility as the dimension within which the patient's lost autonomy as a person in pain or suffering will have to be restituted. The "restitution of the autonomy of the suffering human" ([20], pp. 159, 112), according to Rendtorff, also has to set priorities for public health policy and medical research and research in related sciences and technologies. In this regard, Illhard differentiates between two forms of dialogue between patient and physician. One is "responsibility in reciprocity", the other is "responsibility as a response" ([10], p. 15).

Only the first one is fully compatible with the principle of informed consent as a means towards choosing the correct therapy. The second one cannot be based on informed consent, because the patient has lost part or all of his autonomy by being in pain, desperately in need or for other reasons. Rather, this form of medical responsibility sees the restitution of autonomy (i.e., the precondition of the possibility of informed consent) as a goal to be achieved by treatment, thus making informed consent a goal rather than a precondition for treatment. The task is, as Rendtorff puts it, to "replace absent autonomy, not in order to increase or deepen dependency but in order to eliminate or reduce it." Treating patients was always and will always be a "specific concretization of being a fellow-human, a reflexivity of concrete moral life" ([20], p. 161).

III. MORAL TRIAGE IN MEDICINE

There is a lot of talk about medicine approaching a new age of total triage, i.e., economic triage, because of a shortage in health care professionals, facilities, technologies, etc. Such an uncomfortable situation does not have to occur. Also, I do not think this is the correct description of the situation and challenge of medicine today. There is another form of triage that seems to be more pressing. Still undetected, but growing, it is the undersupply of the capacity to deal adequately with traditional values and principles in new situations of health risks and medical possibilities. This moral triage is a product of the combination of abundant technology available in a society that is diverse in values and personal choices. The decision whether or not medicine will win the battle of the fine art of healing and comforting over the economic, scientific, political or technical diversions and infiltrations will be determined by its capacity to translate and to transform traditional value-related attitudes and responsibility structures into the new world of free and educated citizens with its plurality of value choices and the abundance of technologies and sciences. The prescription Pellegrino recommends against the danger of what we call the moral triage is the same this paper calls for: "in a technologically-oriented society, it is precisely the widespread study of the liberal arts and of values and ethics that is required if we are not to be overwhelmed by our own creations. Isaiah's warning is more pertinent than ever: Their land is full of idols; they worship the work of their hands, that which their fingers have made" ([16], p. 47). We don't have to cite Isaiah. There is an old German saying, putting to rest the overintellectualized epistemological discussions surrounding medicine as well as overpoliticized public health administration: the healer always is right (Wer heilt, hat recht). The new appreciation of the art of healing in an age of technological abundance and abundance of personal choices seems to be not so different from the old times where there was not much technology available and not much personal choice, either. The future of the *Art of Healing* will depend on its capability to reemancipate itself from the procrustean beds of the sciences and the politics, as well as on a skillful art for the transformation and translation of traditional medical authority into the new dimensions of the secularized and technical societies of the future.

Ruhr-Universität,
Bochum, Germany, and
Kennedy Institute of Ethics,
Georgetown University, Washington, D.C., U.S.A.

BIBLIOGRAPHY

1. Ayer, A.J.: 1972, *Language, Truth, and Logic*, Penguin Book, Hammondsworth.
2. Baumgarten, A.G.: 1750, *Aesthetica*, Vols. 1,2, Frankfurt/Oder.
3. Begemann, H.: 1973, 'Emanzipation und Sozialisation als Grundlage und Ziel ärztlichen Handelns und künftiger Krankenhausstrukturen', *Medizinische Klinik* **63**, 553–538.
4. Blohmke, M.: 1983, 'Der ganzheitliche Aspekt der Medizin aus der Sicht der Epide-mologie", *Medizin, Mensch, Gesellschaft* **8**, 72–78.
5. Déscartes, R.: 1641, *Meditationes de Prima Philosophia*, Paris.
6. Engelhardt, H.T.: 1986, 'Bioethik in der pluralistischen Gesellschaft' *Medizin, Mensch, Gesellschaft* **11**, 236–241.
7. Fuchs, J.: 1986, 'Verfügen über menschliches Leben?', *Medizin, Mensch, Gesellschaft* **11** 241–247.
8. Gerok, W.: 1979, *Zur Lage und Verbesserung der klinischen Forschung in der Bundesrepublik Deutschland*, Boldt, Boppard.
9. Gross, R.: 1992, 'Intuition and Technology as Basis of Medical Decision-Making', in this volume, pp. 183–197.
10. Illhardt, F.J.: 1985, *Medizinische Ethik*, Springer, Heidelberg.
11. Jonas, H.: *Das Prinzip Verantwortung*, Suhrkamp, Frankfurt.
12. 'Medicine', *Academic American Encyclopedia*, Vol. XIII, 1980, p. 267.
13. 'Medizin', *Brockhaus Enzyklopaedie*, Vol. XII, 1971, p. 322.
14. Otto, R.: 1929, *Das Heilige*, Klotz, Gotha.
15. Pellegrino E. and Thomasma, D.: 1981, *A Philosophical Basis of Medical Practice*, Oxford University Press, Oxford.
16. Pellegrino, E.: 1986, 'Medical Ethics. Where Is It Going?', *NYU Physician*, Fall, New York School of Medicine, pp. 43–47.
17. Perger, F.: 1983, 'Regulationsdiagnostik zur Erfassung subklinischer Belastungsfaktoren', *Medizin, Mensch, Gesellschaft* **8**, 79–88.

18. Polkert, M.: 1983, 'Die Methoden der chinesischen Wissenschaften als Grundlage einer Ganzheitsmedizin', *Medizin, Mensch, Gesellschaft* **8**, 89–96.
19. Reich, W.: 1986, 'Paradigmen für die Bioethik', *Medizin, Mensch, Gessellschaft* **11**, 231–236.
20. Rendtorff, T.: 1981, *Ethik*, Vol. 2, Kohlhammer, Stuttgart.
21. Roessler, D.: 1971, *Der Arzt zwischen Technik und Humanität*, München.
22. Rickert, H.: 1922, *Die Grenzen der naturwissenschaftlichen Begriffsbildung*, 5th ed., Winter, Heidelberg.
23. Sass, H.M.: 1983, 'Technical Values and Human Values', *Wandlung von Werten und Verantwortungen in unserer Zeit*, Saur, München, pp. 221–236.
24. Sass, H.M.: 1987, 'Philosophical and Moral Aspects of Manipulation and Risk', *Swiss Biotech* 5, (No. 2a) 50–56.
25. Sass, H.M.: 1992, 'Diagnosing the Eleven Month Pregnancy', in J. L. Peset and D. Gracia (eds.) *Ethics of Diagnosis*, Kluwer Dordrecht, pp. 153–162.
26. Schäfer, H.: 1986, 'Gründe und Hintergründe der Kritik an der Medizin', *Medizin, Mensch, Gesellschaft* **11**, 265–273.
27. Schiller, Fr. von: 1795, 'Ueber naive und sentimentalische Dichtung', *Die Horen*, **11**, Tübingen, pp. 43–76.
28. Schuster, H.P. and Weillmann, L.S.: 1981, 'Behandlungsergebnisse internistischer Intensiveinheiten', *Medizinische Klinik* **76**, 419–422, 443–446.
29. Schwanitz, H.J.: 1983, *Homoeopathie und Brownianismus 1795–1844*, G. Fischer, Stuttgart.
30. Siegrist, J.: 1980, 'Koronare Herzkrankheiten: Psychosoziale Aspekte ihrer Prevention', *Fortschritte in der Medizin* **98**, 797–800.
31. Stephens, P.M. and Henry, I.P.: 1977, *Stress, Health and the Social Environment*, Springer, New York.
32. Von Troschke, J. and Schmidt H.: 1983, *Ärztliche Entscheidungskonflikte*, Enke, Stuttgart.
33. Toulmin, S.: 1992, 'Knowledge and Art in the Practice of Medicine', in this volume, pp. 231–248.
34. Uexkuell, Th. von: 1979, *Lehrbuch der psychosomatischen Medizin*, Urban und Schwarzenberg, München.
35. Virchow, R.: 1856, 'Alter und neuer Vitalismus', *Archiv für pathologische Anatomie, 9*.
36. Wieland, W.: 1975, *Diagnose. Überlegungen zur Medizintheorie*, De Gruyter, Berlin.
37. Windelband, W.: 1894, *Geschichte und Naturwissenschaften*, Strassburg.
38. Wolff, H.P.: 1983, 'Hat die Medizin versagt? Resumee einer Generation', *Informationen des Berufsverbandes der Deutschen Chirurgen*, Nr. 7, pp. 120–128.

CORINNA DELKESKAMP-HAYES

IS MEDICINE SPECIAL, AND IF SO, WHAT FOLLOWS?: AN ATTEMPT AT RATIONAL RECONSTRUCTION*

The title of this volume promises "dialogues". Dialogues are an apt manner of presentation where the native complexity of novel topics calls for a multiplicity of perspectives.[1] In contrast to the employment of dialogues as a mere literary device, the "real life" confrontation of different researchers combines the advantage of greater heuristic variety with the disadvantage of greater conceptual confusion. As no one author's mind has arranged the illuminating mutual understandings and misunderstandings between the participants, no one can be sure about understanding them all.

In this essay, I shall try to remedy the latter disadvantage. I shall construe a common conceptual ground on which the authors assembled in this volume, in spite of their heterogeneous interests and their disagreements on special issues, could conceivably all feel at home. In response to the rich variety of subjects, viewpoints and argumentative levels characterizing the authors' very diverse contributions, a coherent account of medicine's place between science, technology, and art will be stipulated.

The difficulty of such an undertaking becomes obvious if one spells out the mutually exclusive positions which the various essays imply. Such oppositions could be reconstructed in many ways, but already one single version will illustrate the point:

1. "Medicine is a science, no longer an art" refers to the transformation of merely empirical or wait-and-see medicine to a discipline favoring rational justification and predictability (Feinstein in [50]).
2. "Medicine is still an art, not a science" insists on the individual nature of patients which cannot simply be subsumed under general laws ([20], notwithstanding [36]).
3. "Medicine is a technology, no longer an art" regards the increasing controllability and effectiveness of patient care [59].
4. "Medicine is still an art, not a technology" emphasizes the personal side of truly efficient health care and the intersubjective aspect of the physician-patient relationship [53].
5. "Medicine is a science, not a technology" highlights the importance of purely scientific research for the eventual improvement of applicable knowledge (Virchow in ([18], [36]).

Corinna Delkeskamp-Hayes and Mary Ann Gardell Cutter (eds.), Science, Technology, and the Art of Medicine, 271–319.

6. "Medicine is a technology, not a science" considers the cost of developing and maintaining both research facilities and clinical equipment, and thus the social asset character and public policy relevance of health care (Fleck in [49]).

Given these (and many other possible) contradictions, the project of determining one single conceptual ground for all the dialogue-partners requires an effort at differentiation. We shall have to distinguish the various respects in which medicine is like or unlike various aspects of science, technology, and art. These respects will be specified by a number of propositions which are devised with a view to general acceptability.

At the outset, it is useful to remember that the interpretation of medicine in terms of science, or of technology, or of art, is not only a matter of hitting upon the appropriate details but also of pursuing certain interests. In the nineteenth century, conceiving of medicine in terms of a "science" implied a program for improving its cognitive presuppositions and for rendering the art of medicine more certain and objective. Similarly, the present-day technological approach aims at improving medicine's effectiveness and efficiency by subjecting practice to technical control. The alotment of medicine to "art", on the other hand, is not motivated by any program for progress. Instead, it serves the conservative interest in medicine's traditional vocation. Or, to put this contrast differently: Subsuming medicine under science or technology is designed to prove medicine's theory or practice to be at least potentially just like other (scientific) theories and (technological) practices, and to encourage present tendencies at realizing this potential. The motive behind calling medicine an "art", however, is to stress the "at bottom" *sui generis* character of medicine and to discourage any radical transformation.

The supposition of such a special status of the "art of medicine" is usually linked with medicine's traditional ethos and with a claim to special professional privileges for the protection of that ethos. While most of the authors of this volume have restricted their deliberations to the structural characteristics of medical theory or practice, i.e. to the "matter of fact" side of medicine, Toulmin [53] and Sass [47] have also addressed its "humane side". In this context even the science and technology of medicine are linked with ethos and professional policy. Moreover, while most of the authors at least implicitly support a technological model, or a reduction of the medical profession to a technological occupation, the latter two provide a more encompassing framework of concepts in which the limitation of those technological components can be understood.

In order, therefore, to secure a common conceptual ground, it must be clarified, to what extent medicine's ethos is indeed operative in medical practice such as to distinguish medicine from other comparable occupations. For this purpose, the concept of "art", which played a somewhat minor role in many of the essays, will be redefined so as to establish the needed link between the structural (matter of fact) characteristics and the ethical as well as social significance (or humane side) of medicine. In addition, it must be clarified why such a difference should necessitate any special protective societal measures.

The conceptual distinctions developed here will thus be divided into two parts. First (A), it will be investigated in what senses not only "science" and "technology" but also a more encompassing concept of "art" can be employed to illuminate just "what medicine is". Second (B), the difficulties attending medicine's ethos will be explored in view of their possible function for supporting claims to protective professional privileges. Here the three title terms will be interpreted in view of medicine's precarious social position.

A. THE CONDITIONS OF MEDICINE

The essays in this volume have been assigned to two separate parts, the first focussing on medical knowledge, the second on medical practice. This established conceptual duplicity distinguishes medicine from each: science, technology, and art (either in the sense of craft, or of fine art). We speak of the practice of medicine as we speak of the practice of an art (a craft), but we do not speak of the practice of a science. On the other hand, we do not speak of "artistic theory" in the sense in which we speak of "medical" and "scientific" theory. As to technology, this term is perhaps still too novel to enjoy any set ways of expression. It does sound odd to speak of the "practice of technology" and we are not very clear what the (less odd) term "technological theory" could stand for.[2]

We do not, however, in common usage separate "engineering theory" from practice in engineering, as we separate both in medicine. Perhaps the reason is that in engineering, both are more intimately connected than they are in medicine and less trivially than in science, where "practicing" mostly amounts to "theorizing" anyway.[3] Thus medicine is an object both "of practice" and "of study"; science is an object of study only; and the arts and crafts — even though they may be learnt — are objects of practice only.

> #1 Or: To claim that medicine is different from science, technology and the arts is legitimate insofar as only medicine incorpo-

> rates both a codified body of knowledge and a codified body of rules for (not merely theoretical) practice.

The term "practice of" medicine or of a craft indicates that there are set rules and ingrown principles for professional performance, and perhaps it is the lack of such rules which accounts for the missing "practice of" science.

> #2 Or: To claim that medicine is an art is legitimate in view of its (not merely theoretical) practice being subject to codified rules.

To be sure, each science establishes a body of methods and even devises a methodology that is far more explicit and rigorous than what medicine or the arts and crafts would employ. But at least in the creative aspects of scientific work, the more important hypotheses have no set way of making themselves known. Perhaps it is this element of invention which also precludes technology from being "a practice" in this regulated sense. Whereas it is a mark of excellence in a scientist and a technologist to come up with novel ideas, in a practicing physician such inventiveness today is deemed suspicious.

Yet, on the other hand, technology, like medicine, is studied and at the same time oriented toward practice in such a way as to "stand between" the arts (as crafts) and the sciences. It is therefore not clear whether the "codified rules"-character and in this sense traditional nature of medicine is in the end sufficient for establishing an essential distinction. This question will have to be decided. If medicine as "art" would be special, medicine as "technology" would not be special at all.

At any rate, performing as a physician (and perhaps as a standard technologist) involves an intertwinedness of knowing and (not merely mental) doing which is different from that involved in performing as either a scientist (and certain special sorts of technologists) or as an artist or a craftswoman. The capacity of performing as a physician (and perhaps as a technologist) can be distinguished from the other capacities already in terms of an analysis of that special intertwinedness. Whatever other capacity-differences may exist, for the desired distinction this limited perspective will be sufficient.

Professional performance is also subject to societal, or objective, conditions. Capacities, by contrast, can be considered "subjective conditions". I shall call that aspect of medicine which can be analyzed in view of the capacities entering into its practice medicine's "matter-of-fact side" (I); medicine's ethical (and thereby also humane) side (or medicine in its full sense) can be analyzed only by including the organizational conditions of that practice, or of medicine as a social institution (II). The first analysis – irrespective of the discovered intertwinedness–will be found to favor a technological understanding of medicine, the second will force us to devise a very special notion of "art".

I. *Medicine as a Capacity*

Three aspects of the intertwinedness of knowing and doing, which characterize medicine as a capacity, can be distinguished: (1) their complementarity; (2) their mutual regard; and (3) their inherence in one another.

1. *The complementarity of knowing and doing.* Capacities can be acquired and they can be realized. In medicine, just as in science, professional capacities are acquired through study.

> #3 Or: To claim that medicine is like a science is legitimate in view of its representing a well-circumscribed academic discipline.

But the study of medicine also involves practice. This practice concerns, first, the learning of certain manual or cognitive skills, such as sewing wounds, employing technical devices, recognizing clinical signs, and evaluating scientific data. Learning here depends on exercising and is similar to the acquisition of an art or of a craft. Even science students have to go through such a narrow training when learning how to set up experiments. In medicine, however, just as in the arts and crafts and unlike in science, acquiring practice also concerns working with "serious" professional issues. Apprentices at masonry are trained in real jobs; likewise medical students shortly after they begin their course of studies deal with "live patients".

In addition, science students have to learn how artificially to set up the ideally simplifying conditions for their experimental practice; medical students, on the contrary, have to train their hands and senses while already dealing with the very real and complex conditions in which their practice will take place. In that sense as well, studying medicine is more like learning a craft, because there too the resistance of an independently shaped material has to be tackled.

> #4 Or: To claim that medicine is an art (in the sense of a craft) is legitimate in view of this (doubly determined) practical training element of their education.

Just as the study of medicine, i.e., the "getting to know" about medical knowledge, needs to be complemented by "doing" medicine, so the latter needs to be complemented by the former. Medical practice is generally thought to imply the need for further study. Medicine, just as science and technology, are developing fields in a sense in which the arts (as crafts) are not. Physicians, scientists, and technologists can perform properly only if they keep abreast of current developments.

> #5 Or: To claim that medicine is like a science or a technology is legitimate in view of the fact that in each case professional performance also requires further study.

To be sure, a baker may also keep informed about new sorts of flours and eating fashions. But there are many jobs she is asked to do which can be satisfactorily completed in the traditional way. Indeed, shoemakers may even take pride in having preserved their ancient handiwork in a sense in which no scientist, technologist, or physician could. The crafts are more conservative than the other disciplines here discussed.

In sum, both study (getting to "know") and practice (in the narrow sense of "doing") enter into both the acquisition and the realization of the capacity of performing as a physician.

2. *The mutual regard of knowing and doing for one another.* It has been noted that medical practice, just as the crafts, obeys rules. Physicians' performance, just as craftwomen's, is judged in view of those rules which function as external standards. In the case of medicine, however, these rules are part of a systematized body of knowledge. Medical practice is oriented towards medical knowledge.

At the same time, this knowledge, insofar as it comprises the needed rules, is also oriented towards practice.

> #6 Or: To claim that medical knowledge is not a science is legitimate in view of its being designed for guiding practice.

That is, the knowing of medical theory includes a know-how about how to do things in medicine, just as that doing is effected with reference to the knowing.

Medicine was subsumed under the "developing" disciplines. This development also concerns medical knowledge. Unlike with crafts, in medicine a merely empirical validity of rules of practice is deemed insufficient. Medical knowledge must be rendered ever more scientific, in order for medical practice to become ever more reliable.

> #7 Or: To claim that medicine is scientific is legitimate in view of this acknowledged goal of the scientification of medical theory and practice.

The acknowledgement of such a goal at the same time encourages the transformation of the art of medicine into a medical technology. Yet even without scientific interest, a similar transformation has occurred in many crafts. Specialization, explicitation, and systematization of practical rules, of knowledge about the material's properties and the possibilities of working with it, have rendered these crafts ever more technical. The result has been a mutual integration of knowledge about how to do things and a way of doing that is exhaustively covered by standardized procedures. Just so, there is in medicine a tendency for complete technological integration of theory and practice. Its art-like dependence on traditional standards, then, has no intrinsic significance. It is to be overcome through scientific and technological progress.

> #8 Or: To claim that medicine is a technology is legitimate in view of the direction of its native progressiveness.

Nevertheless, on closer scrutiny the motivations for that technological transformation are different in both areas. Whereas medicine in many cases needs to develop in order adequately to deal with the jobs it presently has been engaging in, the jobs a craftsman is asked to do can by definition be adequately completed on the basis of the materials, knowledge, and skills available. To ask a craftswoman to do a job that goes beyond her capacities is to ask inappropriately. By contrast, physicians are quite appropriately asked for help not only with hitherto incurable diseases but also with the everyday problems posed by very hard-to-decipher signs and symptoms. Jobs in furniture restoration are selected with a view to the limits of professional competence; physicians' jobs are not. They arise like natural phenomena from the need and suffering presented by (the more afflicted) patients.

Admittedly, the standards of medical practice also provide some guidance concerning what aspects of patient suffering can be medically handled. Still, within the scope of medical responsibilities, limits of capacity have no absolute significance.[4]

In this, medicine also differs from much of technology (except for the tower-of-Babel variety) and agrees with science. For scientists will also quite appropriately undertake the solution of problems which cannot yet be solved.

> #9 Or: To claim that medicine is like science is legitimate in view of both disciplines imposing tasks which go beyond what professionals can deliver.

Admittedly, this difference between medicine and science, on the one side, and the crafts and technologies, on the other, is not a very clear-cut one. Scientists would be considered unreasonable if they would undertake quite obviously impossible tasks, and even physicians will restrict themselves to alleviating patients' sufferings when it is beyond doubt that they cannot be helped. Still, if either a scientist or a physician is unexpectedly successful in either case, such success will justify their merely temporarily unreasonable behavior. At least it will then appear to have been part of their professional duty to try.

To be sure, even locksmiths may feel challenged by a novel problem, and technologists are even asked to investigate ways of realizing what so far had seemed impossible. Yet the craftsman, in taking up such a job, will feel obliged to stick with the possibilities offered by his craft, and the technologist will be assigned certain deadlines for the desired development. Or for both technologists and craftsmen, it would be a disrespectable thing to undertake a task which they cannot deliver in a sense in which this does not hold of scientists and physicians.

On the other hand, the tasks of science and of medicine have different ways of transcending the "possible". Scientists are free to isolate those portions of their problems which promise more immediate success. Physicians, while they may also choose to attend to such isolated portions, still remain confronted with the whole picture of a patient being in pain and danger. As a consequence, whereas the rationality of an understanding gained in science is already tantamount to the solution of a problem, in medicine this rationality is only one, and not even an indispensable, prerequisite to such solution.

Irrespective of this difference, there is in medicine, science, and (creative) technology a certain tolerance for unorthodox solutions which is lacking in the crafts. Even though inventiveness is not a virtue for physicians as it is for scientists and technologists, still, if the physician is successful with his non-standard treatment, he will not be criticized.

All this amounts to saying that for medicine as for science and for only some (more daring) aspects of technology, there exists a separate internal standard for the adequacy of professional performance. This is lacking in the crafts or in standardized technologies. The tasks to be completed by the former are not (exhaustively) defined in terms of what the external standards stipulate. Scientists (and sometimes technologists) must "do justice" to their problems just as physicians must "do justice" to their patients. Where they cannot, they must work at it. By contrast, motivations for technological development incidental to a craft only arise from a regard for new jobs that can thereby be undertaken, or for old jobs that can be completed better, not however for old jobs simply as such.

> #10 Or: To claim that medicine is like a science is legitimate in view of the fact that in both fields internal standards may even suggest different evaluations than those suggested by external standards.

This implies that medicine, quite independently of its ("external-standard"-) striving toward scientific and technological transformation, also involves a certain ("internal-standard") resistance to that rational transformation. Medicine retains a number of merely empirical practices in spite of their implied conceptual backwardness, because they "work". The progressive integration of theory with practice in medicine has a limit wherever the doing is more successful by disregarding the knowing (scientifically).

> #11 Or: To claim that both medicine and some parts of technology involve an art is legitimate in view of the prudent judgement required about how to adjust internal and external standards of adequacy to one another.

As medical knowledge provides an external standard for the adequacy of medical practice, so does medical practice provide an internal standard for the adequacy of medical knowledge: Just as in technology (and unlike in science), so in medicine progress of knowledge is accepted as valid only if it can be "cashed in" by more successful practice. (Of course, a mere promise of future usefulness also counts, cf. note 5.) Admittedly, science, technology, and medicine are alike in sometimes employing experiments as mere attempts, unguided by any systematized knowledge, for the purpose of advancing knowledge. Only the latter two disciplines, however, consider working with properly practical problems as a heuristic means (or a feedback device) not only in advancing knowledge but also in evaluating the relevance of existing knowledge.

> #12 Or: To claim that medicine is also a technology (in the research-oriented sense of the term) and not merely an "applied science" is legitimate in view of this (latter) "experiment"-character of its practice.[5]

On the other hand, it is generally conceded that problems in medicine, even if they do fall within the scope of medical possibilities but have not yet been transformed technologically, sometimes fail in answering to established attempts at solution. That is, not only may a non-(internal-)standard procedure prove successful, but an external-standard procedure may also prove unsuccessful. Medical practice, unlike technological practice (let alone applied science), is essentially uncertain.

> #13 Or: To claim that medicine is not a technology (or an applied science) is legitimate with respect to all those parts of practice which are in this sense merely "experimental".

(It is sometimes said that this inveterate uncertainty renders medicine an "art". But as the practice of a craft, if conducted according to the rules, is supposed to guarantee success, and as in the fine arts uncertainly — as opposed to unpredictability — has no place at all, such a manner of speaking is not helpful.)

As a result of this characteristic uncertainty of medicine, the tolerance for non-scientific useful knowledge, which medicine had shared with technology, in medicine (quite unlike technology) is a source of pervasive professional disagreement between different "schools" of treatment. In medicine, as in religion or economic theory (two other intrinsically uncertain undertakings), "official" ways of proceeding are always accompanied by a vast array of non-standard or "deviant" versions.[6]

> #14 Or: To claim that medicine is not a science, technology, or craft is legitimate in view of the intrinsically contestable character of its professional standards.

This uncertainty is also responsible for the fact that medical knowledge comprises not only what concerns an already established medical practice but also retains large portions of merely theoretical knowledge, the practical utility of which has not yet been established. Such knowledge is preserved and its theoretical completion is pursued in the hope that its applicability will eventually be secured.[7]

> #15 Or: To claim that medicine is neither a science nor a technology is legitimate in view of the tolerance of, medical knowledge not only for scientific as well as unscientific information that is practically useful, but also for scientific information which is only potentially useful.

In other words, medicine's pragmatic orientation differs from that of technology (or the crafts) in that the regard of knowing for doing here has a longer-term orientation.

At the beginning of these deliberations, the fact that there exists a medical knowledge somewhat "apart" from medical practice, and that this distinguishes the medical field from other scientific and technological disciplines, was merely observed. At this point, however, this peculiarity has become intelligible: As medicine's tasks are posed in a different manner than are the tasks of science and technology, it makes sense to preserve a certain independence of theory from practice and vice versa.

3. *The inherence of knowing in doing.*[8] Unlike in science, technology and the arts as crafts, the physician's task also includes a separate job of determining what her task should be. In crafts and in technology, the working materials and the changes to be effected in them are predetermined. The physician, by contrast, must determine by herself what is wrong with her patient- "material" and which changes she ought to effect. Whereas in science, clarifying the nature of a problem is already a step towards solving it, in medicine these tasks can to some extent be separated.

> #16 Or: To claim that medicine differs from science, technology, and the arts is legitimate in view of the separate diagnostic part of a physician's task.

Diagnosis aims at determining what disease the patient has and how it will develop. It can be made to resemble the scientific task, if this is interpreted as recognizing what there is in a specific part of reality and predicting what will happen to it. Still, the diagnostic interest in medicine is usually limited by a regard for what can reasonably be done to help (given the constraints of time, resources, and patient cooperation).

> #17 Or: To claim that the cognitive part of a physician's practical task differs from scientists' cognitive tasks is legitimate in view of the practical limitation of the physician's cognitive interest.

Therefore diagnosing, while in itself a "getting to know" what there is, also aims at getting to know what is to be done, and in particular, what parts of medical knowledge to invoke as direction for doing it. Diagnosing thus not only involves doing things to patients (like applying instruments to his body, or, at the other end of the spectrum, making performative utterances that put him into particular societal roles), but it is itself an instance of (mental) doing that relies on (different parts of) medical knowledge for orientation. The specific intertwinedness of knowing and doing in medicine is most conspicuous here.

Even in technology and the (repair- and maintenance-) crafts, however, some residual elements of diagnostic activity are present. Consumer wishes must first be translated into the conceptual framework of the provider's professional possibilities. Yet in these instances, consumers at least always know that they are in need of help. If professionals would try to tell them, this would be construed as mere self-interested advertisement, not, as in medicine, as a properly professional behavior. Electricians surely may point out to a house-owner that his wiring is not up to safety standards, without being accused of undue advertisement. But such cases of provider-induced demand depend on someone having done a poor job beforehand (or safety standards having changed) — a restriction which does not hold in medicine.

In addition, would-be consumers of medical services may be much more radically ignorant about the kind of help they need than would-be consumers of craftwomen's or technologists' services. While even in these latter areas providers may point out that certain consumer demands cannot be fulfilled, their reasoning will always concern the limits of professional possibilities. By contrast, in medicine patients may be quite radically mistaken about the nature of their problem even where no limits of medicine's possibilities are concerned.

Finally, while even a car mechanic will have to determine what is wrong with a car that is presented as "running poorly", he is at least in a position to assess the grounds of such complaints by himself. He can get into the car and check in what way it is running poorly. By contrast, no physician is able truly to enter into the grounds of patients' complaints; he must gather them from what the patient reports, and from what he himself observes. At the same time, the tasks of diagnosis and therapy are not clearly separated in many courses of physician-patient-encounters. Each of the more dramatic diagnostic or therapeutic measures may change a patient's condition, and as the diagnoses reached are usually hypothetical in nature, the hope of finding them verified or falsified by the results flowing from corresponding therapies is not in all cases realistic.

Hence, not only is the diagnostic task difficult to fulfill, but as a result the specific uncertainty of therapeutic measures noted above is here again compounded: at each stage of a medical intervention, lacking success is in principle compatible with correct diagnosis and treatment or with incorrect diagnosis and treatment, or with any combination crosswise — and the same holds for successful outcomes.

On the ground of these difficulties, the adequacy of a physician's performance is judged in terms of an additional "subjective" standard which medicine shares only with science. Practical (as well as theoretical) training generates an "automatization of competence" or "experience". The hands' dexterity is matched by the mind's intuitive certainty in speedy combination of information and pattern recognition. At least in the routine aspects of such automatized practice, one (almost) does not need to think any more. For the craftsman and the technologist (as well as for the practicing experimenter) routine cases are (usually) themselves recognizable by intuition and indicate a routine way of proceeding as a source of subjective security and objective reliability. The physician, by contrast, when dealing with what intuitively appears as routine, may find himself in a dangerous trap, and may discover his routine way of proceeding a source of error. Hence, while craftsmen and experimenters may in good conscience rely on their intuitive capacities for handling routine cases, physicians — just as practising scientists — are required self-critically to mistrust and doubt that very intuitive understanding, which otherwise is the mark of their professional excellence. One may even conceive it to be a specific professional requirement for physicians (as well as for scientists) to consider their speedy associations and aptness at pattern-seeing as samples of mere "hunches", urging the need to explicate the underlying reasoning.

> #18 Or: To insist that medicine is like a science and unlike crafts and technologies is legitimate in view of this subjective (i.e., attitude-related) standard of adequacy.

The analysis of the peculiar intertwinedness of knowing and doing in medicine has provided a notion of medicine's unique character vis-à-vis science, technology, and the arts. Contrary to common usage, the "art" of medicine was of little help for understanding that uniqueness. It either referred to the craft-aspect of medical practice and thus to its merely undeveloped stage which is waiting for technological transformation. Or it referred to the element of prudent judgement which enters into particular medical and technological assessments concerning the importance of scientific validity versus practical utility. Thus, "art" was conceived as either a preliminary stage of technology or as one of its constituent parts.

This opens the question whether the differences that were found to distinguish medicine from technology are relevant after all. Do current tendencies at further technologization of medicine not promise a progress that should incline us simply to disregard even those remaining differences, which will not eventually be annihilated by that progress anyway?

Medicine has so far been discussed in terms of capacities entering into professional performance, and thus in terms of its subjective conditions. Its objective conditions were defined as the societal frameworks and organizational structures within which medicine as a social institution takes place. In the context of this social nature, questions concerning values must be addressed. The central difference that was so far found to separate medicine from technology (and which is not amenable to annihilation through any progress in the science and technology of medicine) concerned the potential transcendence of medical tasks to medical possibilities. This transcendence implies a reference to medicine's ethos as the particular expression of its value-orientation. The question, then, whether such differences can be disregarded as immaterial bears on the question of how materially medicine's value-implications differ from those of technology.

II. *Medicine as a Social Institution*

Social institutions are collective ways of serving a purpose, or of realizing values.

Science, technology, and the arts are also social institutions. Science realizes epistemological values and this is its main purpose. It also indirectly realizes non-epistemological values. Through the application of scientific knowledge, practical purposes are served, and through providing one general court of appeal for factual questions, social coherence is enhanced. The general supposition that science is "value-free" only emphasizes the fact that the objects of scientific scrutiny are regarded exclusively in view of their epistemological value.

Technology serves the purposes which it is employed for, and these are primarily the practical purposes of having things and enjoying services. It also serves the theoretical purposes of finding out how practical purposes can be better realized. Thus, while primarily pursuing non-epistemological values, technology also realizes a certain selection of epistemological ones.

With the arts, whether conceived as fine arts or as crafts, the emphasis is definitely on the non-epistemological side, and any knowledge sought is restricted very narrowly to either aesthetic or instrumental values or to any combination of both.

In addition, all three kinds of social institutions provide frameworks within which individuals pursue goals such as gain, fame, or immediate satisfaction. Thus, capacities for professional performance are themselves instrumental values in view of such goals.

In what concerns medicine, it clearly fits with the latter ("framework"-)

definition. Just like artists, craftsmen, technologists, and scientists (and at times assuming any one of these roles), doctors work for fun, status, and a living.[9] They also render services for which their expertise is sought. These services always apply medical knowledge and sometimes advance it as well, so they always presuppose and sometimes amount to the realization of epistemological values. One might conclude that medicine as a social institution occupies a space within technology.

> #19 Or: To claim that medicine is a technology is legitimate in view of the classes of values it realizes.

In addition, the values realized by both medical and technological services (or the production of both technological and "health" goods) have a reference beyond individual utilization and towards society as a whole. Just as transportation and communication technologies satisfy not only individuals' needs but enhance economic and social infrastructures, so restored health in many cases not only pleases the individual but also enhances the economic efficiency of a society. In that sense, both medicine and technology serve individual as well as collective goods in a way in which this is at least not quite so true of arts in the sense of crafts such as shoemaking or hairstyling. Accordingly, both are to a large extent publicly funded.

> #20 Or: To claim that medicine is a technology is legitimate in view of the societal bearing of its practice.

Yet there is an obvious difference in medicine's and technology's relation to values. Whereas technology is merely receptive to any values for the realization of which its users employ it, and is oblivious with regard to their nature, medicine has always been considered value-oriented in the sense of carrying specific value-commitments. These are determined by medicine's (explicit or implicit) ethos. Whereas limits in the useability of technology are determined by the limits of technological resources, there are uses to which medical resources could be conceivably put, but which are ruled out by professional standards. Already the Hippocratic version of medicine's ethos excludes certain services even though these might answer consumer needs or physicians' ambition for extending their competence. The standard example for the first is euthanasia, a second is genetic screening. As both show, limits of professional competence are disputed matter, and they change through time. Yet, as such disputes also show, it is deemed an important task for the profession to keep determining the scope and limits of its competence not merely as a matter of fact, but also as a matter of norms.

> #21 Or: To claim that medicine is not a technology is legitimate in view of the normative constraints limiting the use and extension of medical resources.

Yet strictly speaking, at least since the Second World War, it no longer holds that technology is an entirely ethos-free domain. In the last forty years, codes of professional ethos have been designed for many technologies. Take for instance the 1950 code of German engineers [56], or more recent discussions about the fact that "technologists should be socially responsible" ([54], p. 350; see also [30]). The recommended ethical codes center around the double imperative of "doing good" while also "doing no harm", which holds for medicine as well. This double orientation arises from the insight that actions, however they may be well-intended for serving individuals or humanity, cannot be guaranteed to come out only as intended. They may have unforeseen negative consequences which obviate their true purpose. Consumers in the technological as in the medical area are neither individually nor collectively always in a position to determine exactly the kind and extent of services required for their purposes or to have an immediate grasp of these services' quality. The fact that in both areas experts are sometimes invoked who rarely reach unanimous judgements illuminates this point.

> #22 Or: To claim that medicine is like a technology is legitimate in that both require that consumer scrutiny is supplemented by producer obligation.

Still, technological codes differ from medical codes with respect to (1) origin; (2) orientation; and (3) function.

1. *The origin of ethical*[10] *codes.* Technological codes are designed to re-establish a public acceptability ([61], p. 154) which has never been (until recently) doubted in medicine. Technological developments tend to acquire a momentum of their own. They "get out of hand" and become quasi-autonomous factors in changing reality. This problem presently receives much attention in Western democracies (see, for example, [35], p. 287) and has caused some public disenchantment with scientific and technological modernity as a whole. Hence, technological codes must incorporate value choices that are generally accepted within a given society.

To be sure, medical codes can also survive only if they are generally accepted. Already the very notions of health and disease reflect changing societal valuations. Yet, social acceptance of the medicine of any given time does at least not exclusively result from an adaptation to predetermined societal value choices.[11] Instead, it results from a century-old history of mutual influences. Of course, technology also influences a society's value perceptions. Thus, technology in medicine opens up new possibilities of diagnostic and therapeutic intervention, calling for clarification of how established value choices can be rendered applicable or must be adjusted to these new realities. But such technological influences occur merely as a matter of course, and due to a changed order of things. By contrast, the medical profession also defends its professional commitments even against the trends of a society's value preferences (see [10], pp. 7f, 11), or as a matter of intentional advocation.

> #23 Or: To claim that medicine is not a technology is legitimate in view of the intentional and not merely incidental nature of its influencing society's value perceptions.

That is to say that medicine has a cultural identity, or a tradition of value commitments to which it is in turn committed and which limits its capacity to compromise (cf. [31], pp. 3, 22; [14], pp. 150ff; [7]; [8]). In this sense, medicine, unlike technology, imposes a certain devotion toward (what is construed as) a noble past.

> #24 Or: To claim that medicine is not a technology is legitimate in view of its constituting a tradition-oriented vocation.[12]

2. *The orientation of ethical codes.* A central element of that traditional commitment concerns medicine's obligation towards individual patients. Irrespective of any societal goals it is employed for, therapy-oriented medicine is obliged toward individuals. It concerns individuals not merely indirectly and in consequence of affecting people in general, as with technology pursuing the good of anyone as well as the public. Rather, such medicine concerns individuals also directly, as each patient presents herself for medical attention.

> #25 Or: To claim that therapeutic medicine is not a technology is legitimate in view of its concern for individuals.[13]

In addition, this obligation concerns the "good" of those individuals not merely in terms of their expressed consumer wishes (as in the crafts), and not always in a comparatively peripheral manner as does the ethos of technology. It pursues that "good" by reference to medical judgment specifying and correcting these wishes, and sometimes in view of existentially threatening circumstances. Whereas craftsmen and technologists will advise a client against realizing his objectives by arguing that "it won't work", physicians discouraging patient objectives will argue that "it won't help you". While the technological "do no harm" affects individuals either trivially (in demanding conscientiousness of professional performance and thus the avoidance of immediate harm) or indirectly (in demanding that long-term societal consequences should be taken into account so as to avoid merely mediate harm), medicine's ethos responds to individual patients sometimes trusting "themselves" directly and non-trivially into the hands of a physician, who is then responsible for them.

Accordingly, with growing awareness about the psychic factors influencing health status, individual patients come into medical view not only as sick bodies, but as human beings with their thoughts and feelings inseparably attached. The particular difficulty of the diagnostic task in medicine (cf. #16) becomes comprehensible in view of the holistic sense in which the patient's medical "good" must be pursued. The "gnostic" element of "diagnosis" [50] can thus be derived from an intensity of concern that is incompatible with the technological spirit of regarding everything from the perspective of speedy manipulability. What is really wrong with particular individuals must be assessed both on the grounds of their complaints and with a critical reserve vis-à-vis their likely misconceptions of the nature and the seriousness of these complaints.

> #26 Or: To claim that medicine is not a technology is legitimate in view of the necessary proviso that the adequacy of its clients' wishes cannot always be trusted.[14]

Viewing patients thus holistically requires the physician — within the limits set by his profession — to confront them humanely. This is no trivial matter. Addressing someone humanely is ambiguous as to its subjective and objective meaning. In the first case, even though patients may (subjectively) feel addressed personally, the physician may be merely using a psychological technique on them. In the second case, there is a genuine personal involvement on his side. It was Toulmin's [53] main concern to argue that only in the second case does the job get adequately done. One can evaluate the psychic factors that were involved in someone's having fallen ill or muster someone's psychic resources in helping her to get better not by merely translating the technological model into the sphere of psychology,[15] but only by engaging in some honestly personal interaction. Only thus are patients encouraged not only to trust their bodies to the member of a respectable profession but also to open up their problems to a fellow human being. Simply applying psychological categories, simply dispensing psychological advice as one dispenses pharmaceuticals, may be more harmful than helpful. The sense of being personally cared for (rather than being merely taken care of) has in many cases a more beneficial effect on awakening patients' will to get better, and thus presents the hope of greater efficacy of medical services, than an unsympathetic and routine way of spreading cheerful optimism.

> #27 Or: To claim that medicine is not a technology is legitimate in view of the professional acceptability of such humane placebo effects.

Insofar as patients carry their personality into their relationship with their physician, the personal trust that may have been gained in the course of their interaction is in many cases an important prerequisite for the already matter-of-fact success of medical interventions.[16]

> #28 Or: to claim that medicine is not a technology is legitimate insofar as only physicians already carry their merely technical responsibility personally (either individually or as a team) and thus irreplaceably.

3. *The function of ethical codes.* As this personal trust-relationship is a consequence of an obligation towards individuals imposed by medicine's ethos, this ethos already affects the merely matter-of-fact part of medical practice. To be sure, the ethos of both medicine and technology enforces the strict observation of professional standards which are incorporated in the respective practices. But beyond this merely trivial enforcement-function, the obligations imposed by such codes have in each case a different function.

In technology, they affect only the beginning of a production process or its end: only the decision to deliver or not to deliver such a thing or service at all, and whether to make it useable for certain purposes or not. In medicine, these obligations also affect the "productive process" itself. Whereas society's trust in ethos-bound technology is divided between a regard for the technical care taken and a regard for the consideration of indirect consequences, society's (as well as individual patients') trust in ethos-bound medicine concerns both regards at once, or concerns the indirect-consequences consideration as operative in the care-taking itself.

That is to say, technologists can achieve a technically good performance irrespective of the social-consequences regard required by their ethos, whereas physicians cannot achieve a technically good performance irrespective of the individual-well-being-regard required by their ethos. Or in technology, the technical quality of professional activities can be considered apart from their ethical interpretation, whereas in medicine no surgery, for example, can legitimately be considered "well done" if it does not further the patient's health. (The fact that surgery is indeed often so considered constitutes no counter-example: Jokes work precisely through denying the obvious.)

> #29 Or: To claim that medicine is not a technology (nor, for that matter, an art or craft) is legitimate in view of the ethical impregnation of its practice.

It has already been remarked that in medicine, unlike in the other fields, professional standards are contested. One of the additional reasons for this fact lies in the well-established vagueness of fundamental terms guiding medical action, such as "health" and "disease". This vagueness in turn bears witness to a further peculiarity of all ethos-based practice, namely, that it lacks guidelines about how to limit the extent of the obligations imposed. Due to the specific impregnation of medical practice with ethical obligations, this unclarity affects medicine more decisively than technology.

Thus, in particular, as medicine's ethos places no limits on the extent to which the patient's "good" should be medically pursued, it becomes understandable why only medicine's tasks were found to carry the potential of transcending medicine's capacities. It also becomes clear now, why only in medicine the internal standards of adequacy (geared towards helping the patient) should often not be satisfied by what satisfies the external standards (of acting in accordance with accepted rules), and why, finally, medicine was considered to be an essentially developing field for reasons that have nothing to do with making additional tasks possible but arise from the unlimited requirements carried by many of the tasks at hand.

The fact that medicine was found to resemble science in these three regards can also now be understood. Just as every aspect of medical practice is intrinsically directed by its ethos, so every aspect of the scientific practice of researching and theorizing is directed by the (equally unlimited) ethos of science.

Looking back at these considerations, it seems as though the similarities which medicine as an institution carries to technology are after all less ponderous than the dissimilarities. Although both disciplines may be ethos-oriented, still the origin, orientation, and function of these ethoi differ. Rather more like lawyers, ministers, and teachers, physicians belong to a profession that collectively imposes a personal sort of practice. It therefore makes sense to revert once again to the concept of the "art" of medicine and to investigate how such a concept would have to be understood if it were to illuminate what uniquely distinguishes medicine.

The concept of "art" or "the arts" has so far been used mostly with regard to the crafts, that is, to a collective practice which, other than medicine and technology (and for that matter science), realized societal values merely indirectly. Also it was supposed that while crafts today are still sometimes practiced by individuals rather than within corporations, they, other than medicine, are not essentially dependent on the personal side of these individuals. I may prefer my gardener to yours, but the job gets done (pretty much) by either.

Yet the concept of "art" also denotes the fine and the applied arts, where at least the former, in producing "goods of culture", also realizes societal values directly. In addition, both areas of human activity make up for their lack of generally regulated practice by having (at least in the West and perhaps excluding the Middle Ages) a "personal" orientation. A work of (applied) art — not merely irrespective of any standards and rules that may have been set for it but through their very mastery — mirrors (among many other things) the artist's self. Similarly, through the very mastery of (generally valid) professional standards, physicians' personal care relationship to patients also mirrors (among many other things) their selves.

To be sure, with medicine such reflective self expression is never a prime purpose as it sometimes has been with the fine arts. Like with the applied arts, it is rendered subordinate to functionality. Nor do the arts have any standardized ethos that could compare with that of lawyers, ministers, and physicians. They do, however, along with the crafts and in contrast to law and religion, imply a regard for handiwork or practical skill that is also cultivated by medicine.

Hence, perhaps it makes sense to utilize the multi-faceted meaning of the term "art" as an opportunity to select those elements of meaning that fit with what distinguishes medicine from technology.

> #30 Or: To claim that medicine (as a social institution) is an art is legitimate in view of the "professional organization" (as implied in "craft"), the "person-relatedness" (as implied in the fine and applied arts), and the handiwork-aspect (implied in both).

B. THE PRECARIOUSNESS OF MEDICINE

The many and fundamental differences which separate medicine's ethos from that of the purely technological occupations do not, however, account for a further difference which concerns professional policy: The medical profession claims societal privileges, while technology leaves its fate to the market.

Claiming special rights is either a sign of special strength or of a special weakness. In the first case, it simply reflects an existing power structure and is conceptually uninteresting. In the second case, the claim must be defended on conceptual grounds.[17] The "science", "technology", and "art" of medicine will now be considered (I) with a view to illuminating why medicine's ethos should motivate the profession's perception of a special vulnerability, which could (II) render its claims intelligible.

I. *The Difficulty of Medicine's Ethos*

What has so far been observed makes it clear that medicine is uniquely dependent on the functioning of its ethos.

In the beginning of this essay, medicine was likened to a craft because both are oriented towards external standards of adequacy. Unlike the crafts, however, medicine was also found to involve internal standards, and, in opposition to both sorts of "objective" standards, a "subjective" standard of cognitive self-criticism was stipulated as well. In the subsequent examination of medicine as a social institution, it was proposed that, unlike science, technology and the arts, medicine's ethos-based obligations permeate already the merely technical parts of professional performance. The internal adequacy of professional performance in medicine must, therefore, also be judged in terms of these obligations, or in terms of an additional subjective standard of ethical adequacy.

Thus, in the crafts as in technology, the objective standards permit an exhaustive control of professional performance. In medicine, the internal standards contained therein are inseparably linked with subjective standards, the compliance with which cannot be so controlled. To be sure, many aspects of medical practice are accessible to objective control. Even though it may often be practically difficult[18] there is nothing in principle impossible about assessing the diagnostic and therapeutic measures that were taken on the basis of a patient exhibiting certain symptoms. But it is in principle impossible objectively to assess the care that was taken in translating the patient's complaints into the language of those symptoms, in encouraging the patient to be explicit about her complaints, or in furthering her overall well-being through the considerate manner in which medical measures were taken.

Like with religion and education, and unlike technology, the adequacy of professional practice in medicine is not in its fine but sometimes decisive points open to outside scrutiny. Unlike religion and education and like some technological employments, however, medicine involves substantial bodily (and thus potentially irremediable) risks for its consumers. These risks are not restricted to the technical part of patient management. Treatment faults also sometimes arise from a failure of having taken seriously what the patient says (cf. note 14). Hence, unlike all the other comparable fields of societal practice, the quality of medical services both needs to be externally controlled and remains to a considerable extent inaccessible to such controls. Subjective, i.e., ethos-directed self-control, seems therefore particularly important.

At the same time, a closer examination reveals that such self-control is very difficult to achieve. Ethical obligations do not come with earmarked limits. Neither ethical standards in general nor the subjective standards of adequacy in medicine can therefore simply be "applied" for either guiding or evaluating actions. Compliance with and criticism with regard to such standards involves a judgement about the situational context, the sophisticated interpretation of which will usually reveal conflicting value-obligations.

> #31 Hence: To claim that medicine, quite irrespective of any technological progress that can be expected in the future, involves an art in the sense of a personal judgement is legitimate in view of the insufficiency of merely external standards and of the necessity of prudentially evaluating how to comply with those subjective standards, on which the possibility of meeting medicine's internal standard of adequacy depends.

This is not to say that compliance with medicine's ethos is left entirely to doctors' whim. Their compliance can be examined. To be sure, such examination, whether by the agents themselves or by an onlooker, involves an interpretation, which may be subjective in reflecting individual bias. Its general logic, however, can be stated in view of three value-conflicts which are characteristic of medicine[19] and which may be centered around three levels on which the term "art of medicine" is ethically significant: (1) "art" as one of the objects of medicine's ethos; (2) "art" as emphasizing the personal element of medicine and thus as the ground of physicians' personal responsibility; and (3) "art" as a personal asset the use of which the ethos is designed to restrain.

1. *Conflicts between the obligations to the art and to the patient.* Medicine's ethos imposes obligations not only towards the patient but also towards the preservation and perfection of "the art".

> #32 Or: To claim that medicine is an art is also legitimate insofar as medicine's ethos renders the further cultivation of medicine a matter of individual physicians' concern.

Already with respect to mere preservation, the interests of patients, especially in training-hospitals, must sometimes be balanced against the need of educating young physicians. The justification for such a balancing rests on the fact that these patients also enjoy their trained physicians' skills which were acquired through a training that involved previous patients. Subjecting present patients to such a balancing procedure thus means asking them to pay back a debt they owe to former patients. (There are, of course, additional justifications. Patients in training-hospitals are usually supposed to profit from more advanced medicine, and from the supervision of more celebrated experts than other patients.)

With respect to the perfection of the art of medicine, a similar conflict of interests arises and a similar justification holds for the general possibility of utilizing patients in medical research. As the established aspects of their treatment profit from previous such utilizations, one must weigh the conditions under which these patients can be asked to contribute their own part to future improved treatments. (No very clear lines can here be drawn between non-therapeutic and therapeutic research, because even in the latter case, as for instance in randomized clinical trials, medical judgement concerning individual cases is put aside for a while so as not to compromize the scientific results.)

Given the potential experiment-character of all medical practice, a further similar conflict involves physicians' personal commitment to their art and thus (for example) to pursuing the diagnostic task beyond what can reasonably be expected to make a difference for therapy.

In a more indirect manner, this conflict is aggravated through the view towards science (#7) and technology(#8). Both promise long-term progress in the art of medicine, while their immediate utility for the patient at hand is sometimes dubitable.

The danger of disregarding obligations to the patient then arises, first, from the likely psychological deformation resulting from an exclusive regard for science. This not only tends to reduce physicians' empathetic capacities but also tempts them into considering patients merely as cases (cf. [53]). The very existence of a value conflict is thus suppressed. This danger arises, second, from the discrimination against outsider-systems as unacceptable by scientific standards, even though these may permit help to certain otherwise untreatable segments of the patient population. Professional policy justifying society-induced limits on what is reimbursed by state-funded insurances through a determination of what counts as "acceptable" (i.e., "scientific") medicine may thus again obscure the underlying conflict.

Moreover, the ideal of technologization encourages the regard for those isolated and technical aspects of patient care for which the ideal has come closest to realization, but which sometimes stand in the way of furthering overall patient well-being. (A very drastic example is the indiscriminate use of life-preserving technology: its indubitable contribution to advancing the art contrasts with sometimes dubitable benefits for patients, see [22] and, more generally ([34], [62], [4], [9], [28].)

Where these special aspects are rendered subject to special controls and where the results of these controls influence physicians' economic standing, this tends to alter their own perception of their practice and thus the practice itself. In some medical disciplines, physicians have even come to replace their medical by a properly technological ethos. Again, while instituting such controls is surely a help in improving the art of medicine, this may also diminish attention to what particular patients really need.

> #33 Or: To claim that medicine is neither a science nor a technology is legitimate insofar as advances in scientific knowledge and technical manageability constitute only necessary, not sufficient, conditions for advancing the goals set by medicine's ethos.

2. *Conflicts between physicians' and patients' view of patients' "good"*. The obligation toward patients implied that the physician must take on a per sonal responsibility for their well-being (as far as his profession is concerned).

The respects in which this care is a medical concern are determined by the professional standards valid at any particular time (and changing only gradually by virtue of medicine's traditionalism). Yet sometimes several of these respects call for conflicting practical decisions, and sometimes any one of these respects opens alternative ways of medical intervention. In these cases, professional standards are often of little help. In some cases, they encourage courses of action which may not further patients' overall well-being as effectively as non-standard ways of proceeding. In other cases, they are simply insufficient for deciding about the right course of action. In some cases, the physician ought to do more, in other cases, less, than what those standards require, and in most cases, there exists a certain free space for personal preferences.

In addition, these standards are influenced not only by scientific and technological progress but also by organizational structures, habits, and accidents determining the ways in which medical services are rendered, as well as by economic incentives. Questions concerning how long patients should stay at the hospital are answered differently depending on the existence of outpatient hospital services and on the manner of reimbursement of costs.

In sum, acting in accordance with professional standards is not always sufficient, and in some cases not even necessary for improving individual patients' medically relevant well-being. Moreover, just as physicians' view of what they should do for a patient is influenced by many external factors compromising their legitimate authority, so the patient views his own "medical well-being" in a more encompassing context of life plans and personal idiosyncrasies. Hence, there exists a potential of conflict between physicians' appropriately paternalistic role and their equally respectable technological provider function.

Solutions to the conflicts previously noted in connection with medicine's double orientation towards its art and its patients could be conceived in terms of a balancing of opposing interests. The present conflict, to the contrary, calls for a clarification of priorities between medical and non-medical values. As a precondition to this, within the area of medical decision-making, cases in which physicians must take responsibility for patients (and may use their professional authority or persuasive skills in opposing patients' wishes or securing their consent) must be distinguished from cases in which patients may or must take responsibility for themselves.

> #34 Or: To claim that medicine is an art is legitimate not only in view of the persuasive skills required in the pursuit but also in view of the prudent judgement required for distinguishing the limits of acceptable medical paternalism.

Unfortunately, appeals to the "art" character of medicine have often served to ward off patient or public participation in decisions concerning the rendering of medical services. In this context, the inscrutable intricacies of "art" (in the technical sense) are adduced to prove outsiders incapable of deciding on medical matters. Here, different meaning components carried by the concept of the "art of medicine" are used to obscure the fact that there exist illegitimate assumptions of medical authority, or that genuine value conflicts are possible.

The difficulty of separating adequate patient management from appropriate patient respect is aggravated by the influence of science and technology on the practice of medicine. Utilizing scientific data encourages the assumption of an authority in knowing. Using technological devices furthers a self-confidence in doing. Both carry the temptation not only of unduly extending the realm of professional authority, but also — again — of obscuring from view existing value conflicts.

This difficulty affects not only individual physicians' dealings with individual patients, but equally the societal relevance of their care. The sick role carries social advantages. Legal responsibility, claims to social compensation, rehabilitation, and pensions are all decided upon by medical practitioners as experts. Decisions concerning reductions or extensions of the realm of professional competence and on the institutional setup of medical services also rely on physicians' judgement. In all these matters, the societal framework (in which the case-to-case value conflicts described above take place) gets either implicitly or explicitly affirmed or modified. Here, as well, certain assumptions on the extent and limits of professional authority may conflict with what is socially accepted.

On closer look, however, these cases differ substantially. In the former context, a "medical situation" is presupposed: In consulting a physician, patients concede a certain amount of expert authority. Society, on the other hand, when "engaging" the medical profession's advice on policy issues, reserves the expert role to its own officials. Hence, whereas in the former case (of determining individual patients' "good") the required clarification of value-priorities may be reached through conceiving patient respect as a side constraint on physician responsibility (a term used by Engelhardt in [16]), in the present context (of determining how to secure collective actual and potential patients' "good") that clarification requires an advocation of the profession's values. Or whereas in medical situations the relevance of medicine's ethos is presupposed, and the problem concerns only its considerate application, in policy struggles merely the fact of medicine's ethos-orientation is presupposed, and the problem concerns the extent to which, or the sense in which, its policy-implications can be rendered generally acceptable.

Here as well, the natural momentum carrying scientific and technological assumptions of medical authority beyond their proper mark presents the danger of the profession losing sight of existing value conflicts. Or, more precisely, in the individual just as in the collective version of these conflicts the justificatory function of science (knowing objectively) and technology (doing effectively) increases the possibility of improper medical paternalism. Yet, the true basis of that paternalism itself is still medicine's ethos. Hence if that ethos misleads physicians into adducing their individually unlimitable professional responsibility in defense of claims to equally unlimited professional authority and independence, then the social utility of that ethos is reduced.

By contrast, viewing medicine as a whole on the model of technology, that is, available on public and private consumer markets, has an undoubtedly salubrious impact. It functions as a balance against professional arrogance, recalling the importance of professional performance as opposed to the profession as vocation. It thus encourages a more sober evaluation of the costs and benefits involved in medical services. It accomplishes this goal mainly on the collective level, but indirectly also on the individual level, helping patients and the public to recognize their own share in the value decisions that enter into medical transactions.

> #35 Or: To claim that medicine is a technology is legitimate to the extent that this serves as a counterweight against exaggerated conceptions of medical responsibility and authority.

Yet, on the other hand, a mere reduction of medicine to a technology disregards the fact that this model makes sense only as means for better assessing the existing value conflicts between professional obligations and private and public consumer wishes. In disregarding the legitimate grounds of medical paternalism and value advocation, it denies the very existence of such conflicts.

> #36 Hence: To claim that medicine is merely a technology is illegitimate if this confuses the need for conflict resolution with a license for unlimited utilization.

3. *Conflicts between physicians' obligation to the patients and to themselves.* Medicine's ethos has not only a positive function in imposing certain values as obligatory, but also a negative function in placing restraints on the pursuit of certain disvalues. It not only requires a regard for the art of medicine (first value conflict) and an art in personally meeting obligations (second value conflict), but it is also necessitated by the personal-asset-character of medical capacities, or by the possibility of abusing the "art of medicine" in a merely instrumental sense.

As a limit for the personal utilization of that asset, medicine's ethos implicitly and traditionally has imposed a certain altruism on physicians. It has thereby opposed a strictly market-rational behavior. Even though some medical specialties offer opportunities for wealth, medicine is construed as something other than a mere business.[20]

The most important reason for this construction refers to the fact that patients sometimes are "in need" of a help which only professional physicians can provide. Moreover, whereas medicine shares with aviation or food processing technology professional responsibility for customers' lives, patients have usually (except perhaps in some sorts of plastic surgery, for example) not freely chosen their risk. They had (within the limits our culture sets for "rational behavior") "no alternative" than to trust their physician in a dramatic sense. Airplane travelers and canned food consumers may have had "no alternative" in a merely practical sense. Travelers would have to invest only time and money in order to avoid flying, and hungry individuals could in principle turn to private gardening and preserving. By contrast, for cancer patients no investment of energy, time, and funds promises a comparable hope for success. As a response to such relatively enforced helplessness in medicine, it is, for example, deemed unprofessional to set fees according to the intensity of clients' needs, and advertising is discouraged. In addition, it is understood that doctors are prepared sometimes to render charity services.

Whereas the respects in which patient well-being is a medical concern are determined by professional standards, the extent to which medical care should be offered in each of those respects is left undetermined by medicine's ethos. It is limited by professional standards in ways that make it difficult ever to feel comfortable in applying them. Of course, for each category of patients, there are certain diagnostic and therapeutic measures clearly indicated. Yet, there also exists a gray zone of possible interventions where the physician's judgement is required. "Experienced physicians" are supposed to know how far they should enter into each patient's problems. Yet as the problems present themselves concretely, such matters are hard to decide, and "experience" may simply have hardened physicians against the discomfort attending these uncertainties.

Stated as such, this is of course rarely a practical problem. But even in the most private of practices, physicians allocate scarce resources, with the prime resource being their own time and energy. Hence, quite irrespective of any difficulties involved in adjusting standards to ethos, physicians must balance their own interests against those of their patients, and stated in this way the problem becomes practical indeed.

The conflict of interests here is not merely a matter of possibility, as in the first conflict, where the two-fold obligation towards the art and the patient only sometimes involves opposing claims, or as in the second conflict (individual version), where patients and physicians often have no trouble agreeing. Instead, the present conflict characterizes each medical situation in general.

Care for himself may concern the physician either personally or in his capacity for serving other patients. Hence, unless a physician has one patient only, there exists no situation in which his care for one patient does not (in more or less relevant ways) reduce his time and psychic strength to care for others. The conflict here is compounded of a conflict between different obligations both imposed by medicine's ethos (towards this patient and towards other patients) and a conflict between ethos-required altruism and natural egoism.

Conflicts of these sorts require a correspondingly compound balancing. The physicians' interest lies both in securing the economic means for comfortable subsistence and in cultivating or developing their technical skills, thus securing not only personal professional excellence and monetary gains but also the conditions for other patients receiving effective help and charity services.

Moreover, physicians' personal as well as professional self-interest is implicitly also involved in the first two conflicts noted above. What promotes the "art of medicine" often also promotes someone's career, and sometimes physicians' understanding of "legitimate professional authority" hinges upon their shunning the pains of a more extensive communication with patients or the public. Since in all three cases physicians' merely egoistic interests can always be construed as "other-patients' interests", a clear recognition of the psychological forces at work in the balancing is easily avoided. Or in all three cases of ethos-induced value conflicts, egoistic motivations can be clouded by recourse to the ideals of that ethos.[21] Hence, doing the conflict-resolving part right requires an extraordinary reflective impartiality.

> #37 Or: To claim that medicine is an art is legitimate insofar as fairly meeting the standards of medicine's ethos presupposes a personal expertise in ethical self-criticism.

In addition, the native uncertainty of truly beneficial outcomes that hampers some medical practice (unless that practice is restricted to the merely technological solving of detail problems) is hard to reconcile with a truly caring attitude. One may suppose, therefore, that concentration on conscientiousness and ethical intentions is an important help for physicians in securing their own emotional stability. But where reference to one's ethos is already needed for psychological support, it becomes especially difficult to expose that support to the threat of self-criticism. Hence not only for the sake of hiding illegitimate egoism, but also for the sake of supporting the required altruism will physicians feel impelled to neglect that reflective aspect of their art.

Medical practice has been found not to be exhaustively accessible to objective controls. Partial controls, when employed without a view to their merely partial significance, were found to be potentially counterproductive. A more encompassing interpretation of medical practice requires reference to its ethos. Yet its ethical evaluation hinges on value conflicts which appeared difficult to resolve. Medicine offers particularly rich opportunities for ethical self-deception. Hence, even a concept of the art of medicine which was tailored to the needs of such an encompassing evaluation was not ultimately helpful in determining the conditions of medical practice's overall adequacy.

II. *Claims to Professional Privilege*

Medicine as a social institution has been found to occupy a particularly precarious position *vis-à-vis* the obligations imposed by its ethos. This precariousness also affects "science" and "technology", each of which appeared to be both necessary and misleading for an understanding of medicine. It affects the "art" component as well, because the various meanings this term can assume in the medical context were not only found to be both helpful and harmful for understanding medicine's special nature, but also turned out to be less than helpful for evaluating medical practice.

Claims to professional privilege sometimes indicate an awareness of a profession's particularly vulnerable position. The two questions which remain to be answered now are: 1. Does the understanding of that position developed here render medicine's professional claims more intelligible? 2. In what sense does the intelligibility thus gained provide a better ground for supporting these claims?

1. *The rationale behind the claims.* The oddness of the medical profession's claims for social privileges can be described in three respects: (a) the demand for security from economic pressures; b) the demand for freedom from outside intrusion; and (c) the oblique manner in which these demands are put forward.

a) An awareness of medicine's potential for self-deception explains why economic security (and social status) play such a major role in the profession's public relation policy.

Medical practice presents not only the possibility of great benefit but also of (involuntary) great harm. Economic pressures (as well as self-esteem) are supposed to influence both the extent and the quality of medical care. They thereby influence the chances of benefits and harms. Physicians are in their very orientation towards their ethos easily misled when trying impartially to assess such influences in their own practice. Outside controls are no sufficient safeguards. Hence, it is reasonable to try (among other courses of action, such as including ethics courses in medical education or instituting peer review committees in hospitals) to reduce the efficacy of such economic pressures (and status uncertainties). At least the more blatant temptations for unethical behavior can thereby be avoided.

b) The perception of a decreasing scope of physicians' medical responsibility through an increase in public regulation of medical services explains why professional independence is such a major issue in the profession's self-representation.

1) Recent developments in Western societies have placed medicine under increasing public scrutiny. Increased technological effectiveness and costliness of established services and increased technological possibilities for new costly services demand a societal re-assessment of the attendant risks and benefits. At the same time, a greater disparity of value priorities within present-day societies disrupts the uniformity of such assessments. As this difficulty also affects the relation of physicians to patients, regulations have been instituted for the protection of both. These limit, for example, the use of patients in research, thus alleviating one aspect of the first potential value conflict (between obligations to the art and to the patient).

These regulations also require patient consent to risky diagnostic and therapeutic interventions. While this is intended to alleviate the second potential conflict (between physicians' and patients' view of patients' "good"), it in effect sometimes interferes with sound medical judgement about the advisability of detailed patient information. The exercise of legitimate medical authority may thereby be impeded. Thus, a new potential conflict between physicians' legal obligations to comply with such regulations and their medical obligation to act according to medical judgement and thus on behalf of the patient has arisen.

2) In addition, with increased public funding of medical services and increased cost to those services, the third already compound conflict between

patient- and physician-interest is rendered still more complex. To be sure, this conflict is alleviated insofar as in such systems the load of caring for the poor is taken off individual physicians' backs. But at the same time, arbitration between patients' and physicians' interest (whether merely for themselves or on behalf of other patients) must also include a regard for particular (potential and actual) patient communities with their combined medical and financial interests.

This may have different consequences in different societal contexts. In a system or in areas which allow for growth of medical expenses, physicians will sometimes find their personal interest along with that of their patients (in expensive services) opposing that of the paying community, or sometimes their personal interest (in providing such services) opposing that of the patients (where "doing less" would be more beneficial) and of the paying community. In a system or in areas where expenses must be kept steady and where there are ceilings set on medical services, additional factors tending to compromise the adequacy of medical judgement and care are introduced. (See, for example, [11].)

Since "cost explosions" in public health systems have everywhere given rise to cost containment policies, the second alternative is slowly becoming generally effective. Physicians' responsibility is here restrained in view of funding agencies' averaging standards on quality and quantity of medical services. Such standards are rendered effective through linking the physician's economic self-interest with conformity. It thus again becomes a matter of personal charity whether a physician simply complies or whether he takes his responsibility for individual patients seriously. In that sense, nothing much has changed.

The difference lies in a shift of emphasis that introduces an additional source of self-deception. The physician's perception of his allocative function is changed from a matter of ethos-related value conflict to a matter of course. In consequence, since funding agencies' standards are developed and enforced by medical professionals themselves, they figure as "standards of adequate care" while in fact defining only "standards of minimal care". Hence it is easy (and lies in their self-interest) for physicians implicitly to replace the dictate of medical responsibility by the dictate of societal norms. Again they find their professional performance surreptitiously transformed into a technological one.

c) The perceived professional responsibility for securing the conditions of patient trust explains why professional self-representation in medicine is such a cloudy matter. Objectively speaking, patients' trust would have to concern physicians' capacity and willingness adequately to resolve the three value conflicts described above. Yet, if the existence of these conflicts and hence the constraints under which medicine is practiced, along with the difficulties for physicians even adequately to recognize these constraints, were a matter of common knowledge and would therefore become subjectively operative, this might destroy the very condition for patient trust which the profession is obliged to preserve. Patients would rationally have to consider not only those constraints and difficulties, but also the fact that many aspects of actual unfair dealing are rarely discoverable either by themselves or even by other experts. They would feel at a loss on whether to prefer one physician who complies with their stated wishes or another who (perhaps) more conscientiously refuses. The more a particular medical situation is shaped by patient helplessness, emotional anguish, and immediate need, the more would such a consideration reduce the patient's capacity of trusting and his willingness to accept troublesome advice. It would thereby tend to diminish his chances of being successfully treated.

Quite admittedly, all of these complicating factors are to some extent present in all medical systems of today. But the very fact that the profession has already suffered a considerable loss of public confidence renders understandable any attempt to avoid that which would make matters worse. Accordingly, professional representatives have preferred to remain silent on the problems involved in medicine's ethos, even if this may again increase the risk of their deceiving themselves and the members of their profession as to the existence of such problems. And hence it is also understandable that they do not support their claims for economic security and independence by the sound arguments theoretically available and that they avoid marketing their ethos openly and in the style of enlightened technologists.[22] Their custom of resisting any clear-cut business interpretation of their professional activity and of merely improving the profession's public image by invoking generally accepted values (such as scientific authority, technological expertise, and a devotion to their art which includes the cultivation of humanitarian principles of helping the needy and the like) makes therefore sense as well.[23]

Technologists openly engaging in a business need not fear exposing the difference between short-term gains and long-term costs of unethical industrial developments. They can afford to admit that they must "sell their ethos" in order to make up for the short-term losses threatening their taking responsibility for long-term consequences. By contrast, the medical profession quite sensibly resorts to the merely oblique "marketing" device of offering its societal usefulness in exchange for demanded privileges.

As a result of the three preceding arguments, the unusual aspects of physicians' claim to professional privileges can be understood as grounded in the equally unusual constraints of their professional practice. A closer look at these same arguments makes it, however, obvious that this gain in insight is not easy to reconcile with any desire for support.

2. *The acceptability of the claims.* The main impediment has turned out to be not merely the anachronism involved in medicine's guild-like self-representation. It lies, rather, in the aggravating influence which the profession's necessarily oblique manner of claiming privileges exerts on the very tendency for self-deception which in turn had necessitated that obliqueness. Cultivating a public image of the profession's ethical trustworthiness in defense of privilege claims will strengthen physicians' already professionally obligatory habit of appearing trustworthy. But while external appearances sometimes enhance internal reality, the assumption of such appearances also generates a corresponding self-image, and thereby hinders the ethically required reflective impartiality. If existing value conflicts tend to be hushed up, their existence tends to be forgotten. In this sense, medicine's specific professional self-representation, while it is designed to secure social advantages, and thus to reduce any more urgent need for ethical self-deception, in effect increases the opportunity for such deception.

The account of medicine sketched in this essay was to provide a common conceptual ground for the various positions that were taken in the essays of this volume. This account has been centered around a concept of "art" which was specifically defined for the purpose of giving a more encompassing view of what is unique about medicine in spite of its scientific and technological orientation. This uniqueness was linked with medicine's special ethos, which in turn was found to depend on certain societal conditions for its realization. Securing these conditions is the object of a professional policy that was also found potentially to defeat its own purpose.

Where does this leave us? One possibility is to reject the account of medicine's special "artfulness" altogether and to favor the technological transformation of the medical endeavor as a whole. It seems, after all, that technological medicine can succeed where traditional medicine must fail: in the consistent marketing of its ethos. The sort of "corporate medicine", to which the technological model lends itself, allows for that very candidness about the economic constraints of medical practice which the profession has traditionally shunned.

Yet, on closer scrutiny, the matter is not so simple. What the profession traditionally was unable to "market" is medicine's ethos in its full sense, imposing limitless obligations. What corporate medicine could succeed in marketing is restricted to its technology-ethics components, and thus to a mere subset of medicine's traditional commitments. Neither the extent of individual physicians' charity and altruism nor that "personal" aspect of professional care which goes beyond mere friendliness can in good conscience be endorsed within a consistent market medicine: The first would defeat corporations' economic purpose (which rests on separating business from charity), the second is inaccessible to that objective evaluation by independent consumer agencies which is required to secure the (much more down-to-earth) fairness of such medicine.

The question about how to respond to the discovered inconsistency thus involves a meta-ethical decision: When should one prefer a superior (and superior to what degree?) good (i.e., medicine's ethos in its full sense), the realization of which is risky (and risky to what extent?), to a safer inferior good? This is a difficult question and no universal agreement on the implied rating of values and probabilities can be hoped for.

The picture of medicine drawn up here, while offering no escape from this difficulty, at least has proposed a more comprehensive view of what is there to gain or lose on each side of the alternatives available. Whichever side one chooses, one should at least be clear about the price.[24]

Underneath this value issue, the preceding deliberations also have served the conceptual purpose of inquiring into "science", "technology", and "art" in medicine. The proposed 37 propositions (or 40, counting the notes) illustrate that there are many different senses in which traditional (non-reduced) medicine either can or cannot be described by either of these terms. The list of these propositions is not claimed to be complete. But it supports a very simple summary in terms of three observations:

1. There exist meanings of "science", "technology", and "art", and there exist aspects of medicine such that under these aspects medicine is in fact "a science", "a technology" and "an art".
2. There does not exist any meaning of "science", "technology", or "art" such that medicine as a whole could be denoted by any one of these terms. Rather, medicine can only be described in terms of a network of relations between the various aspects of activities which are denoted by the various meanings of these terms.
3. Even though medicine's uniqueness as defined by this network encourages using the term "art", still there exist meanings of "science" and "technology" which do not account for this uniqueness or even strictly

apply to any of the aspects of medicine, but which it makes sense to apply in order to counter-balance against the vicissitudes implied in medicine's interpretation as "art". For medicine is not only placed in between whatever the three title terms variously denote, but it is placed precariously among them.

International Studies in Philosophy and Medicine
Freigericht, Germany

NOTES

* My understanding of knowing and doing in medicine has been greatly furthered by discussions with H. T. Engelhardt, Jr.

[1] "Any question of philosophy . . . which is so obscure and uncertain, that human reason can reach no fixed determination with regard to it, if it should be treated at all, seems to lead us naturally into the style of dialogue and conversation" ([27], p. 128).

[2] To be sure, none of these distinctions is very rigorous. We do speak of a "practicing scientist" (even though his practice consists mostly of theorizing) as we speak of a "practicing physician", but not of a "practicing painter" or a "practicing plumber". This similarity within a dissimilarity arises perhaps from the "professional" connotation of "practice". There are people who have the capacity to perform as a physician or scientist and, even though they are not practicing physicians or scientists, still remain professionally related to their field. They are medical researchers or teachers (cf. [38]), for instance, or they merely apply science. By contrast, it is hard to conceive of a way in which a seamstress who does not practice sewing could still be professionally related to her field. (She could, of course, restrict herself to the economic management of a sewing business or teach sewing or politically represent the seamstress' association, but these options are also open to the analogous cases in medicine or science. The point is that physicians and scientists have additional professional ways of being non-practicing.)

[3] This is why it is a bit inconvenient to distinguish the "theoretical" from the "practical" sciences by stating that only the latter establish norms for practice (cf. [58], p. 32f). After all, the "theoretical sciences" as well contain certain norms about how to set up theories, how to devise proofs and how to decide questions of scientific relevance. And, after all, theorizing is also (a predominantly mental) "doing".

[4] Even where nothing can be done to better a patient's condition, physicians characteristically feel impelled to keep trying, if only *ut aliquid fiat* (see [31], p. 13; [37] p. 35).

[5] In the Introduction, an interpretation of medical practice as an "applied science" was distinguished from a praxological interpretation of medical knowledge. In the first case, the knowledge guiding the practice was gained in the course of theoretical studies pursued (to a certain extent) for epistemological purposes; in the second case that knowledge was developed for immediate practical utilization. Given the experimental character of all medical practice for medical knowledge proposed here, it

makes sense to bracket even the "applied science" interpretation of medicine by its larger praxological context.

6 In Germany, for instance, numerous journals endeavor to establish the scientific respectability of such outsider systems. Examples are [1] or [13] (see also [31], p. 79ff; [51], [39].

7 In different systems of medical knowledge, praxis-orientation may thus be secured in more or less direct ways. A good example is found in von Engelhardt's essay [18]: Cellular pathologists and bacteriologists differed not only as to the internal versus external causes of disease, but also, more fundamentally, as to the function of medical theorizing. The bacteriologists restricted their theoretical interests in etiology to devising means for effective therapy. They believed this restriction to be more beneficial for practice than Virchow's insisting on the (not only political) freedom of research. At the other end of the spectrum, even the conditionalists, who most consistently sacrificed etiological interests to a positivist functionality, argued for the long-term utility of their conceptual frameworks. Preventive medicine, contriving manners of bridging the gap between psychic and physical phenomena, and presenting an antidote against thoughtless physicians, were adduced by them. They also claimed that securing scientific conformity is ultimately even more beneficial for practice than the monocausalists "merely superficial" identification of the goals of medical research with those of therapy.

Similarly, in Moulin's [36] historical account of immunology, it becomes clear that immunological explanations were not primarily devised in the context of practical problem-solving. Instead, within medicine, a vague notion of immunity had led a rather fruitless life for a long time, until progress in molecular biology, pursued quite independently, in the end revealed ways for new practical utilizations.

8 Of course, as Sassower [48] has shown, there is also an inverse inherence of doing in knowing, which characterizes science in a similar way in which it influences technology. Yet Sassower's suggestion to replace the separate concepts of science and technology by one single "technoscience" (and to cover medicine with this new heading) makes sense only with regard to the mental habitus of the professional. Sassower can indeed show that scientists are not pure theoreticians (in the Greek sense of the term) but that they, instead, technically manipulate the very reality which they scientifically analyze. In what concerns the purpose of scientific and technological analyses of reality, however, the established distinction between a primary pursuit of insight and a primary pursuit of practical concerns remains unshaken.

9 Such a "business" interpretation of medicine is incompatible with accounts which construe medicine as an exclusively or even mostly ethical endeavor. These accounts, which are hard to reconcile with obvious facts about the economic and professional implications of practicing medicine and which could therefore at most amount to the statement of an ideal, are usually supported by reference to the "helping people in need" aspect of medical practice. But apart from the fact that this aspect covers only a very small amount of actual doctors' dealings with patients and that other (mundane) occupations also help people in various needs, a more fundamental root of such thinking seems to lie in a not very differentiated view of values and human actions. Thus, Thomasma and Pellegrino [52] start out with the observation that evaluations of medical practice concern not scientific correctness. From this, they conclude that such

observations must concern moral goodness. For them, just because medicine is at bottom practical and not theoretical, it must at bottom be ethical. That is, just because human practice is in principle always open to ethical investigation, it is taken to be in principle ethically relevant. The difference between what immediately affects the subsistence or liberty of a person and thus touches unconditional moral duties, and what merely concerns other matters and thus does not by itself necessitate any ethical scrutiny is thereby obscured. (Cf.: Wear's [57] and Hucklenbroich's ([26], p. 71) criticism of this position.)

A similar identification of values or norms with ethical values or norms seems to lie at the bottom of Sadegh-zadeh's "deontological" interpretation of medicine as a practical science or as a "praxology". He does not consider regarding medicine's practical rules or norms as (at least in many cases) merely instrumental ([46], p. 12f) in view of people's merely contingent wishes.

[10] It is a bit inconvenient that only one adjective ("ethical") exists to match the two nouns "ethics" and "ethos" which, in this essay, are frequently opposed to one another. To revert to "ethos-related" would be clearer but also clumsier. For our present purpose, it is, however, sufficient to let the context determine which meaning is intended.

[11] This relative autonomy of medicine's ethical commitments does not preclude the possibility of policy makers taking advantage of their existence and invoking them for quite extraneous purposes. Evidence for this can be found in (for instance) a 1983 report on future perspectives of societal development, issued by the government of Baden-Württemberg (quoted in [29]). Still, the fact that something can be thus abused for other ends does not imply that its proper purpose may be reduced to them.

Another problem for medicine's relative autonomy arises from medicine's increasing technologization, which presents increasing opportunities for political instrumentalization not just of medicine's ethos, but of medicine itself (see [6]).

[12] One way of conceiving the grounds for this special status of medicine consists in assigning special value-implications to the basic concepts of health and disease. The supposition of such value-implications is compatible with the fact that both the intention of "health" and "disease" and the value connotations of these concepts change drastically over time. This supposition, moreover, does not imply that health is an absolute value or disease an absolute disvalue. Not all individuals value alike, nor does the same individual value consistently throughout his life. Nevertheless, most people at most times prefer to be freed from their more troublesome ailments.

Accordingly, for a physician it is not as neutral a matter whether he takes favorable or disfavorable action against diseases (such as spreading them or curing them) as it is for a technologist who may one day serve the public by constructing atomic power plants and another day serve the same (or another) public by (gently) decomposing what he constructed.

> Or, #24a: to suppose that medicine's basic concepts, unlike those employed in scientific or technological knowledge, carry a normative commitment (of some sort) makes sense for explaining medicine's peculiar professional status.

This supposition also helps conceptually to distinguish within medical knowledge between contributions of merely auxiliary sciences and medical science proper: only the second involves the concepts of "health" and "disease" and thus a normative commitment. (Accordingly, two senses in which practical medicine can be said to "apply science" or to be "an applied science" become distinguishable as well: Insofar as only data from the auxiliary sciences are utilized, the purpose of that utilization is added from without. Insofar as medical science proper is utilized, at least the general direction of its purpose is already determined. In the first case, the application concerns only facts; in the second, it encompasses a norm.)

This reference to a "norm" calls to mind other attempts at distinguishing medicine as a "practical science" from the "theoretical" or "natural" sciences. In the Introduction, Hucklenbroich's work was cited as an example. Yet his distinction remains ambiguous as to whether only subject matters (with their factual versus normative implications) or also manners in which these subject matters are conceived (descriptive versus value-endorsing) should provide the decisive criterion (see [26], p. 56). On p. 61, "statements about actions" are alotted to "action theory" (i.e., practical science), so Hucklenbroich seems to endorse the sole-subject-matter criterion. On p. 62, "statements about functions" are called "hybrid" because they contain an evaluative element alongside their descriptive one, which in turn seems to endorse the manner-of-conception criterion. Both cases differ in that norms or values enter at different levels of consideration: Where "actions" are the subject matter of scientific statements, "purposes", and thus the values envisaged by these purposes, are included as well. By contrast, where "function" in view of "health" and "disease" is used as a descriptive category about subject matters of scientific statements, values occur in a not merely descriptive but an endorsing manner. Hence it seems preferable to distinguish, top line, between theoretical and practical sciences. The former serve (mostly) epistemological, the latter (mostly) non-epistemological values. The former may (one down on the left side), deal either — as natural science — with natural, or — as social science — with social phenomena, such as actions and values, but always in a descriptive manner. Practical science (one down on the right side) may deal with its phenomena either in a value-neutral manner, and can then be called "technological", or in a manner that includes value commitments, such as medical science.

The basis of all these distinctions, that "health" and "disease" carry value implications, is acknowledged by many authors (see, for example, [45]) and disputed by some. Thus Goosens [19] takes these basic concepts of medical knowledge to be like the merely factual concepts of the theoretical sciences (or of technology, so we may add). Value considerations, in his view, enter only on the level of practical application (or in that part of more narrowly clinical knowledge which guides such application). As a reason for his opposition to the "normativist" thesis, he adduces two generally recognized facts about good medical practice: first, not all diseases are treated by physicians in a way in which, if diseases constituted disvalues, (he thinks) one would expect; second, whenever a disease is not treated on the grounds that treating it would not further a particular patient's overall well-being, this is not done in the sense of one duty "overriding" another, as this would have to be (he thinks), if the normativist thesis were true.

Yet Goosens' reasons are not convincing. Engelhardt's normative account of "health" and "disease" stipulates that the presence of a disease (for example) enjoins a physician to action aiming at removing this disvalue, but subject to two side-constraints: one (in ordinary cases), the patient's consent, the other, that such action should do no harm, or that it should further the patient's overall well-being (see [16]). Hence, one can account for the two generally recognized facts about good medicine which Goosens has quoted, without endorsing his position (which in effect likens medicine to a technology). Moreover, on closer scrutiny, Goosens' position makes it even difficult to account for those facts in a way that is consistent with present-day professional practice. Goosens stipulates that the value decisions entering at the level of medical practice center around the goal of "serving Humanity". Yet, as in postmodern societies, no consensus exists about what exactly it is that serves humanity in each particular situation, this definition must be left either to the providers or to the consumers of medical services. In the first case, an unacceptable medical paternalism would ensue; in the second case, consumer wishes would come to determine the scope of professional competence.

Nor can, on the basis of Goosens' approach, either one of these undesirable consequences be avoided through recourse to the sort of consent-stipulation introduced by Engelhardt. The reason is that Goosens, unlike Engelhardt, interprets medicine's value-implications exclusively in ethical terms. These implications therefore ground duties which would be hard to balance off against opposing claims. But such an interpretation is not at all necessary. (On the confusion of extrinsic and intrinsic ethical relevance of human actions in general, see note 9.) For Engelhardt, the disvalue of having a disease can be grasped in aesthetic and instrumental terms. Only in rare and extreme cases of threat to life or permanent bodily integrity can medical help be construed as a moral duty proper. In most cases, by contrast, medical efforts (which at the other end of the continuum may merely serve to reduce a nuisance) are open to multiple utility assessments. Therefore, instituting the two side-constraints makes sense only in the context of his normativist but non-moralist interpretation.

> Or, # 24b, to suppose that medicine's basic concepts, like those of science, technology, and the arts (as crafts), carry no specifically ethical normative commitment makes sense for explaining the side constraints under which the pursuit of non-ethical goals in medicine stands.

Thus, identification of all values with moral values lies at the bottom not only of purely ethical but also of purely technological interpretations of medicine. In the first case (cf. note 9), the value-implications of "health" and "disease" are acknowledged (but conceived in exclusively ethical terms), in the second case (Goosens) they are denied (because conceiving them in exclusively ethical terms is recognized as untenable). This latter view also motivates attempts at reducing particular professional codes such as that of medicine to those generally valid ethical principles which are also appealed to in technological codes (for a recent example see [58], cf. also p. 289f of this essay), thus again construing medicine itself as a technology. The non-ethical obligations imposed by medicine's ethos (such as the obligation to further the art of medicine and to improve patients' conditions even where helping them is not an ethical duty in the strict sense) are here not recognized in their own right. That is to say, wherever medicine's ethos is ethically reduced in a manner either sympathetic or unsympathetic to its special status, either is the domain of the "ethical" expanded in such a way as to deprive it of practical relevance or the specific non-ethical obligations of medicine are neglected.

As a result, the difference between the grounds on which medical and technological codes are erected can now be summarized. The value implications of social institutions lend themselves to description on two different levels: first, with regard to the point in the structural set-up of those institutions at which the values become operative; second, with regard to the kinds of these values. In the present context, the most important division between kinds of values (second aspect) is between moral and non-moral. Only some moral values in some philosophical accounts constitute unconditional duties, and in all philosophical accounts they rank higher than most (and in our context: all) of the other ones. Medicine is in only few instances intrinsically related to moral duties; in most cases it serves — like technology — any values its consumers and providers may wish to realize. Unlike in technology, however, these values in medicine (first aspect) are linked with the basic concepts around which the underlying knowledge is oriented. They do not enter merely at the level of practical (and arbitrary) application. This latter point explains why medicine's ethos, unlike technology's, is specifically medical, whereas the former point explains why medicine's ethos is professional and differs from private ethics.

[13] There is no need to interpret this obligation as such in (privately) ethical terms. If the profession's ethos is based on the (mostly) aesthetic or instrumental disvalue of (having) diseases, and if diseases (strictly speaking) are linked with patients' psycho-physical individuality, then already the professional obligation for fighting diseases will impose a regard for that individuality.

[14] Of course, not sufficiently trusting in patients' own perceptions is also a source of diagnostic error, see, for example [46], p. 2ff: [42].

[15] On the economic instrumentalization of physicians' "humanity", see [60], p. 269f, 275.

[16] Thus, while there may, in some philosophical accounts of private ethics, exist a primary obligation to confront all humans "personally" in the second (strong) sense of genuinely investing oneself, the professional obligation described here is neither a primary one nor does it extend to an unlimited genuineness. Instead, it is both derivative from the obligation to pursue patients' good effectively, and it is limited in view of the physician's personal involvement. After all, whereas in privately human intercourse one person's offering confidential information on private matters will normally oblige the other to reciprocate (to a certain extent), no patient who has been talking about his personal troubles could legitimately expect his physician to do the same. (In fact, such behavior would even be considered unprofessional.)

This is, of course, not to deny that principles valid in private ethics also enter into the professional requirements for physicians' humaneness. The generally valid ethical principle that one should not be unfair, or take advantage of those who are placed in a position of weakness, is valid for professional codes as well. Thus, even insofar as dentists can do their work effectively without attending to their patients' worries and feelings, and even where no concern for staying in business would prompt them to be gentle (as in socialized medicine), still the generally human requirement of "humanity" remains in force. The professional principle of "doing no harm" is wide enough to encompass this requirement.

None of all this, finally, addresses the quite different problem of treating patients "as persons" in the strict sense of respecting their liberty. The requirement of such respect also holds universally. It is an issue for professional medicine in particular, because medical legitimate paternalism sometimes conflicts with respecting patients' wishes. This problem, which was already alluded to in the context of #26, will be discussed on p. 295f.

[17] Actually, the matter is more complicated than this juxtaposition of possibilities suggests. Some authors (see, for example ([24], p. 132f)) have derived the strength of the medical profession's political position from the very fact that, even though it lacks any power structure to speak of, it can conceptually construe any threat to physicians' status as a threat to the conditions of proper medical services. It can thus mobilize (potential) patient support for opposing such threats.

From the perspective of the profession, on the other hand, the possibility of this very construction itself is taken as evidence for the profession's particularly weak position. This weakness does not concern political or societal status as such but the dependence of good, or more precisely ethos-guided, medicine on the general acknowledgement of that status. So, what is rendered as mere sophistry by outsider analysts can also be taken seriously as indicating a genuine problem (and thus a weakness). It is this latter route, I think, which promises further insight into what is special about medicine and about the possibilities of its conceptual representation.

[18] For the difficulties involved in controlling even the technical aspects of health care, see [32], and the discussions in [33], [37], [55].

[19] In the literature on "medical ethics", ethical problems in medicine are usually derived from conflicts between the special rights and obligations imposed by medicine's professional ethos and the ethical obligations valid for humans in general. (A good example is the issue of medical confidentiality.) In the present context, however, the problems involved already in medicine's ethos itself are addressed. Thus, solutions cannot be expected from a simple separation of the private from the professional portions of physicians' obligations or from a reduction of the latter to the former. Rather, such solutions require a more thorough assessment of the complexity of value implications which characterize various aspects of the professional setting.

[20] Present-day tendencies in Western countries encouraging the establishment of group practices for physicians have given fresh impulse to reflection on medicine's ethos in relation to the economic aspects of being a physician. See, for example [41].

[21] To give just one example: Arguments for the desirability of group practice are backed up by reference to the resulting increase in free time for the physician as giving him a chance to keep up with current research and thus to improve the quality of his patient care ([25]; [40], p. 65).

[22] The devious manner in which the medical profession defends its claims to social privileges has been noted by many observers. It is usually interpreted as a deceptive device for hiding collectively egoistic interest behind a veil of collectively professed altruism, that is, as a merely strategic maneuver (see, for example ([4]; [5], p. 234f; [15]). What is lacking in such accounts is a fair appraisal of the specific constraints under which physicians' professional policy must operate in order to meet those very demands of professional responsibility which are also supposed to justify that policy. Unlike technologists, physicians would weaken their case by trying to inconsiderately strengthen it.

[23] A good example for this strategy is [23], and here especially [21], where the demand for prolonged medical training, while in fact also reducing professional com-

petition for established physicians, is defended exclusively in terms of securing a high standard of patient care.

[24] Looking at present-day tendencies in technical and societal developments, one might wonder whether such "choice" has not already been rendered obsolete. The demands of scientific and technological progress shape ever greater areas of medical practice. Many physicians — especially those of the more technical disciplines such as dentistry, ophthalmology, radiology, and anaesthetics — have come to adjust their ethos to that prevalent in the technological occupations. Traditional aspects of Hippocratic medicine such as caring for the aged and the chronically ill, comforting the dying, but also more recent areas of medical competence such as helping people to better cope with their bodily impairments, are being delegated to paramedical personell or returned to the private sector. What is left for medicine is being increasingly restricted to what is manageable in the narrow sense. Such tendencies are supported by the view that traditional medicine is no longer affordable. Rising costs and rising risks have reduced physicians' economic security and professional independence. The still considerable momentum of modern social policy erodes traditional value structures. In Western Europe, medicine is increasingly exposed to bureaucratic controls, whereas in the United States these controls are increasingly replaced by market forces. In each case, however, an enhanced reliance on objectifiable results favors the technological transformation of medicine.

Of course, not all modern tendencies are uncritically accepted today. With regard to our use of natural resources, postmodern disenchantment takes on multiple forms. It points to the direct harms resulting from polluted environments as well as to the indirect harms that ensue from the disappearance of natural species. It points, in other words, to the vulnerability of what is newly recognized as a cosmic arrangement of mutual dependencies.

The question that arises from the juxtaposition of a more encompassing traditional and a reduced technological account of medicine is, whether such ecologically-oriented sensitivities could be extended to the sphere of social institutions as well. To be sure, the social order of life differs from that of the natural. It is human-made, and thus leaves no room for antecedent trust in any cosmic arrangement. But even the specific forms of that life are not human-made in the sense of simple fabrication. They have a way of growing over time as well, sometimes using up centuries, and gradually taking on quite non-intended, quasi-natural characteristics. Among such in-grown forms of social life are arts and ethoi. Their subsistence depends on specific societal conditions. Thus, they are equally vulnerable to public policy or social engineering interference with these conditions as are the natural species with regard to technological interference with their ecological niches.

One of the side effects of this analysis has been the recognition that — in the case of medicine — the endorsement of a special ethos and the claim to a special artfulness are more than mere ideological strategies. It thus opposes (for example) Baier ([2], [3]), who reduces medicine's proclaimed commitment to its ethos (just as that of the technological occupations to their's) to a device for obscuring the pursuit of merely mundane group interest, and concludes that corporate organizations of professionals already as such constitute threats to private citizens' liberty (cf. [44]). Medicine's ethos is seen exclusively as a means for improving physicians' image in the eyes of the public, or even more unsympathetically ([43], pp. 160ff, 240f.)

Yet, even apart from the friendlier picture of medicine developed here, the presuppositions underlying such attacks are weak. Baier not only simply identifies ethos with ethics (see especially [3]), but also ethics with its Kantian interpretation. This is why he denounces the employment of appeals to values for issues of professional policy as illegitimate. He does not seem to consider that a corporate commitment to values can be effectively abused for shielding mundane policy goals only if there is in fact a generally acknowledged difference between the two. One can, after all, pass off interest for ideals only if the effectivity of ideals is generally believed. Such belief, however, in turn encourages the very effectivity it presupposes. It thus constitutes itself an element of social life that is worth preserving against adverse societal change. For such belief encourages specific value commitments and — in the very impasses which the unlimited nature of these commitments introduces — a specific culture of human excellence.

Thus, the meta-ethical choice imposed above should perhaps be rephrased so as to respond to a postmodern ecological awareness. It then involves no longer a simple preferring of one sort of goods to the other, but instead an imperative to render compatible what is valuable on each side. It is, after all, perhaps not as necessary as it appears on the basis of our conceptual juxtaposition to submit medicine as a whole to a technological transformation. Perhaps one can employ technology as a means of improving medical services, without abandoning one's view of these services to a wholesale technological ideology — just as creative artists employ many crafts without defining themselves as craftsmen. Perhaps medical knowledge and practice could be made to profit from the superior technological model of practical rationality, while organizational and economic determinants of that practice could be kept responsive to a more traditional ethos.

Nor does societal value pluralism, just because it precludes any particular generally accepted limits to medicine's technological availability (as in abortion or euthanasia), imply the impossibility of a general agreement on the need for limits *per se*, and hence on the ethos-based dignity of the art of medicine (as Baier ([2]; [3], p. 146) seems to think). Engelhardt's metaethical account of procedural fairness (see especially [17]) is designed to secure conditions for general acceptability even within value-pluralistic societies. In that case, individual or group providers of medical services would have to publicize their understanding of that ethos and thus account within medicine for the existing differences in value priorities. But of course, as a consequence the question whether such ethos-variability could better be pursued within or without the network of professional organizations presently employed (cf. [12]) is again thrown open.

> #38 Or: To still tolerate for a while medicine's claim to "artful uniqueness" is perhaps legitimate as a condition for that more careful (meta-) "technology assessment" which concerns the risks and benefits involved in medicine's potential technologization.

BIBLIOGRAPHY

1. *Allgemeine Homoeopathische Zeitung*, Heidelberg.
2. Baier, H.: 1987, 'Benötigen wir eine Ethik der Medizin? Der Freiraum der Ärzte zwischen Moral, Politik und Recht', in L. Bress (ed.), *Medizin und Gesellschaft. Ethik, Ökonomie, Ökologie*, Springer Verlag, Berlin, p. 131ff.
3. Baier, H.: 1988, 'Gibt es eine Ethik des Sozialstaats?', in G. Gäfgen, (ed.), *Neokorporatismus im Gesundheitswesen*, Nomos Verlagsgesellschaft, Baden Baden, p. 231ff.
4. Bauch, J.: 1980, 'Technisierung, Gewinn oder Verlust an medizinischen Kompetenzen?', *Medizin, Mensch, Gesellschaft* **5**, 241ff.
5. Bauch, J.: 1982, 'Zwischen Gesinnungsethik und Verbandszwang', *Soziale Welt* **33** (2), 221f.
6. Bauch, J.: 1985, 'Probleme der Finalisierung der Medizin', *Medizin, Mensch, Gesellschaft* **10**, 207ff.
7. Bergmann-Kraus, B.: 1981, 'Freie Berufe im Sozialstaat', *Medizin, Mensch, Gesellschaft* **6**, 28ff.
8. Bergmann-Kraus, B. and Schuller, A.: 1986, 'Wertewandel bei freien Berufen — am Beispiel des Zahnarztberufs', *Medizin, Mensch, Gesellschaft* **11**, 112ff.
9. Borgers, D.: 1986, 'Für eine ganzheitliche Perspektive in der Anwendung medizinischer Technologien', in *Medizin und Technologie, Argument Sonderband 141*, Argument Verlag, Berlin, pp. 49ff.
10. Bourmer, H.: 1975, 'Zur Einführung; Gedanken zum Selbstverständnis des Arztes am 75. Jahrestag des Hartmannbundes', in Schadewaldt H. *et al.* (eds.), *75 Jahre Hartmannbund*, Verband der Ärzte Deutschlands, Bonn-Bad Godesberg, pp. 6ff.
11. Brenner, G.: 1988, 'Das Gesundheitsreformgesetz aus der Sicht des ambulanten kassenärztlichen Bereichs', *Medizin, Mensch, Gesellschaft* **13**, 75ff.
12. Butterfass, E.: 1986, 'Freie Ärzteverbände — Gewerkschaften oder Berufsverbände?', *Medizin, Mensch, Gesellschaft* **11**, 30ff.
13. *Curare, Zeitschrift für Ethnomedizin und transkulturelle Psychiatrie*, Braunschweig.
14. Deppe, H.-U.: 1987, *Krankheit ist ohne Politik nicht heilbar*, Suhrkamp Verlag, Frankfurt.
15. Deppe, H.-U. *et al.* (eds.): 1987, *Medizin und Gesellschaft, Jahrbuch 1, Ärztliches Behandlungsmonopol und ambulanter Sicherstellungsauftrag*, Campus Verlag, Frankfurt.
16. Engelhardt, H.T., Jr.: 1986, *The Foundations of Bioethics*, Oxford University Press, New York.
17. Engelhardt, H.T., Jr.: 1986, 'Bioethik in der pluralistischen Gesellschaft', *Medizin, Mensch, Gesellschaft* **11**, 236ff.
18. Engelhardt, D. von: 1992, 'Causality and Conditionality in Medicine Around 1900', in this volume, pp. 75–104.
19. Goosens, W. K.: 1980, 'Values, Health and Medicine', *Philosophy of Science* **47**, 100ff.
20. Gross, R.: 1992, 'Intuition and Technology as the Bases of Medical Decision-Making', in this volume, pp. 183–197.
21. Häussler, S.: 1985, 'Ärzteschaft und Politik', in Hartmannbund — Verband der

Ärzte Deutschlands (ed.), *Medizin, Gesundheit, Politik. Hartmannbund Jahrbuch 1985*, Deutscher Ärzte Verlag, Köln, p. 205ff.
22. Hannich, H.J.: 1988, 'Überlegungen zum Handlungsprimat der Intensivmedizin', *Medizin, Mensch, Gesellschaft* **13**, 238ff.
23. Hartmannbund — Verband der Ärzte Deutschlands (ed.): 1985, *Medizin, Gesundheit, Politik. Hartmannbund Jahrbuch 1985*, Deutscher Ärzte Verlag, Köln.
24. Henke, K.: 1988, 'Funktionsweisen und Steuerungswirksamkeit der konzertierten Aktion im Gesundheitswesen', in G. Gäfgen (ed.), *Neokorporatismus im Gesundheitswesen*, Nomos Verlagsgesellschaft, Baden Baden, p. 113ff.
25. Henrard, G.D.: 1979, 'Das Prinzip der ärztlichen Ethik in freiberuflichen Gruppenpraxen', in Friedrich Thieding Stiftung (ed.), *Internationaler Kongress für Gruppenmedizin 4, Dokumentation Deutsch*, Hartmannbund — Verband der Ärzte Deutschlands, Bonn-Bad Godesberg, p. 55ff.
26. Hucklenbroich, P.: 1981, 'Action Theory as a Source for Philosophy of Science' *Metamedicine* **2** (1), 55–73.
27. Hume, David: 1971 (1779), *Dialogues Concerning Natural Religion*, The Bobbs-Merrill Company, Inc., Indianapolis.
28. Jordan, J. and Krause-Girth, C.: 1986, 'Technologische Entwicklung der Medizin aus psychosomatischer Sicht', in *Medizin und Technologie, Argument Sonderband 141*, Argument Verlag, Berlin, p. 69ff.
29. Kühn, H.: 1987, 'Die rationalisierte Ethik', in *Medizin, Moral und Markt, Jahrbuch für kritische Medizin 2*, Argument Verlag, Berlin.
30. Lenk, H.: 1982, 'Verantwortung und technische Macht', in F. Rapp and P.T. Durbin (eds.), *Technikphilosophie in der Diskussion*, Friedrich Vieweg und Sohn, Braunschweig, p. 187ff.
31. Lüth, P.: 1986, *Medizin in unserer Gesellschaft*, VCH Verlagsgesellschaft, Weinheim.
32. McDermott, W.: 1977, 'Evaluating the Physician and his Technology', *Daedalus*, 135ff.
33. *Medizin, Mensch, Gesellschaft* **5**(1): 1980.
34. Meifort, B.: 1988, 'Prävention und Gesundheitsförderung', *Medizin, Mensch, Gesellschaft* **13**, 180ff.
35. Mitcham, C.: 1980, 'Philosophy of Technology', in P.T. Durbin (ed.), *The Culture of Science, Technology and Medicine*, The Free Press, New York, N.Y., p. 282ff.
36. Moulin, A.-M.: 1992, 'The Dilemma of Medical Causality and the Issue of Biological Individuality', in this volume, p. 53 – 162.
37. Münnich, F.E.: 1984, 'Kosten- und Allokations-Wirkungen des technischen Fortschritts im Gesundheitswesen', in F.E. Münnich and K. Öttle (eds.), *Beiträge zur Gesundheits-ökonomie, Vol. 6: Ökonomie des technischen Fortschritts in der Medizin*, Bleicher Verlag, Gerlingen, p. 22ff.
38. Narr, H.: 1989, 'Betragspflicht zur Ärztekammer', *Hessisches Ärzteblatt* **7**, 397f.
39. Oepen, I.: 1987, 'Unkonventionelle medizinische Methoden; Anbieter, Spielregeln, Konventionen', in L. Bress (ed.), *Medizin und Gesellschaft. Ethik, Ökonomie, Ökologie*, Springer Verlag, Berlin, p. 164ff.

40. Parsons, A.H.: 1979, 'Überlegungen zur gruppenärztlich orientierten Primärversorgung', in Friedrich Thieding Stiftung (ed.), *Internationaler Kongress für Gruppenmedizin 4, Dokumentation Deutsch*, Hartmannbund-Verband der Ärzte Deutschlands, Bonn-Bad Godesberg, p. 63ff.
41. Payne, D.: 1979. 'Das Prinzip der ärztlichen Ethik in Gruppenpraxen', in Friedrich Thieding Stiftung (ed.), *Internationaler Kongress für Gruppenmedizin 4, Dokumentation Deutsch*, Hartmannbund — Verband der Ärzte Deutschlands, Bonn-Bad Godesberg, p. 59ff.
42. Piertkin, R.: 1968, 'Fehldiagnose in der Klinik' *Therapiewoche* **6**, 226.
43. Rauskolb, C.: 1976, *Lobby in Weiss*, Europäische Verlagsanstalt, Frankfurt.
44. Reich, W.T.: 1986, 'Paradigmen für die Bioethik', *Medizin, Mensch, Gesellschaft* **11**, 231ff.
45. Rothschuh, K.E.: 1972, 'Der Krankheitsbegriff', *Hippokrates* **43**, 3ff.
46. Sadegh-zadeh, K.: 1983, 'Medizin als Ethik und konstruktive Utopie', *Medizin, Ethik, Philosophie* **1**, 1ff.
47. Sass, H.-M.: 1992, 'Medicine — Beyond the Boundaries of Science, Technologies, and Arts', in this volume, p. 259–270.
48. Sassower, R.: 1992, 'Technoscience and Medicine', in this volume, pp. 219–228.
49. Schäfer, L.: 1992, 'On the Scientific Status of Medical Research: Case Study and Interpretation According to Ludwik Fleck', in this volume, p. 23–38.
50. Spicker, S.F.: 1992, 'Intuition and the Process of Medical Diagnosis: The Quest for Explicit Knowledge in the Technological Era', in this volume, p. 199–210
51. Strik, W.: 1987, 'Schulmedizin und Naturheilkunde, Konkurrenten oder Partner', in L. Bress (ed.), *Medizin und Gesellschaft. Ethik, Ökonomie, Ökologie*, Springer Verlag, Berlin, pp. 157ff.
52. Thomasma, D.C. and Pellegrino, E.D.: 1981, 'Philosophy of Medicine as the Source for Medical Ethics', *Metamedicine* **2** (1), 5ff.
53. Toulmin, S.: 1992, '*Knowledge and Art in the Practice of Medicine*: *Clinical Judgement and Historical Reconstruction*', in this volume, pp. 231–249.
54. Tribe, L.H.: 1981, 'Towards a New Technological Ethic', in T.J. Kühn and A.L. Porter (eds.), *Science, Technology and National Policy*, Cornell University Press, Ithaca, New York, pp. 347–355.
55. Überla, K.K.: 1982, 'Möglichkeiten und Grenzen der Qualitätsbeeinflussung in der Medizin', *Medizin, Mensch, Gesellschaft* **7**, 115ff.
56. Verein deutscher Ingenieure: 1950, *Bekenntnis des Ingenieurs,* single page, no publisher, Düsseldorf.
57. Wear, S.: 1981, 'Nuancing the Healer's Art', *Metamedicine* **2** (1), 27–30.
58. Wieland, W.: 1986, *Strukturwandel in der Medizin und ärztliche Ethik*, Carl Winter, Heidelberg.
59. Wieland, W.: 1992, 'The Concept of the Art of Medicine', in this volume, pp. 165–181.
60. Wilhelm, J. and, Schneider W.: 1987, 'Das Geschäft der Niederlassung und sein Publikum, in H.-U. Deppe *et al.* (eds.), *Medizin und Gesellschaft, Jahrbuch 1, Ärztliches Behandlungsmonopol und ambulanter Sicherstellungsauftrag*, Campus Verlag, Frankfurt, p. 239ff.
61. Zimmerli, W.C.: 1982, 'Prognose und Wert: Grenzen einer Philosophie des Technology Assessment', in F. Rapp and P.T. Durbin (eds.), *Technikphilosophie*

in der Diskussion, Friedrich Vieweg und Sohn, Braunschweig, p. 139ff.

62. Zöckler, C.E.: 1984, 'Fortschrittsglaube in der Medizin', in H. Kleinsorge and C.E. Zöckler (eds.), *Fortschritt in der Medizin, Versuchung oder Herausforderung*?, TM Verlag, Hameln, p. 17ff.

NOTES ON CONTRIBUTORS

Mary Ann Gardell Cutter, Ph.D., is Assistant Professor, Department of Philosophy, University of Colorado, Colorado Springs, Colorado, U.S.A.

Corinna Delkeskamp-Hayes, Ph.D., is Director of European Programs, International Studies in Philosophy and Medicine, Freigericht, Germany.

H. Tristram Engelhardt, Jr., Ph.D., M.D., is Professor, Department of Medicine, Baylor College of Medicine; Professor, Department of Philosophy, Rice University; and Member, Center for Ethics, Medicine, and Public Issues, Houston, Texas, U.S.A.

Dietrich von Engelhardt, Ph.D., is Professor, Institut für Medizin und Wissenschaftsgeschichte, Medizinische Universität Lübeck, Lübeck, Germany.

Anne M. Fagot-Largeault, Ph.D., M.D., is Professor, Department d'Histoire et Philosophie de la Medecine, University of Paris XII and C.N.R.S., Paris, France.

Rudolf Gross, M.D., M.D.h.c., is Professor of Medicine and former Head of Department of Medicine, University of Cologne, Köln, Germany.

Eric T. Juengst, Ph.D., is Program Director, Ethical, Social, and Legal Program, National Center for Human Genome Research, National Institutes of Health, Bethesda, Maryland, U.S.A.

Reidar Krummradt Lie, M.D., Ph.D., is Director, Center for Medical Ethics, University of Oslo, Oslo, Norway.

Ray Moseley, Ph.D., is Program Director, Medical Humanities Program, Department of Community Health and Family Medicine, University of Florida College of Medicine, Gainesville, Florida, U.S.A.

Ruth M. Walker Moskop, M.A., Ph.D., is Visiting Assistant Professor of English, East Carolina University, Greenville, North Carolina, U.S.A.

Anne Marie Moulin, M.D., Ph.D., is Professor, Centre National de la Recherche Scientifique, Paris, France.

José Luis Peset, M.D., is Professor, Centro de Estudios Históricos del Consejo Superior de Investigationes Científicas, Duque de Medinaceli, Madrid, Spain.

Hans-Martin Sass, Ph.D., is Professor, Institut für Philosophie, Ruhr-Universität Bochum, Germany, and International Scholar and Director,

Corinna Delkeskamp-Hayes and Mary Ann Gardell Cutter (eds.), Science, Technology, and the Art of Medicine, 321–322.

European Professional Ethics Program, The Kennedy Institute of Ethics, Georgetown University, Washington, D.C., U.S.A.

Raphael Sassower, Ph.D. is Associate Professor and Chair, Department of Philosophy, University of Colorado, Colorado Springs, Colorado, U.S.A.

Lothar Schäfer, Ph.D., is Professor, Philosophisches Seminar, Universität Hamburg, Hamburg, Germany.

Stuart F. Spicker, Ph.D., is Professor Emeritus, University of Connecticut Health Center, Farmington, Connecticut; and Visiting Professor, Department of Community Medicine, and Member, Center for Ethics, Medicine, and Public Issues, Baylor College of Medicine, Houston, Texas, U.S.A.

Stephen Toulmin, Ph.D., is Henry Luce Professor, Center for Multiethnic and Transnational Studies, University of Southern California, Los Angeles, California, U.S.A.

Nelly Tsouyopoulos, Ph.D., is Professor, Institut für Theorie und Geschichte der Medizin, Wilhelms-Universität Münster, Münster, Germany.

Wolfgang Wieland, Ph.D., is Professor, Philosophisches Seminar, Universität Heidelberg, Heidelberg, Germany.

INDEX

Philosophy and Medicine

1. H. Tristram Engelhardt, Jr. and S.F. Spicker (eds.): *Evaluation and Explanation in the Biomedical Sciences*. 1975 ISBN 90-277-0553-4
2. S.F. Spicker and H. Tristram Engelhardt, Jr. (eds.): *Philosophical Dimensions of the Neuro-Medical Sciences*. 1976 ISBN 90-277-0672-7
3. S.F. Spicker and H. Tristram Engelhardt, Jr. (eds.): *Philosophical Medical Ethics: Its Nature and Significance*. 1977 ISBN 90-277-0772-3
4. H. Tristram Engelhardt, Jr. and S.F. Spicker (eds.): *Mental Health: Philosophical Perspectives*. 1978 ISBN 90-277-0828-2
5. B.A. Brody and H. Tristram Engelhardt, Jr. (eds.): *Mental Illness*. Law and Public Policy. 1980 ISBN 90-277-1057-0
6. H. Tristram Engelhardt, Jr., S.F. Spicker and B. Towers (eds.): *Clinical Judgment: A Critical Appraisal*. 1979 ISBN 90-277-0952-1
7. S.F. Spicker (ed.): *Organism, Medicine, and Metaphysics*. Essays in Honor of Hans Jonas on His 75th Birthday. 1978 ISBN 90-277-0823-1
8. E.E. Shelp (ed.): *Justice and Health Care*. 1981 ISBN 90-277-1207-7; Pb 90-277-1251-4
9. S.F. Spicker, J.M. Healey, Jr. and H. Tristram Engelhardt, Jr. (eds.): *The Law-Medicine Relation: A Philosophical Exploration*. 1981 ISBN 90-277-1217-4
10. W.B. Bondeson, H. Tristram Engelhardt, Jr., S.F. Spicker and J.M. White, Jr. (eds.): *New Knowledge in the Biomedical Sciences*. Some Moral Implications of Its Acquisition, Possession, and Use. 1982 ISBN 90-277-1319-7
11. E.E. Shelp (ed.): *Beneficence and Health Care*. 1982 ISBN 90-277-1377-4
12. G.J. Agich (ed.): *Responsibility in Health Care*. 1982 ISBN 90-277-1417-7
13. W.B. Bondeson, H. Tristram Engelhardt, Jr., S.F. Spicker and D.H. Winship: *Abortion and the Status of the Fetus*. 2nd printing, 1984 ISBN 90-277-1493-2
14. E.E. Shelp (ed.): *The Clinical Encounter*. The Moral Fabric of the Patient-Physician Relationship. 1983 ISBN 90-277-1593-9
15. L. Kopelman and J.C. Moskop (eds.): *Ethics and Mental Retardation*. 1984 ISBN 90-277-1630-7
16. L. Nordenfelt and B.I.B. Lindahl (eds.): *Health, Disease, and Causal Explanations in Medicine*. 1984 ISBN 90-277-1660-9
17. E.E. Shelp (ed.): *Virtue and Medicine*. Explorations in the Character of Medicine. 1985 ISBN 90-277-1808-3
18. P. Carrick: *Medical Ethics in Antiquity*. Philosophical Perspectives on Abortion and Euthanasia. 1985 ISBN 90-277-1825-3; Pb 90-277-1915-2
19. J.C. Moskop and L. Kopelman (eds.): *Ethics and Critical Care Medicine*. 1985 ISBN 90-277-1820-2
20. E.E. Shelp (ed.): *Theology and Bioethics*. Exploring the Foundations and Frontiers. 1985 ISBN 90-277-1857-1
21. G.J. Agich and C.E. Begley (eds.): *The Price of Health*. 1986 ISBN 90-277-2285-4
22. E.E. Shelp (ed.): *Sexuality and Medicine*.
    Vol. I: Conceptual Roots. 1987 ISBN 90-277-2290-0; Pb 90-277-2386-9

23. E.E. Shelp (ed.): *Sexuality and Medicine.*
Vol. II: Ethical Viewpoints in Transition. 1987
ISBN 1-55608-013-1; Pb 1-55608-016-6
24. R.C. McMillan, H. Tristram Engelhardt, Jr., and S.F. Spicker (eds.): *Euthanasia and the Newborn.* Conflicts Regarding Saving Lives. 1987
ISBN 90-277-2299-4; Pb 1-55608-039-5
25. S.F. Spicker, S.R. Ingman and I.R. Lawson (eds.): *Ethical Dimensions of Geriatric Care.* Value Conflicts for the 21th Century. 1987
ISBN 1-55608-027-1
26. L. Nordenfelt: *On the Nature of Health.* An Action- Theoretic Approach. 1987
ISBN 1-55608-032-8
27. S.F. Spicker, W.B. Bondeson and H. Tristram Engelhardt, Jr. (eds.): *The Contraceptive Ethos.* Reproductive Rights and Responsibilities. 1987
ISBN 1-55608-035-2
28. S.F. Spicker, I. Alon, A. de Vries and H. Tristram Engelhardt, Jr. (eds.): *The Use of Human Beings in Research.* With Special Reference to Clinical Trials. 1988 ISBN 1-55608-043-3
29. N.M.P. King, L.R. Churchill and A.W. Cross (eds.): *The Physician as Captain of the Ship.* A Critical Reappraisal. 1988 ISBN 1-55608-044-1
30. H.-M. Sass and R.U. Massey (eds.): *Health Care Systems.* Moral Conflicts in European and American Public Policy. 1988 ISBN 1-55608-045-X
31. R.M. Zaner (ed.): *Death: Beyond Whole-Brain Criteria.* 1988
ISBN 1-55608-053-0
32. B.A. Brody (ed.): *Moral Theory and Moral Judgments in Medical Ethics.* 1988
ISBN 1-55608-060-3
33. L.M. Kopelman and J.C. Moskop (eds.): *Children and Health Care.* Moral and Social Issues. 1989 ISBN 1-55608-078-6
34. E.D. Pellegrino, J.P. Langan and J. Collins Harvey (eds.): *Catholic Perspectives on Medical Morals.* Foundational Issues. 1989 ISBN 1-55608-083-2
35. B.A. Brody (ed.): *Suicide and Euthanasia.* Historical and Contemporary Themes. 1989 ISBN 0-7923-0106-4
36. H.A.M.J. ten Have, G.K. Kimsma and S.F. Spicker (eds.): *The Growth of Medical Knowledge.* 1990 ISBN 0-7923-0736-4
37. I. Löwy (ed.): *The Polish School of Philosophy of Medicine.* From Tytus Chałubiński (1820–1889) to Ludwik Fleck (1896–1961). 1990
ISBN 0-7923-0958-8
38. T.J. Bole III and W.B. Bondeson: *Rights to Health Care.* 1991
ISBN 0-7923-1137-X
39. M.A.G. Cutter and E.E. Shelp (eds.): *Competency.* A Study of Informal Competency Determinations in Primary Care. 1991 ISBN 0-7923-1304-6
40. J.L. Peset and D. Gracia (eds.): *The Ethics of Diagnosis.* 1992
ISBN 0-7923-1544-8
41. K.W. Wildes, S.J., F. Abel, S.J. and J.C. Harvey (eds.): *Birth, Suffering, and Death.* Catholic Perspectives at the Edges of Life. 1992 ISBN 0-7923-1547-2

42. S.K. Toombs: *The Meaning of Illness*. A Phenomenological Account of the Different Perspectives of Physician and Patient. 1992 ISBN 0-7923-1570-7
43. D. Leder (ed.): *The Body in Medical Thought and Practice*. 1992 ISBN 0-7923-1657-6
44. C. Delkeskamp-Hayes and M.A.G. Cutter (eds.): *Science, Technology, and the Art of Medicine*. European-American Dialogues. 1993 ISBN 0-7923-1869-2
45. R. Baker, D. Porter and R. Porter (eds.): *The Codification of Medical Morality*. Historical and Philosophical Studies of the Formalization of Western Medical Morality in the Eighteenth and Nineteenth Centuries, Volume One: Medical Ethics and Etiquette in the Eighteenth Century. 1993 (forthcoming) ISBN 0-7923-1921-4

KLUWER ACADEMIC PUBLISHERS – DORDRECHT / BOSTON / LONDON